Lecture Notes in Operations Research

Lecture Notes in Operations Research is an interdisciplinary book series which provides a platform for the cutting-edge research and developments in both operations research and operations management field. The purview of this series is global, encompassing all nations and areas of the world.

It comprises for instance, mathematical optimization, mathematical modeling, statistical analysis, queueing theory and other stochastic-process models, Markov decision processes, econometric methods, data envelopment analysis, decision analysis, supply chain management, transportation logistics, process design, operations strategy, facilities planning, production planning and inventory control.

LNOR publishes edited conference proceedings, contributed volumes that present firsthand information on the latest research results and pioneering innovations as well as new perspectives on classical fields. The target audience of LNOR consists of students, researchers as well as industry professionals.

This series is indexed by Research Papers in Economics (RePEc), zbMATH and has been accepted by Scopus.

Xiaolei Xie · Kejia Hu · Guiping Hu ·
Weiwei Chen · Robin Qiu
Editors

AI, Society and Digital Transformation

Proceedings of 2025 INFORMS Conference on Service Science, Oxford, UK, July 1–3, 2025

Springer

Editors
Xiaolei Xie
Department of Industrial Engineering
Tsinghua University
Beijing, China

Kejia Hu
Saïd Business School
University of Oxford
Oxford, UK

Guiping Hu
School of Industrial Engineering
and Management
Oklahoma State University
Stillwater, OK, USA

Weiwei Chen
Department of Supply Chain Management
Rutgers, The State University of New Jersey
Piscataway, NJ, USA

Robin Qiu
Division of Engineering and Information
Science
Pennsylvania State University
Malvern, PA, USA

ISSN 2731-040X ISSN 2731-0418 (electronic)
Lecture Notes in Operations Research
ISBN 978-3-032-13115-7 ISBN 978-3-032-13116-4 (eBook)
https://doi.org/10.1007/978-3-032-13116-4

This Springer imprint is published by the registered company Springer Nature Switzerland AG
The registered company address is: Gewerbestrasse 11, 6330 Cham, Switzerland

Contents

Inverse Optimization in Finite-State Continuous-Time Markov Decision Processes

Nhi Nguyen and Archis Ghate(✉)

Industrial and Systems Engineering, University of Minnesota, Minneapolis, MN 55455, USA
{nguy5655,archis}@umn.edu

Abstract. Inverse optimization involves finding parameter values that would render given values of decision variables optimal. Contrast this with the usual (forward) optimization, where optimal values of decision variables are determined using given values of parameters. Inverse optimization has been studied for models such as linear programs, network flow problems, shortest path problems, integer programs, semi-definite programs, and conic programs. It has also been studied in finite-state, discrete-time Markov decision processes (MDPs), to find rewards or transition probabilities that would render a given policy optimal. This paper considers inverse optimization in finite-state continuous-time Markov decision processes (CTMDPs). Unlike the model-free nonparametric version that is prevalent in discrete-time MDPs, we consider a model-based parametric problem here. That is, rewards are characterized using a finite set of parameters. The goal is to find values of these parameters that would make a given policy optimal. In fact, among all such parameter values, we wish to find those that are closest to some given estimates. We use uniformization to write Bellman's equations of optimality for the forward problem and utilize them to formulate the inverse problem. We computationally illustrate our methodology on a batch manufacturing problem and a queue routing problem.

Keywords: Parametric uncertainty · uniformization · convex optimization

1 Introduction

The goal in inverse optimization (IO) [5] is to find values of parameters that would render given values of decision variables optimal to a mathematical program. This is in contrast with (forward) optimization, where optimal values of decision variables are computed using known problem parameters. IO has been applied to transportation, demand management, production planning, finance, and healthcare.

Ahuja and Orlin [1] applied IO to find costs that would make a given feasible solution optimal to a linear program. They searched for costs that were closest,

X. Xie et al. (Eds.): ICSS 2025, LNOR, pp. 1–14, 2026.
https://doi.org/10.1007/978-3-032-13116-4_1

in weighted absolute sum distance, to a given estimate. They employed complementary slackness to rewrite this problem as another linear program. Since then, IO has been studied for problems such as shortest paths [4], conic programs [10], semi-definite programs [9], network optimization [2], and integer programs [12].

Erkin et al. [7] applied IO to infer rewards that would render a given policy optimal for a discrete-time discounted Markov decision process (MDP). They reformulated this as a linear program. Ghatrani and Ghate [8] applied IO to impute transition probabilities in discrete-time discounted MDPs. That formulation is necessarily bilinear. They proposed sequential convex programming methods for approximate solution of this non-convex problem. These papers pursued a model-free nonparametric approach, where the rewards or transition probabilities were arbitrary real numbers with some side constraints.

We state an IO problem for imputing rewards in discounted continuous-time MDPs (CTMDPs) with finite state- and action-spaces. In our model-based framework, rewards are characterized by a finite set of parameters. Our inverse problem involves inferring the values of these parameters from certain compact sets. In fact, we wish to find parameter values that are as close as possible, in the weighted absolute sum distance, to some given estimates. We first apply uniformization to convert the forward CTMDP into a CTMDP with identical sojourn time distributions across states and actions. We transform this uniformized CTMDP into an equivalent finite-state discrete-time MDP. We then formulate the inverse problem using Bellman's equations for this discrete-time MDP. We identify when this formulation is convex or linear and hence tractable. This methodology is illustrated using computational results on an inverse batch manufacturing problem and an inverse queue routing problem.

2 Forward Problem

A discounted continuous-time Markov decision process (CTMDP) is described as follows (see Sect. 11.5 of [11]). A decision-maker (DM) continuously monitors the state of a system. The set of possible states is $\mathcal{S} \stackrel{\text{def}}{=} \{0, 1, 2, \ldots, n-1\}$. Here, n is a finite positive integer. Immediately after the system completes a transition into any state $s \in \mathcal{S}$, the DM chooses an action $a \in \mathcal{A}_s$. Here, $\mathcal{A}_s$ denotes the set of feasible actions in state s; it is a non-empty subset of $\mathcal{A} \stackrel{\text{def}}{=} \{0, 1, 2, \ldots, m-1\}$. Here, m is a finite positive integer. The system remains in state s for a random duration of time $\tau(s, a)$ (called sojourn time) that is exponentially distributed with parameter $0 \leq \nu(s, a) < \infty$. The system then transitions into some state $s' \in \mathcal{S}$ with probability $p(s'|s, a)$. This sequence of events, wherein a decision is made after completing each state transition, continues forever. To avoid technicalities, we assume that an initial decision is made upon observing the state at time $t = 0$. In our notation, $\nu(s, a) = 0$ means that the system never leaves state s upon choosing action a. We allow for "hypothetical" self-transitions, that is, $p(s|s, a)$ may be positive for some states $s \in \mathcal{S}$ and some actions $a \in \mathcal{A}$. Upon choosing action $a \in \mathcal{A}_s$ in state $s \in \mathcal{S}$, the DM receives a lump sum reward of $\rho_1(s, a)$ and also earns a reward at a rate

$\rho_2(s,a)$ until the next state transition (or equivalently, decision epoch) occurs. Rewards are discounted using a continuous-time discount rate $\gamma > 0$. That is, the present worth of one unit of reward earned t time units in the future equals $e^{-\gamma t}$. Let $r(s,a)$ denote the expected discounted reward that the DM earns until the next decision epoch, for choosing action a upon transitioning into state s. That is, we have (see Eq. 11.5.3 in [11]),

$$r(s,a) = \rho_1(s,a) + \rho_2(s,a)\mathbb{E}\left\{\int_0^{\tau(s,a)} e^{-\gamma t}dt\right\} = \rho_1(s,a) + \frac{\rho_2(s,a)}{\gamma}\mathbb{E}\left(1 - e^{-\gamma\tau(s,a)}\right)$$
$$= \rho_1(s,a) + \frac{\rho_2(s,a)}{\gamma + \nu(s,a)}. \tag{1}$$

A stationary deterministic policy π is a rule that prescribes the DM to choose a decision $\pi(s) \in \mathcal{A}_s \subseteq \mathcal{A}$ upon transitioning into state $s \in \mathcal{S}$. Specifically, the DM's decision does not depend on the history of the process such as how many transitions previously occurred, which states were previously visited, and how much time was spent in those states. Moreover, a single action from $\mathcal{A}_s$ is prescribed in state s, instead of randomizing over actions from $\mathcal{A}_s$. As such, π can be seen as a mapping from $\mathcal{S}$ into $\mathcal{A}$. We denote the finite set of all stationary deterministic policies by Π.

The expected total discounted reward that the DM earns upon implementing a policy $\pi \in \Pi$ over an infinite-horizon, starting in state s_0 at time $t = 0$, is denoted by $J^\pi(s_0)$. In the literature, $J^\pi : \mathcal{S} \to \mathbb{R}$ is called the value function of policy $\pi \in \Pi$. The optimal expected total discounted reward earned over an infinite-horizon, upon starting in state s_0 at time $t = 0$, is given by

$$J^\star(s_0) \stackrel{\text{def}}{=} \max_{\pi\in\Pi} J^\pi(s_0). \tag{2}$$

Consequently, $J^\star : \mathcal{S} \to \mathbb{R}$ is called the optimal value function. A policy $\pi^\star \in \Pi$ is said to be optimal for the CTMDP if it attains the maximum in the above problem simultaneously across all states $s_0 \in \mathcal{S}$. That is, if $J^\star(s_0) = J^{\pi^\star}(s_0)$, simultaneously for all $s_0 \in \mathcal{S}$. The DM's goal is to find such a policy, if one exists.

Let $\mathcal{X} \stackrel{\text{def}}{=} \{(s,a) \in \mathcal{S} \times \mathcal{A} | a \in \mathcal{A}_s\}$ denote the finite set of all feasible state-action pairs.

Remark 1. Since $\mathcal{X}$ is finite, there exists a constant $0 < \nu < \infty$ such that $\nu(s,a) \leq \nu$, for all $(s,a) \in \mathcal{X}$. This also implies that $(1 - p(s|s,a))\nu(s,a) \leq \nu$, for all $(s,a) \in \mathcal{X}$.

Remark 2. Since $\mathcal{X}$ is finite, let $B \stackrel{\text{def}}{=} \max_{(s,a)\in\mathcal{X}} |r(s,a)| < \infty$.

For each $(s,a) \in \mathcal{X}$, let

$$\tilde{r}(s,a) \stackrel{\text{def}}{=} r(s,a)\frac{\gamma + \nu(s,a)}{\gamma + \nu}, \tag{3}$$

and

$$\tilde{p}(s'|s,a) \stackrel{\text{def}}{=} \begin{cases} 1 - \frac{(1-p(s'|s,a))\nu(s,a)}{\nu}, & \text{for } s' = s \\ \frac{p(s'|s,a)\nu(s,a)}{\nu}, & \text{for } s' \neq s, \end{cases} \tag{4}$$

as in Equations (11.5.7) and (11.5.8) in [11]. In light of Remarks 1 and 2, observe that $|\tilde{r}(s,a)| \leq B$, for all $(s,a) \in \mathcal{X}$. One can utilize Remarks 1 and 2 to establish that the CTMDP is equivalent to a discrete-time MDP with state-space $\mathcal{S}$, action-space $\mathcal{A}$, bounded rewards $\tilde{r}(s,a)$, transition probabilities $\tilde{p}(s'|s,a)$, and discount factor $\nu/(\gamma+\nu)$. In particular, several results about such discrete-time MDPs from Chap. 6 of [11] apply. One such result, rephrased from Theorem 11.5.2 of [11], is recalled below.

Theorem 1. The optimal value function $J^\star$ is the unique solution of the system of Bellman's implicit equations

$$V(s) = \max_{a \in \mathcal{A}_s} \left\{ \tilde{r}(s,a) + \frac{\nu}{\gamma+\nu} \sum_{s' \in \mathcal{S}} \tilde{p}(s'|s,a)V(s') \right\}, \ \forall s \in \mathcal{S}. \tag{5}$$

Actions that attain the maxima in (5) form an optimal stationary deterministic policy.

Consequently, the CTMDP can be solved by applying standard techniques such as value iteration, policy iteration, or linear programming, to this discrete-time MDP. Finally, since the absolute value of the reward earned upon each transition of the discrete-time MDP is at most B, the absolute expected total discounted reward earned over an infinite-horizon is bounded above by

$$B + \left(\frac{\nu}{\gamma+\nu}\right) B + \left(\frac{\nu}{\gamma+\nu}\right)^2 B + \ldots = \frac{B}{1 - \frac{\nu}{(\gamma+\nu)}} = \frac{(\gamma+\nu)B}{\gamma}.$$

Specifically, the optimal value function also satisfies $|J^\star(s)| \leq (\gamma+\nu)B/\gamma$, for all $s \in \mathcal{S}$. The process of converting the CTMDP into an equivalent discrete-time MDP is called uniformization (see Sects. 11.5.1, 11.5.2, and Fig. 11.5.3 in [11]).

3 Inverse Problem

In many applications, rewards are characterized by a finite set of parameters. For example, in inventory management problems, they are determined by the selling price per unit, ordering cost per unit, and inventory holding cost per unit. In queuing problems, costs are determined by the waiting cost per customer and the service cost per customer. To accommodate such applications, suppose rewards are characterized by a finite tuple of parameters $\theta = (\theta_1, \ldots, \theta_k) \in \mathbb{R}^k$. To make this explicit, we now denote lump sum rewards by $\rho_1(s,a;\theta)$ and reward rates between transitions by $\rho_2(s,a;\theta)$, so that (1) becomes

$$r(s,a;\theta) = \rho_1(s,a;\theta) + \frac{\rho_2(s,a;\theta)}{\gamma+\nu(s,a)}. \tag{6}$$

We use $\Theta \subset \mathbb{R}^k$ to denote the set of all possible values of parameters θ, and assume that Θ is compact. In applications, this could constitute restrictions such as non-negativity or lower and upper bounds on plausible parameter values as dictated by the physics of the problem. We also assume that, for each $(s,a) \in \mathcal{X}$, the functions $\rho_1(s,a;\theta)$ and $\rho_2(s,a;\theta)$ are continuous in θ over $\mathbb{R}^k$. The reward function $r(s,a;\theta)$ defined in (6) thus inherits this continuity property.

We now state the inverse problem as follows. An estimate of the reward parameters $\hat{\theta} \in \Theta$ is given. Positive weights $\beta_1, \ldots, \beta_k$ are given. A stationary deterministic policy $\hat{\pi}$ is also given. Parameters $\nu(s,a)$ of the sojourn time exponential distributions, transition probabilities $p(s'|s,a)$, and continuous-time discount factor γ are known. We wish to find reward parameters θ with the following properties.

1. Parameters θ must belong to the given set Θ;
2. the given policy $\hat{\pi}$ must be optimal to the CTMDP with rewards $r(s,a;\theta)$, for $(s,a) \in \mathcal{X}$; the given transition probabilities $p(s'|s,a)$, for $(s,a) \in \mathcal{X}$; the given continuous-time discount factor γ; and
3. the sum of weighted absolute differences $\sum\limits_{j=1}^{k} \beta_j |\theta_j - \hat{\theta}_j|$ must be as small as possible.

This inverse problem can be formulated as follows.

$$\text{(P)}\ f^\star \stackrel{\text{def}}{=} \inf_{\theta, V}\ f(\theta) \stackrel{\text{def}}{=} \sum_{j=1}^{k} \beta_j |\theta_j - \hat{\theta}_j| \tag{7}$$

$$V(s) = \tilde{r}(s, \hat{\pi}(s); \theta) + \frac{\nu}{\gamma + \nu} \sum_{s' \in \mathcal{S}} \tilde{p}(s'|s,a) V(s'),\ s \in \mathcal{S} \tag{8}$$

$$V(s) \geq \tilde{r}(s, a; \theta) + \frac{\nu}{\gamma + \nu} \sum_{s' \in \mathcal{S}} \tilde{p}(s'|s,a) V(s'),\ \{(s,a) \in \mathcal{X} | a \neq \hat{\pi}(s)\} \tag{9}$$

$$\theta \in \Theta \subset \mathbb{R}^k \tag{10}$$

$$V \in \mathbb{R}^n. \tag{11}$$

Constraints (8)–(9) ensure that $V \in \mathbb{R}^n$ satisfies Bellman's equations (5) and that actions $\hat{\pi}(s)$, for all $s \in \mathcal{S}$, attain the maxima therein. That is, these constraints imply that V is an optimal value function and policy $\hat{\pi}$ is optimal by Theorem 1. Let $B(\theta) \stackrel{\text{def}}{=} \max\limits_{(s,a) \in \mathcal{X}} |r(s,a;\theta)|$. Thus, for any particular choice of parameters θ, the corresponding optimal value function would satisfy $\frac{-(\gamma+\nu)B(\theta)}{\gamma} \leq V(s) \leq \frac{(\gamma+\nu)B(\theta)}{\gamma}$, for all $s \in \mathcal{S}$. Since $B(\theta)$ is the maximum of a finite number of continuous functions $r(s,a;\theta)$, it is continuous in θ over $\mathbb{R}^k$. Moreover, since Θ is a compact set, the function $B(\theta)$ attains its maximum over $\theta \in \Theta$. We denote this maximum by $B^\star$. Since parameters θ belong to the set Θ, we know that the optimal value function would satisfy $\frac{-(\gamma+\nu)B^\star}{\gamma} \leq V(s) \leq \frac{(\gamma+\nu)B^\star}{\gamma}$, for all $s \in \mathcal{S}$, regardless of the choice of $\theta \in \Theta$. Recall that Θ in constraint (10) is compact. This means that the set of $(\theta, V) \in \mathbb{R}^k \times \mathbb{R}^n$ pairs that are feasible to the inverse problem is compact. Since $r(s, a; \theta)$ is continuous in θ over $\mathbb{R}^k$ and constraints (8)–(9) are linear in $V \in \mathbb{R}^n$, the set of $(\theta, V) \in \mathbb{R}^k \times \mathbb{R}^n$ pairs that satisfy those constraints is closed. Thus, the feasible region is a closed subset of a compact set. It is therefore compact. We assume that the feasible region of (P) is not empty, to avoid a vacuous a problem statement. The objective function of (P) is continuous. The inverse problem therefore has an optimal solution and the "inf" in (7) can be changed to a "min".

The absolute value difference $|\theta_j - \hat{\theta}_j|$ in the objective function can be linearized by using a standard variable transformation [1]. The transformation works as follows. We first introduce non-negative variables x_{+j} and x_{-j}, for $j = 1, \ldots, k$. We then express the difference $\theta_j - \hat{\theta}_j$ of two real numbers as the difference $x_{+j} - x_{-j}$ of non-negative real numbers. Then the absolute value difference $|\theta_j - \hat{\theta}_j|$ in the objective function can be replaced with $x_{+j} + x_{-j}$ without loss of optimality. The reason behind this is as follows. At least one of x_{+j} and x_{-j} must be 0 in any optimal solution. For if both are positive, then the one with the smaller value can be reduced to 0 and the other can be reduced by an equal amount. This decreases the objective function without changing the difference $\theta_j - \hat{\theta}_j = x_{+j} - x_{-j}$. In particular, if $\theta_j - \hat{\theta}_j$ is positive, then x_{+j} is also positive but x_{-j} is 0. On the other hand, if $\theta_j - \hat{\theta}_j$ is negative, then x_{-j} is positive but x_{+j} is 0. As such, x_{+j} is called the positive part of the difference $\theta_j - \hat{\theta}_j$ and x_{-j} is called the negative part of the difference $\theta_j - \hat{\theta}_j$. This motivates the subscripts + and − in our notation. To express this transformation in tuple-format, we write $\theta - \hat{\theta} = x_+ - x_-$, where differences between tuples are interpreted componentwise. With this notation, we consider problem (Q) below instead of (P).

$$(\text{Q})\ f^\star \stackrel{\text{def}}{=} \min_{x_+, x_-, V} \ g(x_+, x_-) \stackrel{\text{def}}{=} \sum_{j=1}^{k} \beta_j (x_{+j} + x_{-j}) \tag{12}$$

$$V(s) = \tilde{r}(s, \tilde{\pi}(s); \hat{\theta} + x_+ - x_-) + \frac{\nu}{\gamma + \nu} \sum_{s' \in \mathcal{S}} \tilde{p}(s'|s, a) V(s'),\ s \in \mathcal{S} \tag{13}$$

$$V(s) \geq \tilde{r}(s, a; \hat{\theta} + x_+ - x_-) + \frac{\nu}{\gamma + \nu} \sum_{s' \in \mathcal{S}} \tilde{p}(s'|s, a) V(s'),\ \{(s, a) \in \mathcal{X} | a \neq \tilde{\pi}(s)\} \tag{14}$$

$$\hat{\theta} + x_+ - x_- \in \Theta \subset \mathbb{R}^k \tag{15}$$

$$x_+, x_- \in \mathbb{R}^k_+ \tag{16}$$

$$V \in \mathbb{R}^n. \tag{17}$$

The objective function in (Q) is linear in x_+, x_-. Constraints (13)–(14) are linear in V. Thus, the computational difficulty in solving problem (Q) may largely depend on (i) the algebraic characterization of the set Θ in constraint (15); and (ii) the form of rewards $\tilde{r}(s, \tilde{\pi}(s); \theta)$ as functions of θ. For instance, if the former is convex and the latter is linear, then problem (Q) will likely be tractable. We illustrate on two applications.

4 Applications

4.1 Batch Manufacturing

This application is adapted from Bertsekas [3]. A manufacturer processes orders in batches. Orders arrive according to a Poisson process with rate $\nu > 0$ per unit time. That is, successive arrival times are independent and exponentially distributed with parameter ν (and mean $1/\nu$ time-units). For each order, there is a cost $c > 0$ per unit time while it remains unfilled. A setup cost of $K > 0$ is incurred immediately upon filling orders (regardless of the size of the batch filled). Costs are discounted using a continuous-time rate of $\gamma > 0$. Upon arrival of a new order, the manufacturer has to decide whether to fill the accumulated batch of orders or to wait for the next order to arrive. The manufacturer is not allowed to accumulate more than $n - 1$ orders, for some integer $n > 1$. That is, whenever $n - 1$ orders are waiting, they must all be filled. The manufacturer's goal is to minimize the expected total discounted costs over an infinite horizon.

We use $\mathcal{S} = \{0, 1, \ldots, n-1\}$ as the state-space representing the accumulated batch size and $\mathcal{A} = \{0, 1\}$ as the action-space of this system. Here, action $a = 1$ corresponds to the decision to fill the accumulated orders, whereas $a = 0$ corresponds to the decision to wait. Therefore, $\mathcal{A}_0 = \{0\}$; $\mathcal{A}_{n-1} = \{1\}$; and $\mathcal{A}_s = \mathcal{A}$, for $s = 1, \ldots, n-2$. If action $a = 0$ is chosen in any state $s = 0, 1, \ldots, n-2$, then a new order arrives after an exponentially distributed amount of time with parameter ν and the system transitions into state $s + 1$ with probability $p(s+1|s, 0) = 1$. If action $a = 1$ is chosen in any state $s = 1, 2, \ldots, n-1$, then the batch is immediately processed, the system clears instantaneously, and a new order arrives after an exponentially distributed amount of time with parameter ν. The system thus transitions into state $s = 1$ with probability $p(1|s, 1) = 1$. Thus, this is a particularly simple system where exponential sojourn times are all identically distributed with parameter ν regardless of the state and action. That is, $\nu(s, a) = \nu$, for all $(s, a) \in \mathcal{X}$. Moreover, transitions are deterministic. We interpret negative rewards as costs. So we have the lump sum reward $\rho_1(s, a) = -K$ and the reward rate $\rho_2(s, a) = 0$ when $a = 1$. On the other hand, we have the lump sum reward $\rho_1(s, a) = 0$ and the reward rate $\rho_2(s, a) = -cs$ when $a = 0$.

Thus, applying (1), we obtain,

$$r(s, 0) = \frac{-cs}{\gamma + \nu}, \text{ for } s = 0, 1, \ldots, n-2, \text{ and} \tag{18}$$

$$r(s, 1) = -K, \text{ for } s = 1, \ldots, n-1. \tag{19}$$

This yields, from Remark 2, that

$$B = \max\left\{\frac{c(n-2)}{\gamma + \nu}, K\right\}. \tag{20}$$

Since exponential sojourn time distributions are identical across states and actions in this case, formulas (3)–(4) imply that $\tilde{r}(s, a) = r(s, a)$ and $\tilde{p}(s'|s, a) =$

$p(s'|s,a)$, for all $(s,a) \in \mathcal{X}$. Putting everything together, Theorem 1 yields that the optimal value function $J^\star$ is the unique solution of the following system of Bellman's equations:

$$V(0) = \frac{\nu}{\gamma+\nu}V(1), \tag{21}$$

$$V(s) = \max\left\{\underbrace{\frac{-cs}{\gamma+\nu} + \frac{\nu}{\gamma+\nu}V(s+1)}_{a=0}, \underbrace{-K + \frac{\nu}{\gamma+\nu}V(1)}_{a=1}\right\}, \text{ for } s = 1,2,\ldots,n-2, \tag{22}$$

$$V(n-1) = -K + \frac{\nu}{\gamma+\nu}V(1). \tag{23}$$

Actions that attain the maxima in (22) define an optimal stationary deterministic policy.

We now consider an inverse version of this problem. Recall that rewards in that problem are characterized by parameters $(c,K) \in \mathbb{R}^2$. Suppose the lower and upper bounds on possible values of these parameters are given by $\underline{c} > 0$, $\overline{c} > \underline{c}$, $\underline{K} > 0$, and $\overline{K} > \underline{K}$, respectively. Thus, we have, $\Theta \stackrel{\text{def}}{=} \{(c,K) \in \mathbb{R}^2 | \underline{c} \leq c \leq \overline{c},\ \underline{K} \leq K \leq \overline{K}\}$. This set is compact. It is easy to verify that functions $\rho_1(s,a;(c,K))$ and $\rho_2(s,a;(c,K))$ are continuous in $(c,K) \in \mathbb{R}^2$, for each fixed $(s,a) \in \mathcal{X}$. Thus, the reward function $r(s,a;(c,K))$ inherits this continuity.

In the inverse problem, we are given estimates $(\hat{c},\hat{K}) \in \Theta$ of the two parameters. We are given a particular stationary deterministic policy $\hat{\pi}$. Suppose this policy has an intuitive threshold structure. That is, we are given a positive integer $\hat{n}$, such that $1 \leq \hat{n} \leq n-1$. The policy $\hat{\pi}$ prescribes the manufacturer to process all cumulated orders if $s \geq \hat{n}$ and to wait if $s < \hat{n}$. We wish to find parameters $(c,K) \in \Theta$ that would render this threshold policy optimal, while minimizing $\beta_1|c-\hat{c}| + \beta_2|K-\hat{K}|$, where β_1, β_2 are given positive weights. This problem can be formulated as follows.

$$\text{(P1) } f^\star \stackrel{\text{def}}{=} \min_{(c,K),V} \ f(c,K) \stackrel{\text{def}}{=} \beta_1|c-\hat{c}| + \beta_2|K-\hat{K}| \tag{24}$$

$$V(0) = \frac{\nu}{\gamma+\nu}V(1) \tag{25}$$

$$V(s) = \frac{-cs}{\gamma+\nu} + \frac{\nu}{\gamma+\nu}V(s+1),\ s = 1,\ldots,\hat{n}-1 \tag{26}$$

$$V(s) = -K + \frac{\nu}{\gamma+\nu}V(1),\ s = \hat{n},\hat{n}+1,\ldots,n-2 \tag{27}$$

$$V(n-1) = -K + \frac{\nu}{\gamma+\nu}V(1) \tag{28}$$

$$V(s) \geq -K + \frac{\nu}{\gamma+\nu}V(1),\ s = 1,\ldots,\hat{n}-1 \tag{29}$$

$$V(s) \geq \frac{-cs}{\gamma+\nu} + \frac{\nu}{\gamma+\nu}V(s+1),\ s = \hat{n},\hat{n}+1,\ldots,n-2 \tag{30}$$

$$\underline{c} \leq c \leq \overline{c} \tag{31}$$

$$\underline{K} \leq K \leq \overline{K} \tag{32}$$

$$(c, K) \in \mathbb{R}^2 \tag{33}$$

$$V \in \mathbb{R}^n. \tag{34}$$

This problem is convex and, in fact, can be converted into a linear program using the variable transformation $c - \hat{c} = x_{1+} - x_{1-}$ and $K - \hat{K} = x_{2+} - x_{2-}$, where $x_+, x_- \in \mathbb{R}^2_+$, as described above. We omit further details about this reformulation here.

We solved both the forward and inverse problems using `cvxpy`, a Python-based modeling package for convex optimization [6]. We set the parameters $\nu = 1$, $\gamma = 0.1$, $c = 1$, and $n = 100$, and solved the forward problem for each integer-valued K between 10 and 200 to obtain a set of optimal thresholds n^*. Figure 1a plots the result for the forward problem using blue dots. As c is fixed, the figure shows that optimal thresholds are non-decreasing in K, meaning it is better to wait for more orders to accumulate in the batch before processing as K grows. For each unique value of n^*, we then solved the inverse problem to impute the cost K using estimates $\hat{c} = 1$, $\hat{K} = 100$, and $\hat{n} = n^*$. We forced $c = 1$ in the constraints and imposed that $0 \leq K \leq 200$. Figure 1a shows that the imputed cost for each threshold n^* is the closest point to $\hat{K}$ within the discretized set of costs for which n^* is optimal. This aligns with our objective function to minimize distance to $\hat{K}$. Next, we fixed $K = 100$ to perform sensitivity analysis for the waiting cost c, varying c between 1 and 20 in increments of 0.1, using $\hat{c} = 10$ as the estimate. Figure 1bshows similar behavior, where the imputed cost c is the closest feasible value to $\hat{c}$.

We were also interested in testing the inverse formulation when both c and K are relaxed. We solved the forward problem for each (c, K) pair where $1 \leq c \leq 4$ with discretization size 0.05, and $100 \leq K \leq 400$ with discretization size 5. Each optimal threshold obtained were input into the inverse problem, with estimates $\hat{c} = 2.5$ and $\hat{K} = 250$, and weights $\beta_1 = 1$ and $\beta_2 = 0.01$. Figure 2 shows the discretized space of optimal thresholds with K on the $x-$axis and c on the $y-$axis. Each imputed cost pair was plotted as a point at the corresponding

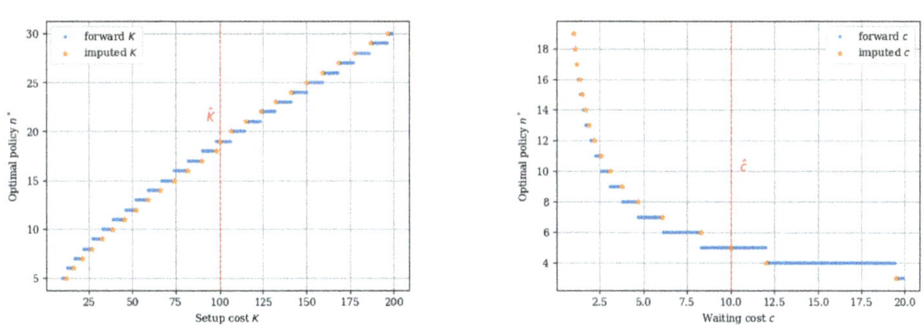

(a) Fixed $c = 1$ and $10 \leq K \leq 200$.

(b) Fixed $K = 100$ and $1 \leq c \leq 20$.

Fig. 1. Sensitivity analysis for optimal thresholds and imputed costs.

(c, K) coordinate. We added some contour lines to aid visualization. Along each contour of (c, K) pairs that yield the same policy, the imputed point lies close to the boundary of the feasible region and tends to match either $\hat{c}$ or $\hat{K}$ when it is possible to do so.

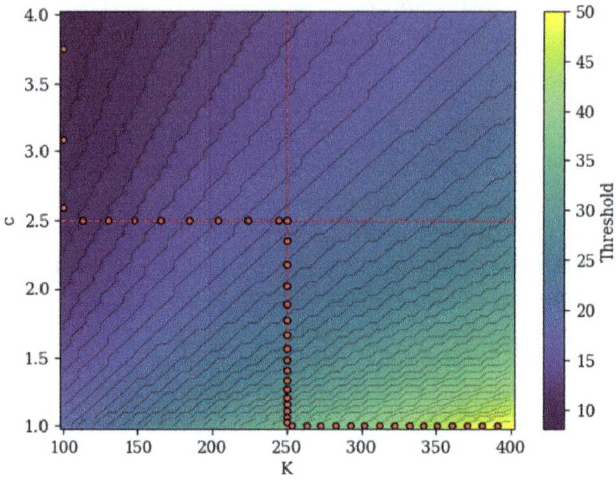

Fig. 2. Optimal thresholds n^* for (c, K) pairs between $(1, 100)$ and $(4, 400)$. The red dots show imputed (c, K) pairs for different thresholds.

4.2 Queue Routing

This application is adapted from Exercise 11.18 of Puterman [11]. In the forward problem (see Fig. 3), customers arrive according to a Poisson process at rate λ. Each arrival must be routed to one of two servers as long as there is room. Queues at each server have a capacity of at most $n-1$ customers. An arriving customer must be rejected at no cost if both queues are at capacity. Service durations are exponential with rates μ_1, μ_2, and costs c_1, c_2 are incurred per customer waiting per unit time, in the two queues. Costs are discounted at a continuous-time rate of $\gamma > 0$. The goal is to minimize the expected total discounted cost over an infinite horizon.

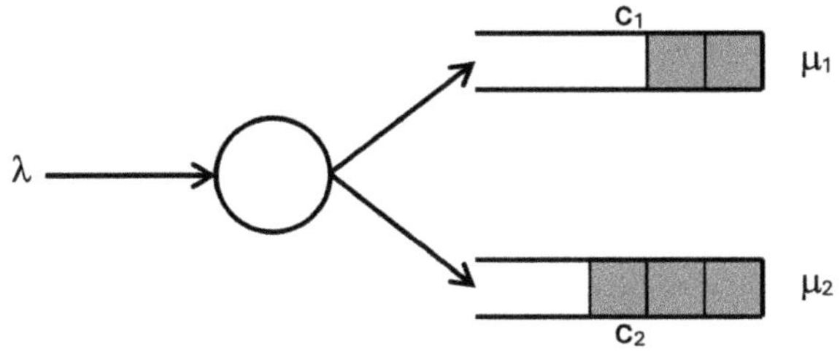

Fig. 3. A queue routing application.

We represent states as triples (b, s_1, s_2), where b indicates whether there's a new arrival to the system, and s_1, s_2 are the numbers of jobs in the two queues. Thus, the state space is $\mathcal{S} = \{(b, s_1, s_2) \mid b \in \{0, 1\}, s_1, s_2 \leq n-1 \text{ and } s_1, s_2 \in \mathbb{N}\}$. The set of all possible actions is $\mathcal{A} = \{1, 2\}$, where action $a = 1$ means route to server 1 and $a = 2$ means route to server 2. Both actions are available in all states with the following exceptions: $\mathcal{A}_{(0,s_1,s_2)} = \varnothing$, $\mathcal{A}_{(1,n-1,n-1)} = \varnothing$, $\mathcal{A}_{(1,n-1,s_2)} = \{2\}$, $\mathcal{A}_{(1,s_1,n-1)} = \{1\}$. Broadly speaking, the transition probabilities capture the natural transitions of this queuing system as an "exponential race" between three processes with rates μ_1, μ_2, and λ. At each point in time, one of three things could happen: server 1 completes service first, server 2 completes service first, or a new job arrives before either server finishes service. This race ends after an exponentially distributed amount of time with rate $\nu(s, a)$. Specifically, $\nu(s, a) \leq \nu = \lambda + \mu_1 + \mu_2$ for all s, a. Thus, we can apply uniformization as described in Sect. 3.

By (4), the uniformized transition probabilities are

$$\tilde{p}(s'|s,a) = \begin{cases} \dfrac{\mu_1}{\lambda+\mu_1+\mu_2}, & s' = (0,(s_1-1)^+,s_2),\ s=(0,s_1,s_2),\ a=0, \text{ or} \\ & s' = (0,s_1,s_2),\ s=(1,s_1,s_2),\ a=1, \text{ or} \\ & s' = (0,(s_1-1)^+,\max\{s_2+1,n-1\}),\ s=(1,s_1,s_2),\ a=2. \\ \dfrac{\mu_2}{\lambda+\mu_1+\mu_2}, & s' = (0,s_1,(s_2-1)^+),\ s=(0,s_1,s_2),\ a=0, \text{ or} \\ & s' = (0,\max\{s_1+1,n-1\},(s_2-1)^+),\ s=(1,s_1,s_2),\ a=1, \text{ or} \\ & s' = (0,s_1,s_2),\ s=(1,s_1,s_2),\ a=2. \\ \dfrac{\lambda}{\lambda+\mu_1+\mu_2}, & s' = (1,s_1,s_2),\ s=(0,s_1,s_2),\ a=0, \text{ or} \\ & s' = (1,\max\{s_1+1,n-1\},s_2),\ s=(1,s_1,s_2),\ a=1, \text{ or} \\ & s' = (1,s_1,\max\{s_2+1,n-1\}),\ s=(1,s_1,s_2),\ a=2. \end{cases}$$

The continuous-time reward rate is linear in the queue size immediately following a routing decision. Since there are no lump-sum rewards, the uniformized reward functions are given by

$$\tilde{r}((1, s_1, s_2), 1) = \frac{-c_1(s_1 + 1) - c_2 s_2}{\gamma + \lambda + \mu_1 + \mu_2} \quad \text{for } 0 \le s_1 < n - 1, 0 \le s_2 \le n - 1,$$
$$\tilde{r}((1, s_1, s_2), 2) = \frac{-c_1 s_1 - c_2(s_2 + 1)}{\gamma + \lambda + \mu_1 + \mu_2} \quad \text{for } 0 \le s_1 \le n - 1, 0 \le s_2 < n - 1,$$
$$\tilde{r}((0, s_1, s_2), \varnothing) = \frac{-c_1 s_1 - c_2 s_2}{\gamma + \lambda + \mu_1 + \mu_2} \quad \text{for } 0 \le s_1, s_2 \le n - 1,$$
$$\tilde{r}((1, n - 1, n - 1), \varnothing) = \frac{-c_1(n - 1) - c_2(n - 1)}{\gamma + \lambda + \mu_1 + \mu_2}.$$

For brevity, we do not include the system of Bellman's implicit equations.

We used $n = 20$, $\gamma = 0.5$, $\mu_1 = \mu_2 = 1$, $\lambda = 1$, $c_1 = 1$ and $c_2 \in \{1, 1.005, 1.05, 1.4\}$ to solve four forward problem instances. A monotone structure of the corresponding four optimal policies is shown in Fig. 4. For each fixed s_1 and increasing s_2, once the policy switches from one queue to another, it should stay the same for all larger values of s_2—similarly for each fixed s_2 and increasing s_1. We see that with service rates being equal, an increase in c_2 results in the policy strongly favoring server 1, only choosing server 2 when s_1 is much larger than s_2.

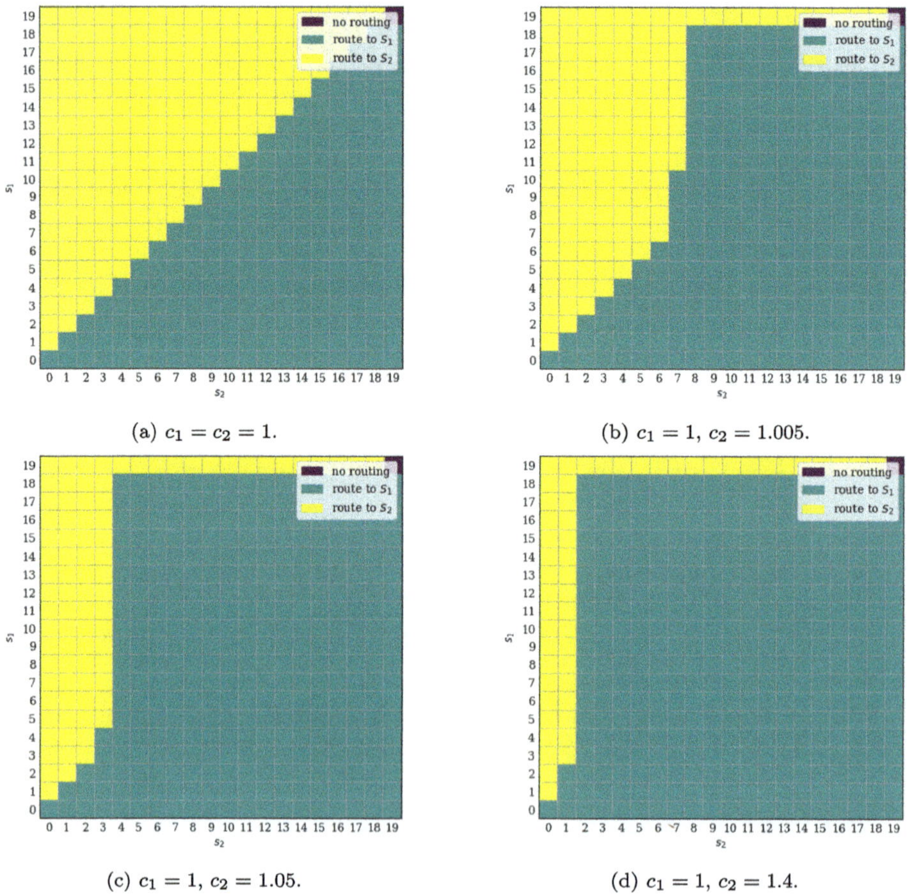

(a) $c_1 = c_2 = 1$.

(b) $c_1 = 1$, $c_2 = 1.005$.

(c) $c_1 = 1$, $c_2 = 1.05$.

(d) $c_1 = 1$, $c_2 = 1.4$.

Fig. 4. Optimal routing policies for fixed c_1 and different values of c_2.

Our inverse problem is to find c_1, c_2 that would render a given routing policy $\hat{\pi}$ optimal. The set $\Theta \in \mathbb{R}^2$ requires these parameters to belong to given intervals $[c_{1L}, c_{1H}]$ and $[c_{2L}, c_{2H}]$. We used the four optimal policies obtained from our forward problems as input $\hat{\pi}$ to the inverse problem. For the inverse problem, we used $\hat{c}_1 = 1$ and $\hat{c}_2 = 0.5$ as estimates, and in fact forced $c_1 = 1$. We let $0.5 \leq c_2 \leq 1.5$. The imputed costs c_2 were 1, 1.0046, 1.049, and 1.267, respectively, for the four policies $\hat{\pi}$. Re-solving with a larger estimate, $\hat{c}_2 = 1.5$, led to imputed c_2 values of 1, 1.005, 1.071 and 1.405. We plugged the imputed values back into the forward problem and verified that they indeed still result in the same policy.

5 Conclusions and Future Work

We presented an inverse optimization method for imputing reward-defining parameters in finite-state CTMDPs. We first applied uniformization to write Bellman's equations of optimality for the forward problem, and then utilized them to formulate the inverse problem. We identified when this problem would be convex or even linear (after applying a standard variable transformation). We illustrated this methodology on two classic applications in service science. Our computational results suggest that in both applications, the inverse formulation can find reward parameters closest to the given estimates.

Inverse optimization comes up in two contexts: system identification and response assurance. System identification problems come up in natural systems whose parameters are difficult to estimate. A classic example is the problem of estimating geophysical properties of a terrain based on observed paths of seismic waves. Response assurance problems arise in the design of human-made systems. A familiar example is that of finding tolls that would induce a particular traffic pattern. Our work can be applied to response assurance problems where the underlying system is modeled as a CTMDP. Results of inverse optimization may be utilized interactively within a feedback loop with a domain expert. For instance, if a domain expert is implementing a particular policy, but does not believe the optimal parameter values it induces via inverse optimization, then this provides the domain expert an opportunity to interactively re-examine both the quality of their policy and the appropriateness of their initial parameter estimates.

It would be interesting to extend this work to countable-state CTMDPs and semi-Markov problems.

Acknowledgments. This research was funded in part by the United States National Science Foundation grant CMMI #2449675.

References

1. Ahuja, R.K., Orlin, J.B.: Inverse optimization part I: linear programming and general problem. Oper. Res. **49**(5), 771–783 (2001)
2. Ahuja, R.K., Orlin, J.B.: Combinatorial algorithms for inverse network flow problems. Networks **40**(4), 181–187 (2002)
3. Bertsekas, D.P.: Dynamic Programming and Optimal Control, 4th edn. Athena Scientific (2017)
4. Burton, D., Toint, P.L.: On an instance of inverse shortest path problems. Math. Program. **53**(1–3), 45–61 (1992)
5. Chan, T.C.Y., Mahmood, R., Zhu, I.Y.: Inverse optimization: theory and applications. Oper. Res. **73**(2), 1046–1074 (2025)
6. Diamond, S., Boyd, S.: CVXPY: a python-embedded modeling language for convex optimization. J. Mach. Learn. Res. **17**(83), 1–5 (2016)
7. Erkin, Z., Bailey, M.D., Maillart, L.M., Schaefer, A.J., Roberts, M.S.: Eliciting patient's revealed preferences: an inverse Markov decision process approach. Decis. Anal. **7**(4), 358–365 (2010)
8. Ghatrani, Z., Ghate, A.: Inverse Markov decision processes with unknown transition probabilities. IISE Trans. **55**(6), 588–601 (2023)
9. Ghatrani, Z., Ghate, A.: Inverse optimization in semi-definite programs to impute unknown constraint matrices. Comput. Oper. Res. **168**, 106681 (2024)
10. Iyengar, G., Kang, W.: Inverse conic programming with applications. Oper. Res. Lett. **33**(3), 319–330 (2005)
11. Puterman, M.L.: Markov Decision Processes: Discrete Stochastic Dynamic Programming. Wiley (2005)
12. Schaefer, A.J.: Inverse integer programming. Optim. Lett. **3**(4), 483–489 (2009)

Predicting Instability at Home and in Foster Care, Challenges and Opportunities

S. Ayca Erdogan(✉), Nafiseh Saberi, Afreen Chaus, and Egemen Ilkimen

Industrial and Systems Engineering Department, San Jose State University, San Jose, CA 95126, USA
ayca.erdogan@sjsu.edu

Abstract. The primary goal of foster care systems is to help foster children achieve the permanency stage through reunification with their family, adoption, or another suitable arrangement in the shortest possible time. Accomplishing this goal is challenging, both in terms of finding the best permanent option for foster children and providing them with a healthy, safe, and stable environment during their temporary foster care episode. Sometimes permanency planning decisions do not work as intended, leading to additional removals from home. Moreover, some children experience multiple placement settings during their out-of-home care. This paper provides an overview of the foster care ecosystem and its challenges, and how digital transformation can improve system performance to benefit its stakeholders and society. This paper also presents prediction models to examine factors associated with foster children's high number of removals from home, as well high number of placement settings during their current foster care episode. The analysis indicates that the child's age, race, ethnicity, clinical diagnoses, history of adoption, adoption age, circumstances associated with the child's removal, caretaker family structure, and location based on state and rural-urban category have a relationship with the child's placement instability in and out of care.

Keywords: Foster care · placement instability · child welfare · machine learning

1 Introduction

Foster care is a temporary service provided for children who are removed from their biological families due to abuse, neglect or other safety concerns that made children's environment unsafe to live in. Main goal is to find the best permanent homes for foster children as soon as possible. For the majority of cases, the goal is reunifying children with their parents after the prompted removal reasons are rectified, or placing them with relatives for adoption purposes in case of parental rights termination. There are other options to achieve permanency like being adopted with nonrelatives, emancipation, and guardianship when kinship adoption is not possible. Despite efforts to achieve the desirable plans in the shortest possible time, a significant number of children spend long periods in care awaiting adoption or another permanent arrangement.

Foster care is a network that consists of various components and stakeholders. The ecosystem has legal, social, financial, health and educational components as pictured

X. Xie et al. (Eds.): ICSS 2025, LNOR, pp. 15–25, 2026.
https://doi.org/10.1007/978-3-032-13116-4_2

in Fig. 1. Each component and stakeholder need to communicate and collaborate with others in order to follow through legal mandates and policies to keep the children safe. The system faces numerous complex challenges that impact the well-being of children and the effectiveness of the system itself. One major issue is placement instability, where children are frequently moved between homes due to mismatches with foster families or lack of proper training and support. This is compounded by a shortage of foster families, particularly for teens, sibling groups, and children with special needs, often due to caregiver burnout and inadequate support. Caseworkers are typically overburdened, managing high caseloads that lead to delays in services and high staff turnover, which disrupts continuity of care. Many foster children also struggle with unmet mental health needs stemming from trauma, abuse, or neglect, and without adequate therapeutic resources, behavioral issues often go unaddressed. Those who age out of the system often do so without life skills or support, facing higher risks of homelessness, unemployment, and incarceration. Systemic inequities further complicate outcomes, as children of color and marginalized youth are disproportionately represented and may face discrimination or instability. Efforts toward family reunification are hindered by under-resourced birth families and sluggish service delivery, while legal and bureaucratic delays, including court backlogs and policy inconsistencies, slow down the process of achieving stable, permanent placements. Together, these challenges highlight the need for comprehensive reform and greater investment in support systems for both children and caregivers.

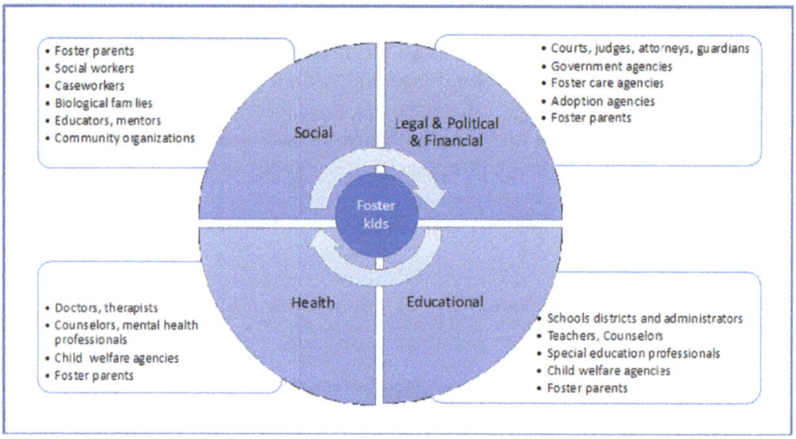

Fig. 1. Foster care ecosystem and stakeholders

Analytics and digitalized systems can play a crucial role in addressing many of the challenges within the foster care system by enabling smarter, more efficient, and data-driven approaches. Predictive analytics can help identify families at risk and allow for early interventions, potentially preventing the need for removal in the first place. Digital platforms can improve placement stability by matching children with foster families based on compatibility factors like experience, cultural background, and specific needs. These systems also assist in managing caseworker caseloads, reducing burnout, and ensuring timely responses through real-time alerts and task tracking. Integrated data systems can enhance coordination among healthcare, education, and social services, ensuring that children's mental health and support needs are met more holistically. By tracking outcomes over time, agencies can measure program effectiveness and allocate resources more strategically. Digital tools also increase transparency and accountability across the system, while analytics can help identify youth at high risk when aging out of care, enabling targeted support in areas such as housing, employment, and life skills. Overall, service science principles and methodologies, along with these technologies offer the potential to create a more proactive, responsive, and human-centered foster care system to achieve value co-creation.

This paper presents an application case on utilizing the data, and analytics techniques to gain insights on two types of instability for children in foster care. First, repeated removals from home to be placed in foster care, second, changes in placement settings while in foster care. This paper presents statistical techniques and machine learning to process a nationwide data and build prediction models for instability. The paper is organized as follows: We provide brief information on instability types in Sect. 2, and relevant literature in Sect. 3. Section 4 presents exploratory analysis of data and initial findings. Section 5 summarizes data pre-processing and the methodology used for model building, and finally, Sect. 6 presents the results and future directions.

2 Multiple Removals from Home and Instability While in Foster Care System

While children are in out-of-home care, they can be placed in different settings such as a relative or nonrelative foster home, institution, group home, pre-adoptive home, or supervised independent setting (Fig. 2). In most cases, children who stay longer in care have to face multiple placement settings. Although these changes are not considered removals since they take place inside the foster care system, they still provoke instabilities in a child's life and impact it in similar ways as removals.

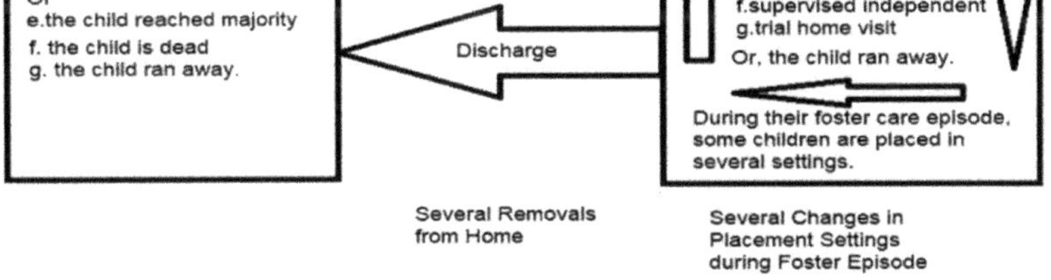

Fig. 2. Instability in permanent home and during foster care placement

Being removed from home even once is a traumatic and frustrating experience for a child and occurring frequent removals has detrimental effects on the child's emotional and social life [1]. Besides, chronic instability in children's placement can arouse severe toxic stress which develops risks of numerous health and social problems [2]. Both types of instabilities are illustrated in Fig. 2. Placement unsteadiness has a deleterious impact on children's safety, permanency, and well-being [3]. Safety is shaken up as a result of maltreatment in a resource home; permanency is postponed due to numerous placements and well-being is influenced in several ways such as mental, behavioral, and cognitive issues. Child development research indicates that children need consistency, predictability, and attachment to a caring adult to develop and succeed. Children in foster care who already have experienced removal trauma, especially need more attention to transition in and out of care. Children in foster care are more likely to develop behavioral problems and additional trauma from multiple placements will increase the risk of these issues [4]. Frequent placements have also been found to prompt deferred permanency outcomes, academic desperate straits, and difficulties to develop meaningful attachments [5].

Placement stability remains to be one of the most significant challenges for foster care and child protection agencies. In this paper, factors related to both types of instability including frequent removals from permanent homes and several temporary placement settings will be analysed. Children experience more placement settings the longer they stay in foster care. There are other factors like child individual characteristics, removal reasons, family structure, foster parent characteristics, and organizational elements that are associated with permanent and placement stability.

3 Literature Review

Various studies record the detrimental consequences of placement instability for foster children from the psychological and health perspectives [1–3, 5]. However, meticulous research on this problematic issue is not sought after as other child welfare outcomes. Furthermore, there is a lack of consistency among the existing research on the placement instability definition. Konijn et al., in their ten multilevel meta-analyses to examine factors that can affect instability of foster care placement included 42 studies, indicate that child behavioral problems, kinship care, and quality parenting are of medium significant effects [6]. They also, reported when children are fostered at a later age, and have experienced maltreatment in their birth family, the risk of instability of a foster care placement increases with more complicated situation when the foster parents are not relatives of the child. According to Connell et al., in their longitudinal study on examining child and case characteristics associated with changes in placement among Rhode Island children in foster care, infants change placement least, and risk increases with child age [7]. They also, stated emergency shelter settings have the highest risk of placement change, followed by nonrelative settings, group home settings, and relative foster care. They added the reasons for removal from the home, the history of previous placements, the interactions between foster care setting and some child characteristics to predict placement changes.

Sattler et al., in their study on the age-specific factors associated with placement instability among foster children in Texas, concluded that the effects of different risk factors varied by placement end reason and across developmental stages [8]. They indicated kinship placement is associated with an increased risk of substandard care disruption. They also, suggested that placements with black children were more likely to end due to placement mismatch, and adolescence age group to be always associated with the greatest increase in risk. O'Neill et al., in their study on the placement stability in the context of child development indicated that behaviour of the child, caretaker experience and age affected placement stability [9]. They also, reported child characteristics to predict placement stability for middle childhood group compared to the early childhood group who do not affected placement stability as much as their caregivers.

This paper provides analysis on instability in and out of care using a nationwide data, and presents machine learning models for predicting instability.

4 Data Source

The data used in this paper (Data set 257; Data Set Title AFCARS Foster Care File, 6-month periods (FY2016A - 2021 A) was obtained from the National Data Archive on Child Abuse and Neglect (NDACAN) and have been used in accordance with its terms of Use Agreement license [10]. The Administration on Children, Youth, and Families, the Children's Bureau, the original data set collection personnel or funding sources, NDACAN, and Cornell University and their agents or employees bear no responsibility for the analyses or interpretations presented in this report. AFCARS Foster Care, 6-month periods data files have entries for each child entered into and continues to be in care.

4.1 Exploratory Data Analysis

Initial explorative data analysis shows that on average in each semi-annual reporting period about 33% of the children in care experience more than two placement settings and about 14% experience 5 or more placement settings in their current foster care episode. Furthermore, on average more than 4% of the children in each reporting period are removed more than 2 times and 0.25% are removed more than 4 times from their home. Although these percentages might seem negligible, the consequences of not resolving this issue are overwhelming. Factors associated with the following three items are studied in this paper.

- High (3 or more times) number of removals from home for foster children
- High (5 or more) number of placement settings in children's current foster care episode

The focus in this report for temporary changes during the foster care episode is on 3 or more removals from home, and 5 or more placement settings in foster care. By comparing each period with the average of all periods, considering the trends, and the desire to study the most recent information to include as much historical data as possible, period FY2020A is chosen as the best representative of the whole data set. Another important reason to choose year 2020 data was the fact that 2020 data is the most recent data set before observing the impacts of COVID-19 pandemic. Period FY2020A started on October first, 2019, and ended on March 31st, 2020. Periods FY2020B and FY2021A might be influenced by COVID-19 shelter-in-place orders in most states and should be analysed considering the impacts of seeking health-related safety actions during those periods. One should note that, AFCARS 6 months data set includes records for each child when there is a periodic review or an update on the case of the child such as placement change. While there are variables such as demographic information of the child which does not change form record to record, some variables get updated such as number of removals, or total number of placement settings.

Figure 3 shows plots from initial exploratory analyses of data set FY2020A. According to data, the children who were removed from home more than 3 times were prominently from California, Florida and Ohio. While most of the removal reasons for these children included parental related causes such as neglect, drug and alcohol abuse, and inability to cope; child behaviour problems were also among the frequent causes of removals from home. Among highly removed children with clinical diagnosis, disability and emotional disturbance were leading cases.

Figure 4, shows the distribution of number of placement settings for the children in foster care. While a great majority of the children have less than 4 placement settings, there exist cases with a large number of placement setting which indicates an alarming level of instability. Data analysis shows that, after being removed, most of the children are placed with nonrelative or relative foster homes. About 41% of children are placed in nonrelative foster homes regardless of their number of placement settings. However, compared to all kids in foster care, the percentage of kids placed with kins decreases from 30% to 11% for children with higher instability, and percentage group home, institution, and supervised independent living increases. As the age distribution of children shows higher ages with high number of placements, placements other than relative foster homes can be expected for older kids.

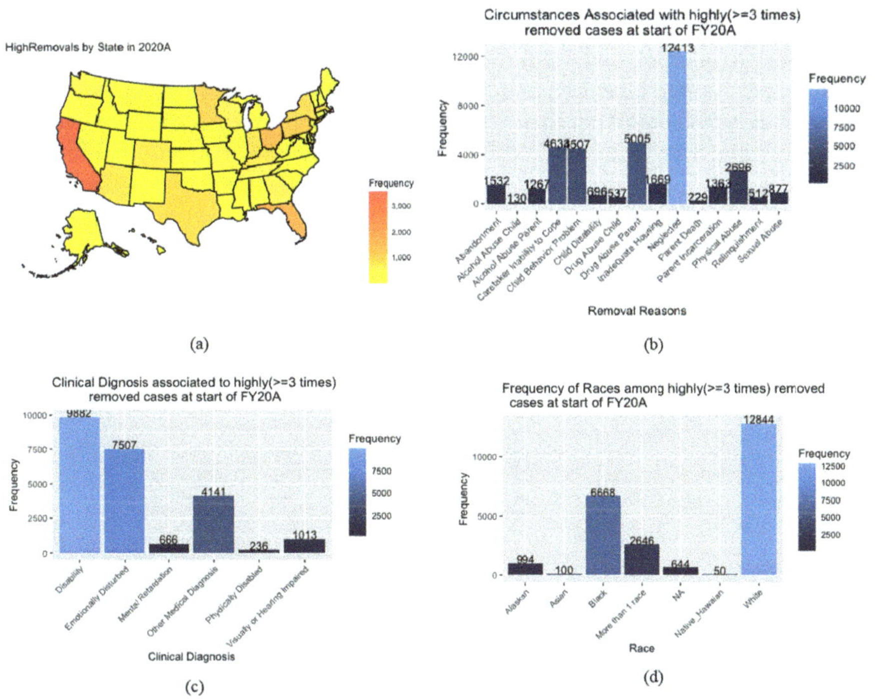

Fig. 3. Exploratory data analysis of children highly (> = 3 times) removed from home (a) High removals by state, (b) High removal causes, (c) Clinical diagnosis of highly removed children (d) High removals by race

4.2 Data Preprocessing

The data is cleaned and prepared for building machine learning models to predict high removals from home, and high number of placements in foster care. Two derived binary response variables for high removal and high number of placement indicators are added to the data set. For reduction of features, variables with more than 50% missing data are eliminated. In addition, eight variables related to funding resources, eight variables regarding organizational dates, and eleven variables that are derived variables are removed. Pearson correlation is used to eliminate highly correlated numerical variables, and Chi-square test is used to evaluate categorical variables' relationship with the two response variables separately to remove variables with no relationship to the response variables. Nine categorical variables are removed from the data set for predicting high removals, and no categorical variables are removed for predicting high number of placements. Eighty percent of the data is used for training, and 20% for testing accuracies of the models.

5 Methodology

We used two machine learning algorithms to train the predictive models for foster children instability. Logistic regression is a statistical model that predicts the probability of a binary outcome (such as yes/no or success/failure) based on one or more input variables.

It assumes a linear relationship between the inputs and the log-odds of the outcome, making it easy to interpret and useful when the relationship is relatively straightforward. Random forests, on the other hand, are an ensemble learning method that builds multiple decision trees and combines their results to improve accuracy and reduce overfitting. Each tree is trained on a random subset of the data and features, and their combined predictions lead to more robust and accurate classifications. While logistic regression offers simplicity and interpretability, random forests tend to perform better on complex datasets with nonlinear patterns or interactions between variables.

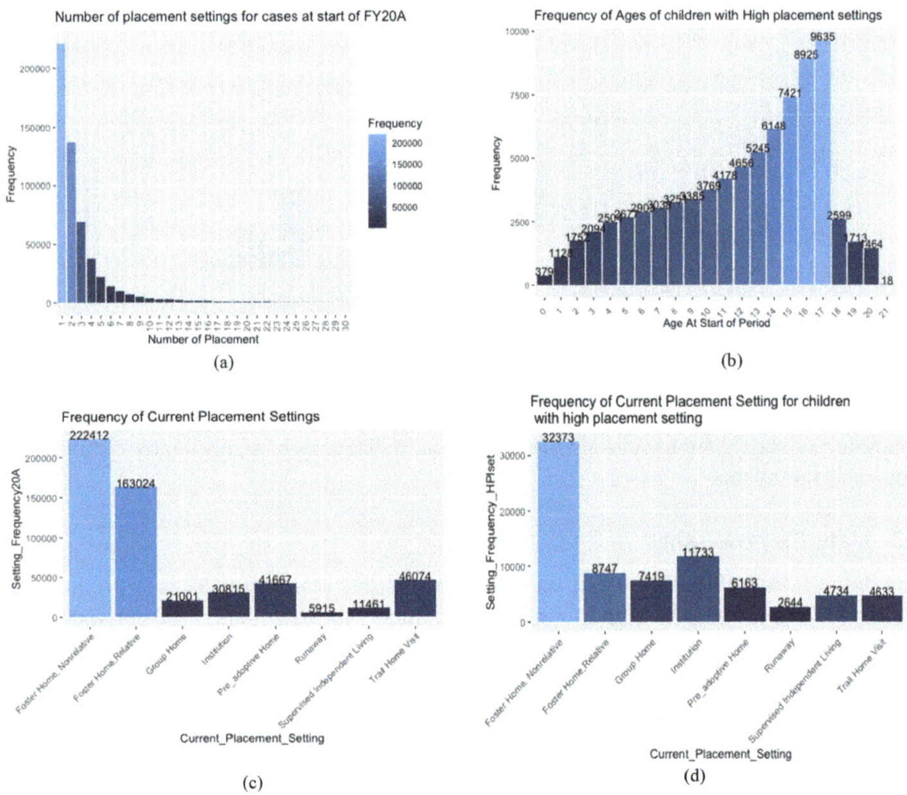

Fig. 4. Exploratory Analysis of foster children with high number of placement settings (> = 5) (a) Number of placement settings in foster case, (b) Age at start of period, (c) Placement settings in foster care (all children), (d) Placement settings in foster care (with high number of placement settings)

R programming language is used for data analyses, visualization, and machine learning [11]. Synthetic minority oversampling technique (SMOTE) is used for addressing the class imbalance problem. Instead of simply duplicating existing minority class examples, SMOTE generates **synthetic samples** by interpolating between existing minority class instances and their nearest neighbours. This helps create a more balanced dataset without overfitting, as it introduces slight variations in the new data points. By expanding the representation of the minority class, SMOTE allows machine learning models to better learn patterns associated with the underrepresented group, leading to improved predictive performance and fairness, especially in scenarios where imbalanced data could cause models to be biased toward the majority class.

6 Results and Conclusion

After data cleaning and feature reduction, 240,112 observations and 59 variables are used in the analyses. Results from four machine learning models, using logistic regression and random forest algorithms for two response variables, high removals, and high placements are presented in Table 1. These prediction models are first trained using the original data set which was imbalanced, and then new models are built after applying SMOTE and balancing the data. Accuracy, precision, recall and F1 scores of all models are presented in Table 1. Due to the importance of detecting high risk children, a low threshold probability of 0.04 is used for logistic regression models. Based on the results, the models for predicting both children with high number of removals, and predicting children with high number of foster placements have high accuracy but demonstrated low recall using the imbalanced data. After applying SMOTE, recall rates as well as F1 scores improved (Table 2).

Table 1. Performance of prediction models with imbalanced data set

		Accuracy	Precision	Recall	F1 Score
High Removals	Logistic Regression	95.79%	50.50%	24.04%	0.33
	Random Forests	96.25%	75.27%	16.80%	0.26
High Placements	Logistic Regression	79.34%	33.37%	76.22%	0.46
	Random Forests	90.78%	70.21%	37.33%	0.49

Table 2. Performance of prediction models with balanced data set

		Accuracy	Precision	Recall	F1 Score
High Removals	Logistic Regression	88.89%	90.00%	86.00%	0.88
	Random Forests	98.09%	99.61%	96.35%	0.98
High Placements	Logistic Regression	75.72%	91.30%	74.02%	0.82
	Random Forests	97.30%	99.22%	97.09%	0.98

Figure 5 shows the variable importance plots for predicting high removals, and high placements based on random forest algorithms, respectively. According to these plots, age of first foster care taker, removal date, age at last removal, length since last removal, neglect, parent's drug abuse, race, location based on state and rural/urban category are among important variables to predict high removals, whereas original caretaker age, current placement setting, and case goal (such as reunification, long term foster care, adoption) are additional important variables for predicting high placement settings during foster care. In general, mostly parent related factors such as abuse, neglect, or having a single male parent affect high removals, whereas both foster parent and child related factors such as behaviour problems, clinical diagnosis, and being male affect high number of placements in foster care. In addition, kids placed with relatives, and pre-adoptive families are less likely to have high number of placements compared to kids that are placed with non-relatives or at institutions. Results of such machine learning models can inform and support decision making to implement early intervention protocols such as mental health and housing assistance to support families before removal of high-risk kids. On the other hand, identifying high-risk kids for instability within foster care can inform individualized intervention plans, motivate for better matching systems, and provide high-quality tailored training for foster parents.

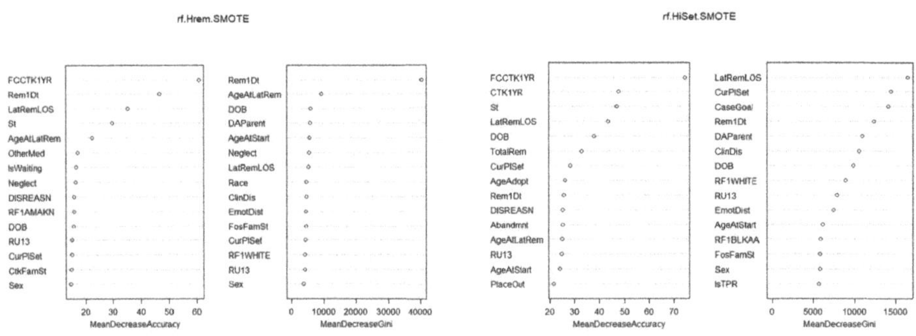

(a) (b)

Fig. 5. Variable important plots from random forest models for predicting (a) high removals (b) high number of placements

This study highlights the pressing issue of placement instability within the U.S. foster care system, emphasizing the dual challenges of repeated home removals and high numbers of placement settings during care episodes. Using nationwide data and machine learning techniques, we identified key factors—such as child age, race, clinical diagnoses, removal reasons, and caretaker characteristics—that correlate with these instability patterns. Our predictive models, particularly after balancing class distributions with SMOTE, demonstrate strong potential in identifying children at elevated risk for instability. These findings not only offer a data-driven approach to understanding foster care dynamics but also provide a foundation for developing targeted interventions and policies.

Moving forward, integrating predictive analytics into child welfare systems could support more informed decision-making, timely interventions, and ultimately, better permanency outcomes for vulnerable children. Improving placement matching in foster care requires high-quality, comprehensive, and timely data—but several persistent challenges hinder this goal. First, **data fragmentation** across agencies (e.g., child welfare, healthcare, education, juvenile justice) makes it difficult to form a holistic view of a child's needs. These silos limit coordination and lead to missed opportunities for matching children with the most suitable caregivers. Additionally, **inconsistent data entry practices** and **missing or outdated information**—such as incomplete health records, behavioral assessments, or caregiver profiles—can result in suboptimal placement decisions.

Acknowledgements. The data used in this publication were made available by the National Data Archive on Child Abuse and Neglect and have been used with permission. Data from the Adoption and Foster Care Analysis and Reporting System (AFCARS) were originally reported to the Children's Bureau. Funding for the project was provided by the Children's Bureau, Administration on Children, Youth and Families, Administration for Children and Families, U.S. Department of Health and Human Services. The receiver of the original data, the funder, the Archive, Duke University, Cornell University and their agents or employees bear no responsibility for the analyses or interpretations presented here.

References

1. Lawrence, C.R., Carlson, E.A., Egeland, B.: The impact of foster care on development. Dev. Psychopathol. **18**(01) (2006). https://doi.org/10.1017/s0954579406060044
2. Fisher, P.A., Mannering, A.M., Van Scoyoc, A., Graham, A.M.: A translational neuroscience perspective on the importance of reducing placement instability among foster children. Child Welfare. **92**(5), 9 (2013)
3. Rubin, D.M., O'Reilly, A.L.R., Luan, X., Localio, A.R.: The impact of placement stability on behavioral well-being for children in foster care. Pediatrics. **119**(2), 336–344 (2007). https://doi.org/10.1542/peds.2006-1995
4. Jacobsen, H., Bergsund, H.B., Wentzel-Larsen, T., Smith, L., Moe, V.: Foster children are at risk for developing problems in social-emotional functioning: a follow-up study at 8 years of age. Child Youth Serv. Rev. **108**, 104603 (2020). https://doi.org/10.1016/j.childyouth.2019.104603
5. Burns, D., Espinoza, D., Adams, J., Ondrasek, N.: California's Students in Foster Care: Challenges and Promising Practices. Learning Policy Institute (2022)
6. Konijn, C., et al.: Foster care placement instability: a meta-analytic review. Child Youth Serv. Rev. **96**(96), 483–499 (2019). https://doi.org/10.1016/j.childyouth.2018.12.002
7. Connell, C.M., Vanderploeg, J.J., Flaspohler, P., Katz, K.H., Saunders, L., Tebes, J.: Changes in placement among children in foster care: a longitudinal study of child and case influences. Soc. Serv. Rev. **80**(3), 398–418 (2006). https://doi.org/10.1086/505554
8. Sattler, K.M.P., Font, S.A., Gershoff, E.T.: Age-specific risk factors associated with placement instability among foster children. Child Abuse Negl. **84**, 157–169 (2018). https://doi.org/10.1016/j.chiabu.2018.07.024
9. O'Neill, M., Risley-Curtiss, C., Ayón, C., Williams, L.R.: Placement stability in the context of child development. Child Youth Serv. Rev. **34**(7), 1251–1258 (2012). https://doi.org/10.1016/j.childyouth.2012.02.018
10. Children's Bureau Administration on Children, Youth and Families.: AFCARS Foster Care File, 6-month periods (FY2016A - 2022B) [Data set]. National Data Archive on Child Abuse and Neglect (NDACAN). https://doi.org/10.34681/yjxr-zz92 (2023)
11. CRAN.: The Comprehensive R Archive Network. R-project.org, https://cran.r-project.org/ (2019)

Shaping the Future of Work in the AI Era: Business Model Innovation in Digitalisation for Decent Work and Economic Growth

Chen Ye[1] and Dequn Teng[1,2](✉)

[1] Institution for Manufacturing, University of Cambridge, Cambridge CB3 0FS, UK
{cy333,dt517}@cam.ac.com
[2] Cambridge Judge Business School, University of Cambridge, Cambridge CB2 1AG, UK

Abstract. The rise of Artificial Intelligence (AI) and digital technologies is reshaping work, creating opportunities and challenges related to Sustainable Development Goal 8 (SDG 8: Decent Work and Economic Growth). AI enables business model innovation (BMI), yet less research examines how specific BMIs shape work practices toward SDG 8 goals. This research uses practice theory and insights from 56 semi-structured interviews with 34 companies to explore work quality challenges (SDG 8), AI-driven BMIs reshaping work, and resulting pathways. We identify challenges like 'complexity in scaling up' capabilities, highlight BMI designs such as 'Gen-AI as a service', and detail pathways like 'realised customisation'. A theoretical framework shows how AI-enabled BMI influences work practices and pathways, impacting Decent Work (SDG 8) attainment.

Keywords: Business model innovation · Future of Work · AI · Decent Work · Economic Growth

1 Introduction

AI and digitalisation are transforming work, shifting focus beyond economics to employment quality, conditions, and sustainability. 'Decent Work' (SDG 8) is a key benchmark for evaluating technology's contribution to desirable work futures [1]. Achieving this requires integrating considerations like gender equality (SDG 5) [2] and understanding technology's impact on sustainable employment [3]. This paper examines challenges and aspects to realise desirable work futures aligned with SDG 8.

AI and digitalisation present opportunities, like supporting ICT entrepreneurship via digital transformation, and challenges, like job insecurity and declining mental well-being [4]. Digitalisation creates jobs and introduces new challenges, especially via digital labour platforms [5]. Policy frameworks ensuring decent work in the AI era are crucial, considering post-COVID shifts and the need for strategies addressing automation's impact on job creation and transformation [6].

Firms must innovate their fundamental BMIs to respond to AI-driven shifts. BMI, enabled by digital tech and capabilities [7], helps organisations redesign work, adapt structures, and navigate future work challenges. It facilitates AI integration, enhances

X. Xie et al. (Eds.): ICSS 2025, LNOR, pp. 26–38, 2026.
https://doi.org/10.1007/978-3-032-13116-4_3

performance, contributes to sustainable goals like decent work [8], and maintains agility [9].

This paper explores how firms use AI-driven BMI to navigate the future of work, focusing on challenges and pathways related to SDG 8. It reviews literature on the future of work, SDG 8, and digital BMI. It then presents methodologies, findings on work challenges, AI-driven BMI opportunities, and subsequent pathways. The discussion synthesizes findings with literature to argue for contributions. Finally, limitations and future research are concluded.

2 Literature Review

2.1 Theoretical Lens: Practice Theory for Understanding Work Transformation

This research adopts a Practice Theory perspective, common in Strategy-as-Practice (SAP) [10, 11], to understand how AI and digital BMI reshape work. This lens examines how actual work practices are performed and reconstructed in response to AI and new BMI strategies [12]. Instead of viewing technology as an external force, it explores how individuals actively use, adapt to, and are shaped by new technologies and models daily. It highlights the recursive relationship between human actions and context [13], offering a detailed view of how work practices evolve and contribute to outcomes like decent work. This approach emphasizes practical relevance [14] and the ongoing nature of work transformation.

2.2 The Future of Work in the AI Era

The work landscape is transforming due to AI and digital technologies. This "Future of Work" involves shifts in jobs, skills, structures, and employment nature. Key debates centre on AI's potential for task automation (causing job displacement concerns) [15] versus its ability to augment human capabilities and create new roles. Studies show AI automates routine tasks, while complex, non-routine tasks requiring human skills like creativity and critical thinking remain, suggesting job restructuring over elimination. This transformation relies on digital infrastructure like cloud computing [16], enabling scalable AI and distributed work models. Increased digital reliance necessitates robust cybersecurity.

This shift demands workforce adaptation via upskilling and reskilling. Demand grows for technical AI skills and complementary human skills (collaboration, communication, adaptability) [17]. Bridging the "AI skills gap" via diverse education and lifelong learning is crucial. The AI era sees evolving work models, especially remote and hybrid arrangements accelerated by AI tools for collaboration and monitoring. Managing distributed teams requires attention to cybersecurity, culture, and well-being. Critical examination focuses on job quality, equity, and governance, including algorithmic management, data privacy, AI bias, and worker autonomy impacts.

2.3 Work-Related Challenges in the AI Era (SDG 8 Perspective)

Navigating the future of work requires addressing significant challenges to achieve Decent Work (SDG 8). Integrating policies supporting decent work within AI-driven economies is crucial. Traditional policies may fail as automation transforms tasks. Operationalising 'decent work' in policy is crucial for assessing AI's impact on labour conditions and ensuring safety nets [19]. Policies must tackle socio-economic issues, including vulnerable populations, using interdisciplinary approaches.

Employment dynamics are shifting, affecting skill demands and job structures globally [4]. Research explores technology's influence on employment, impacts of work types (e.g., temporary work), and factors like foreign investment. Comparative analyses show varying policy impacts on labour markets, stressing factors like gender equality for decent work outcomes globally [2]. Sustainable and inclusive work futures face challenges, requiring alignment of AI-enabled growth with environmental sustainability and inclusive social policies (e.g., for aging populations) [3]. Inclusive innovation systems and socially responsible design are key. Educational institutions' role in fostering skills for inclusive innovation is crucial.

AI's impact on labour markets and skills is a central concern [20]. Digital technology aids sustainable development but poses inequality challenges and requires new skills [21]. Digitalisation transforms professions, demanding readiness for Industry 4.0 [22] and new competencies. Aligning AI-driven growth with societal and worker well-being is critical. This needs integrated approaches considering entrepreneurial accessibility, supportive institutions, diverse population integration, and education's role. Achieving SDG 8 demands multi-dimensional strategies linking economic performance, AI-driven labour shifts, environmental sustainability, and worker well-being [11].

2.4 The Role of BMI in Shaping Work Futures

BMI is a critical strategy for firms in the AI-driven future [23]. Firms use BMI proactively to reshape operations, value propositions, and work environments [24]. Digitally enabled BMI helps firms adapt to market and workforce shifts [9]. In the future of work context, BMI involves rethinking value creation [25, 26], impacting roles, skills, structures, and experiences. Literature on mission-oriented innovation, stakeholder value [27], sustainable models [8], and digital entrepreneurship touches upon how new models influence work. This paper focuses explicitly on BMI as the mechanism for translating AI and digital potential into work practice changes, balancing economic goals with SDG 8.

2.5 AI-Enabled BMI Designs and Their Impact on Work

The AI era necessitates evolving business models leveraging technology, reshaping work itself [7]. Examining these BMI designs reveals implications for tasks, skills, and work organization. Service-dominant business models emphasize customization and value co-creation [28], often requiring employees with strong customer interaction, adaptability, and problem-solving skills, shifting work towards knowledge-intensive service encounters, enhanced by digitalization through integrated product-service systems [29].

Internet of Things (IoT) enabled business models to leverage real-time data from connected devices [30], which can automate monitoring tasks, enable predictive maintenance (changing maintenance work) [31], create new data analysis roles, and potentially increase performance monitoring or worker surveillance. Multi-sided platforms and digital ecosystems fundamentally change work structures by mediating connections between user groups like gig workers and customers [32], fostering flexibility but raising questions about job security, algorithmic management, and worker rights within platform work [33]. AI-driven business models are central [34], automating routine tasks, personalizing services, optimizing workflows, and augmenting human decision-making, often leveraging scalable cloud platforms. The rise of Generative AI introduces capabilities for content creation, coding, and complex problem-solving, further blurring human-machine task boundaries and demanding new human-AI collaboration skills. Open Innovation models emphasize external collaboration and knowledge sharing, impacting work by requiring more teamwork, external communication skills, and proficiency with digital collaboration platforms. Viewing the business model as a complex system highlights how adopting AI triggers interdependent changes in work processes, roles, and skill requirements [35]. Understanding these designs' implications for work, including the need for trust and robust cybersecurity in platform and data models, is crucial.

2.6 Pathways of BMI: Consequences for Work and Workers

The implementation of AI-enabled BMI leads to various outcomes or pathways with direct consequences for the workforce and work quality. Improved Performance, often manifested as efficiency gains and cost savings from data analytics or automation, presents a complex picture. While potentially reducing tedious tasks, it can also intensify workloads, lead to job displacement if labour cost reduction is primary, or heighten performance monitoring. Enhanced Customer Experiences, achieved through AI-driven personalization or integrated services, often reshape frontline service work, potentially enabling employees but requiring stronger interpersonal and complex problem-solving skills. Increased Competitiveness from successful BMI can enhance firm longevity, benefiting workers via job security, though intense pressure might incentivize further efficiencies impacting wages or conditions. Collaborative Value Creation, associated with open innovation or platform ecosystems [33] enabled by digital technologies, implies changes in work organization, necessitating more teamwork, cross-boundary communication, and adaptability, potentially leading to dynamic work but requiring digital tool proficiency. Adaptability and Flexibility [9], key outcomes enabling firms to respond dynamically, can translate into flexible work arrangements (remote and hybrid), benefiting work-life integration, but firm flexibility might sometimes rely on non-standard employment, potentially increasing worker instability. Analyzing these pathways through their impact on decent work (SDG 8) is essential for understanding BMI's complex role.

2.7 Research Gaps and Questions

While literature explores AI's work impact and digital BMI separately [23], limited empirical research investigates the specific mechanisms through which AI-enabled BMI reshapes work practices and influences SDG 8 outcomes. Studies often focus macro or on specific tech, leaving a gap in understanding holistic BMI adaptations and their concrete work consequences. This research addresses this via: (1) What key work-related challenges (SDG 8) trigger AI-enabled BMI, and what specific AI-driven BMI opportunities (designs) and pathways (outcomes) emerge? (2) How do these AI-driven BMI opportunities and pathways influence challenges associated with achieving SDG 8 in the digital landscape?

3 Methodology

This research employs qualitative methods for in-depth understanding. Respondents can express their views freely. Qualitative research excels at studying intricate problems deeply and exploring new areas. The case study method captures real-world activities via interviews and observations. A multi-case approach analyzes data within and across diverse scenarios, critical for exploring digital BMI.

3.1 Data Collection

From 2022 to 2024, face-to-face interviews were conducted with business directors, BDM, marketing, and strategic managers from various companies. No restrictions on company location, size, or ownership were imposed. Of 52 contacted companies, 34 (65%) participated. The sample included seven micro, ten small, eight medium, and nine large firms. Each had 1–4 informants. Most had adopted digital tech and new models, few created new ones. Non-participants often cited a lack of adoption, busy schedules, or confidentiality. Anonymity was offered. The research team maintained independence. Data validity was ensured using primary (semi-structured interviews) and secondary (archival data like reports, databases) sources. Triangulation increased research quality. Data collection focused on firm challenges, AI-enabled digital BMIs, and perceived impacts on work practices, capabilities, and SDG 8 outcomes.

Fifty-six interviews were conducted (1–4 informants per company) using the protocol in the Appendix. Questions probed challenges (work and workforce related), BMI initiatives (esp. AI), and perceived outcomes (efficiency, job quality, competitiveness). Table 1 shows company and interviewee profiles. Most firms were IT-related or had IT departments interviewed. Locations included UK, US, Germany, France, Denmark, Sweden, Norway, the Netherlands, the Czech Republic, and Israel. All operated in the UK; interviewees were English fluent. Some interviews were recorded with consent; all were transcribed for coding.

Site visits to two manufacturing firms allowed exploration of operational environments. These visits helped understand challenges, implemented BMIs, effects on shop-floor work, and SDG 8 strategies. Dialogues with managers, engineers, and workers preceded visits. Job shadowing helped understand processes and routines. Ethnographic interviews were conducted with randomly selected staff in workspaces. Field notes captured key phrases due to recording restrictions. Memos were drafted within 24 h. This data validated managerial responses and compared them against observed practices. Table 1 details participants.

Diverse internal (consulting reports, scorecards) and external (corporate and industry reports, newspapers, databases like Company House, FAME, Bloomberg, LinkedIn) data were used. Publicly listed firms offered open-source info. SMEs (74% of the sample) required long data acquisition. Third-party reports (Statista, etc.) supplemented interview data. Archival data triangulated interviews, sharpening focus.

3.2 Data Analysis

Table 1. Selected companies and interviewee profile

Index	Industry	Size	Region	Interviewee type
1	Computer Networking Products	2-10	UK	Director
2	Software Development	51-200	Israel	Specialist
3	IT Services and IT Consulting	1,001-5,000	US	Partner
4	Computer and Network Security	11-50	US	Founder
5	Financial Services	11-50	UK	Marketing
6	Real Estate	2-10	UK	Director
7	Design Services	2-10	UK	Events manager
8	Technology, Information and Internet	51-200	UK	Business development specialist
9	Advertising Services	51-200	UK	Manager
10	Telecommunications	501-1,000	Norway	Business development manager
11	Insurance	11-50	US	Head of Marketing
12	Software Development	51-200	Czech Republic	Commercial and business director
13	IT Services and IT Consulting	5001-10000	US	Business development manager
14	Business Consulting and Services	2-10	UK	Director
15	IT Services and IT Consulting	1,001-5,000	US	Business development specialist
16	IT Services and IT Consulting	11-50	UK	Head of business development
17	Software Development	501-1,000	France	Business development specialist
18	Software Development	11-50	UK	Founder and CEO
19	IT Services and IT Consulting	2-10	UK	Business development Consultant
20	IT System Custom Software Development	501-1,000	US	Head of business development
21	Financial Services	51-200	UK	Chief Marketing Officer
22	Software Development	11-50	UK	CEO, founder
23	IT Services and IT Consulting	11-50	Netherland	CEO, CTO, Senior Vice President
24	IT Services and IT Consulting	2-10	UK	Marketing specialist
25	IT Services and IT Consulting	51-200	Denmark	Account manager
26	IT Services and IT Consulting	11-50	UK	Director
27	Financial Services	51-200	US	Sales Manager
28	Telecommunications	11-50	UK	Business development specialist
29	Information Services	2-10	France	Marketing specialist
30	Telecommunications	51-200	UK	Technology and Engineering
31	Software Development	501-1,000	US	Strategic Client Manager
32	Retail	10,001+	UK	Director & Business development manager
33	Motor Vehicle Manufacturing	10,001+	Sweden	Business development manager & Product Manager
34	Automation Machinery Manufacturing	11-50	UK	Business development professionals

Interview and archival data were analysed to answer research questions on work challenges triggering AI-driven BMI, emergent opportunities and pathways, and SDG

8 influence. Gioia et al.'s three-step approach built a theoretical model [36]: (1) Identified 1st-order informant-level categories from data patterns, labelled using informant language. (2) Grouped 1st-order categories into 2nd-order themes, summarized as theoretical dimensions. Generated data structure for findings. (3) Used constant comparison to update coding and framework, analysing across informants, companies, and data sources. Reliability was ensured by independent coding by two researchers, discussing differences until consensus. The method identified practice-based blocks for theory on workforce challenges, BMI impacts on work design, and pathways affecting work quality and efficiency. The resulting framework responds to call for theory beneficial for practice and conceptualisation. Findings are categorised into challenges, opportunities, and pathways.

4 Results

4.1 Finding 1: Key Challenges Hindering Positive Work Futures and Decent Work (SDG 8) in the AI Era

The analysis identified three challenge dimensions blocking firms' navigation of future work and SDG 8 achievement: work quality issues, adaptation barriers, and capability gaps (Fig. 1). 'Rigid and routine tasks' negatively impact engagement and productivity (e.g., non-interactive tools, manual onboarding, monotonous processes). These fuel dissatisfaction and are prime for AI reshaping. 'Labour-intensive & difficult-to-manage tasks' highlight operational burdens (e.g., on-premise call centres, manual data insight extraction, difficult training and monitoring). These point to AI opportunities for improving efficiency, feedback, and job quality.

Barriers prevent the proactive shaping of work futures. 'Lack of measurable valuation' for workspace indicators hinders data-driven optimisation. 'Exploration of BMI in the digital era', specifically lack of awareness of AI's potential, is significant. Managers may overlook AI gains; market shifts create pressure, yet firms struggle to identify AI opportunities. This hinders adaptation.

Capability gaps hinder transformation. 'Complexity in scale-up' barriers implementation of new work models (e.g., traditional solutions lacking cloud and API integration limit scalability; cross-national ops add complexity). 'High cost, and lack of capabilities of internal development' limits access to better tools and processes (lack skills and resources for customization, adopting advanced tech, managing cybersecurity, improving communication). Manufacturing noted costly inspections replaceable by tech. 'Lack of quick coordination and trust' impedes new tech and collaboration adoption. 'Lack of cheap and reliable access' to external capabilities limits leveraging tools hindering digital ecosystem participation. Lack of trust for lenders post-crisis reflects this.

4.2 Finding 2: BMI Opportunities Shaping Future Work Practices and Models

Emerging from challenges, three dimensions of digital BMI opportunities reshape work performance, management, and organization via AI and cloud: enabling remote and flexible work, transforming tasks (automation and intelligence), and redefining boundaries

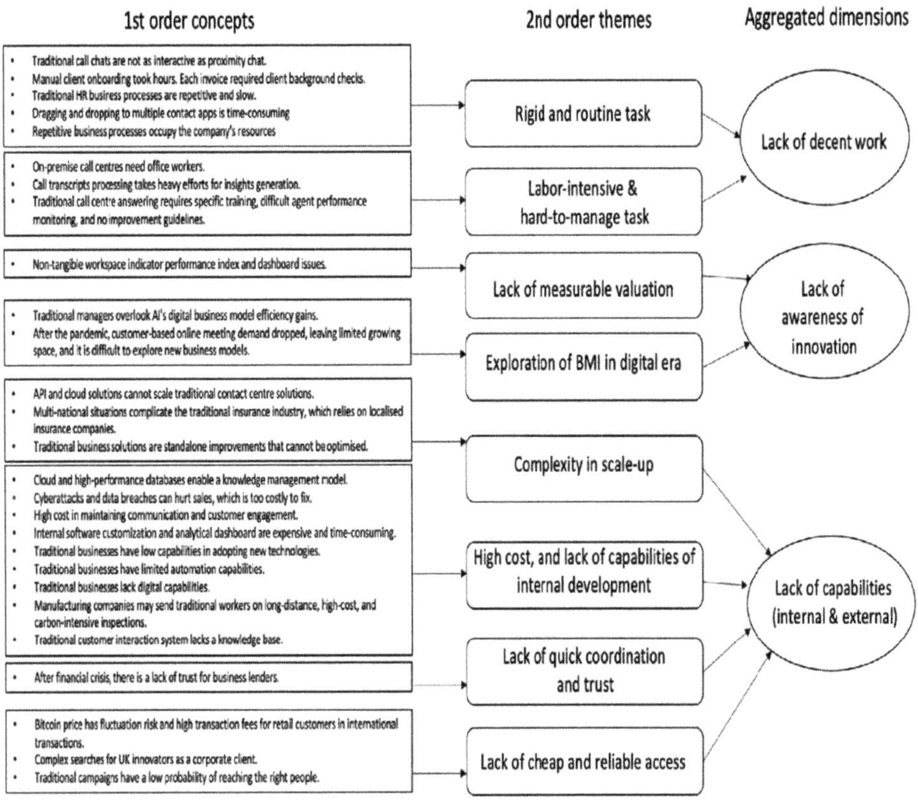

Fig. 1. Data structure of challenges

(partnerships and outsourcing). Figure 2 shows data structure. Digital tech enables BMIs to facilitate remote and flexible work. Cloud solutions reduce the need for physical server proximity. AI support systems improve remote efficiency. Example: extending online communication to enterprise chatbots supports remote customer service. Example: a mobile robot with interactive screen enables remote monitoring and inspection, offering flexibility. 'Work process automation as a service' BMIs impact human tasks. AI and software integration (e.g., RPA) automates routine processes, shifting human work, potentially requiring upskilling. Example firm provides RPA consulting, implementation, AIaaS, training. 'Business intelligence as a service' BMIs augment human analysis. AI and data analytics offer solutions for customer and employee experience, integration, workspace strategy, innovation intelligence, potentially making analysis efficient but raising skill demands. 'Generative AI as a service' significantly impacts knowledge and creative work. Firms deploy marketing chatbots or use GenAI in cloud software for customer service knowledge management, automating and augmenting cognitive labour.

BMIs involving digital partnerships and outsourcing reshape where and by whom work is done. 'Digitalisation outsourcing and consulting service' models shift work externally (e.g., IT solutions for SMEs, streaming and CRM consulting, data exchange

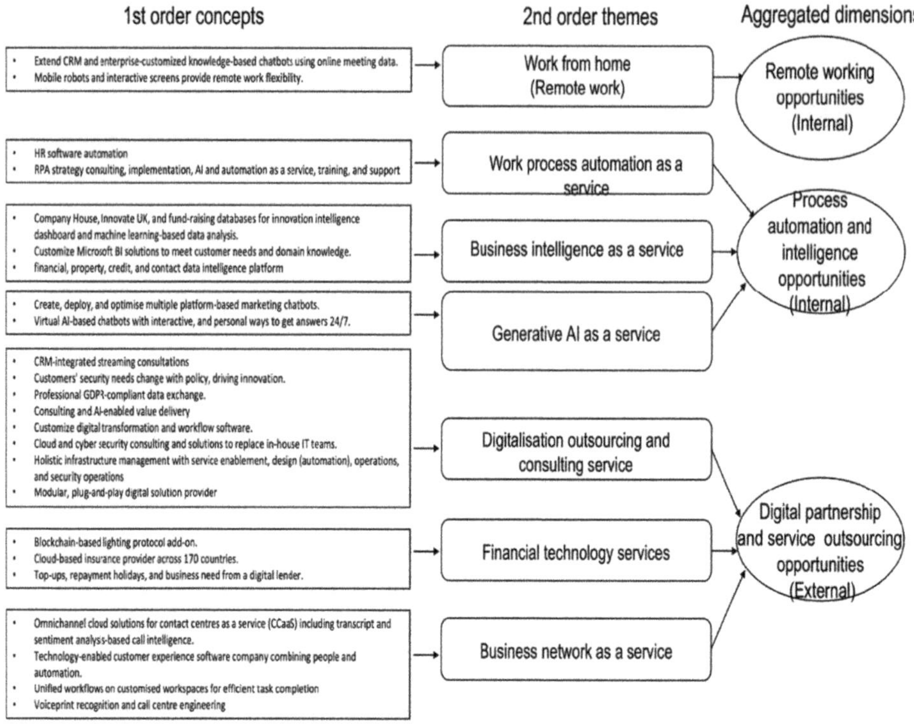

Fig. 2. Data structure of opportunities for BMI

services, custom software, cloud and cybersecurity consulting, infrastructure management). This impacts internal workforce structure and skills. Policy changes can drive innovation here. 'Financial technology services' alter work in finance and customer interactions (e.g., blockchain add-ons, cloud insurance platforms, flexible digital lending). 'Business network as a service' models create integrated platforms restructuring workflows (e.g., contact centre solutions blending human and automation, unified customized workspaces, specialized AI like voiceprint recognition). These change coordination, collaboration, and skill needs. These opportunities show firms leveraging tech and BMI to reshape where (remote and outsourced), what (automated and augmented), and how (platformed and networked) work is done.

4.3 Finding 3: Pathways of AI-Driven BMI and Their Consequences for the Future of Work (SDG 8)

BMI opportunities lead to pathways impacting work practices, experiences, and SDG 8 progress. Categories: efficiency, cost, and convenience improvements; enhanced flexibility and customization; greater reliability and accuracy. Figure 3 details data structure. Cost-effectiveness impacts workforce investment and job structure. Examples: lower transaction fees enabling new models; significant ROI via labour cost and license fee reductions. RPA saves employee time and costs. Cost reduction focus raises job security and wage questions. Efficiency improvements alter work pace and nature. Digital solutions streamline processes. Enhanced communication effectiveness reflects operational efficiency. Gains reduce burden but may raise throughput expectations. Convenience (via integrated solutions) improves employee experience by reducing friction. Comprehensive platforms simplify workflows. Compliance convenience and real-time agent recommendations simplify tasks.

Customisation pathway, enabled by AI-driven BMI, can lead to specialized and engaging roles. Firms realize "customised potential… in AI" or offer "modularised solution[s]", implying shifts needing higher skills and adaptable planning, potentially boosting satisfaction. The flexibility pathway relates to agile practices. "Joint optimisation for flexible workflow" is key. Operational flexibility might translate to adaptable schedules and locations, but needs management to ensure quality and prevent boundary blurring.

4.4 Finding 4: A Framework for AI-Driven BMI Shaping the Future of Work (SDG 8)

The proposed theoretical framework (Fig. 4) links challenges, BMI designs, and outcomes. Future of Work and SDG 8 challenges influence the adoption of AI-Enabled

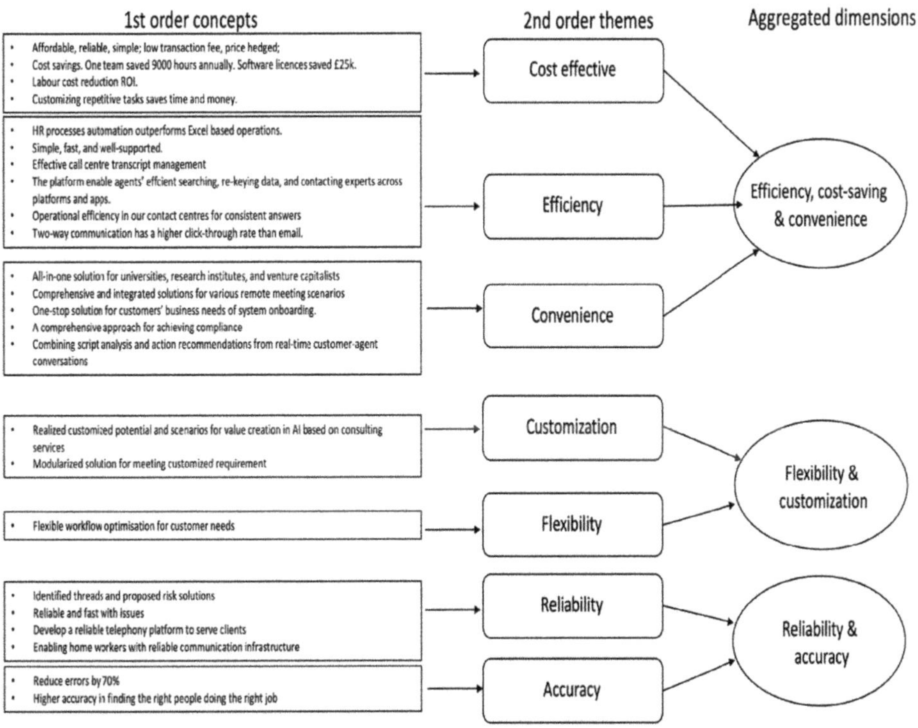

Fig. 3. Data structure of pathways

BMI Designs Impacting Work. Feasibility relies on Enabling Digital Technologies (AI-Centric: AI, Cloud, Cybersecurity). BMI implementation results in BMI Pathways: Work-Related Outcomes. These outcomes are feedback to Address and Shape initial challenges, influencing progress towards SDG 8. Identified challenges and opportunities linked to SDG dimensions: 1) SDG 8.2 (productivity via tech innovation); 2) SDG 8.3 (decent job creation, innovation support, SME growth); 3) SDG 8.10 (access to enabling services).

5 Discussion and Conclusion

This research contributes empirically grounded insights on BMI's role in shaping work transformations in the AI Era. Beyond direct tech impacts, findings show firms strategically leverage AI via BMI to navigate challenges, altering work practices and outcomes. It details AI-enabled BMI's impact on tasks and processes using examples like 'Work process automation,' 'BIaaS,' and ' GenAIaaS', extending literature by showing BMI wrappers. It illuminates BMI's facilitation of new work arrangements like remote work. It highlights workforce capabilities and challenges, echoing AI skills gap literature, with pathways like 'realised customisation' suggesting the need for higher skills. Crucially, it connects shifts to Decent Work (SDG 8), analyzing pathway impacts on work quality and conditions, contributing to the debate on AI's effect on employment outcomes, and highlighting opportunities and risks needing management.

This research proposes an empirically grounded framework (Fig. 4) conceptualizing the interplay between challenges (SDG 8 context), enabling tech (AI-centric), AI-driven BMI designs, and outcomes, and offering a mechanism analyzing how BMI mediates tech's impact. Using Practice Theory, it offers a micro-level perspective showing how new work practices are performed and negotiated, highlighting human agency. It bridges BMI and Future of Work literature, empirically showing BMI as a tool for translating AI and digital potential into tangible work changes relevant to SDG 8.

The predominance of IT firms means findings may not fully generalize; more diverse sector research is needed. The qualitative nature provides rich insights but lacks statistical generalizability; quantitative verification is needed. Future work could include comparative cross-industry studies, quantitative validation of the framework (Fig. 4), longitudinal studies tracking firms adopting specific AI-driven BMIs, and investigation into policy interventions guiding AI-driven BMI towards high-quality work aligned with SDG 8 principles. Building on these, future research could broaden sectoral and geographical diversity, including non-IT industries like healthcare and agriculture, and focus on developing countries where AI adoption may exacerbate inequalities. Quantitative methods, such as structural equation modeling, could validate the framework by operationalizing constructs like BMI designs and pathways. Longitudinal studies over 5–10 years would track dynamic feedback loops, assessing impacts on job engagement and well-being. Policy research could evaluate interventions like upskilling programs and AI ethics guidelines to mitigate risks such as bias in platform BMIs. Interdisciplinary approaches integrating sociology and computer science could develop "responsible BMI" frameworks, ensuring AI supports equitable decent work. Future studies could incorporate emerging AI trends, such as Gen-AI in creative sectors, examining their effects on skill demands. Mixed methods design combining big data analytics with qualitative insights would bridge macro and micro levels. Finally, action research co-creating BMIs with workers could translate findings into practice, fostering inclusive innovation.

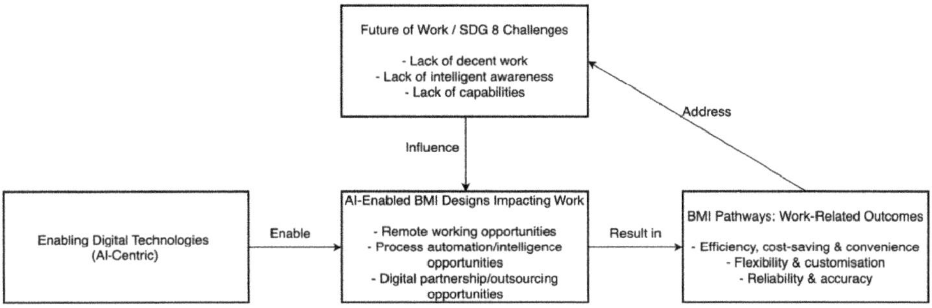

Fig. 4. Proposed theoretical framework

References

1. Chigbu, B.I., Nekhwevha, F.: Exploring the concepts of decent work through the lens of SDG 8: addressing challenges and inadequacies. Front. Sociol. **8** (2023)
2. Rai, S.M., Brown, B.D., Ruwanpura, K.N.: SDG 8: decent work and economic growth–a gendered analysis. World Dev. **113**, 368–380 (2019)
3. Alic, J.A.: Technological change, employment, and sustainability. Technol. Forecast. Soc. Chang. **55**(1), 1–13 (1997)
4. Charles, L., Xia, S., Coutts, A.P.: Digitalization and employment. Int. Labour Organ. Rev. 1–53 (2022)
5. Cusumano, M.A., Yoffie, D.B., Gawer, A.: The future of platforms. MIT Sloan Manage. Rev. 26–34 (2020)
6. Bloom, D.E., McKenna, M., Prettner, K.: Demography, unemployment, automation, and digitalization: Implications for the creation of (decent) jobs, 2010–2030 (technical report). National Bureau of Economic Research (2018)
7. Crittenden, A.B., Crittenden, V.L., Crittenden, W.F.: The digitalization triumvirate: how incumbents survive. Bus. Horiz. **62**(2), 259–267 (2019)
8. Parida, V., Sjodin, D., Reim, W.: Reviewing literature on digitalization, business model innovation, and sustainable industry: past achievements and future promises. Sustainability **11**(2), 391 (2019)
9. Verhoef, P.C., et al.: Digital transformation: a multidisciplinary reflection and research agenda. J. Bus. Res. **122**, 889–901 (2021)
10. Whittington, R.: Strategy as practice. Long Range Plann. **29**(5), 731–735 (1996)
11. Albertsen, R.R., et al.: Strategizing together for a better world: institutional, paradox and practice theories in conversation. J. Manage. Inq. 10564926231210238 (2023)
12. Jarzabkowski, P.: Strategy as practice: recursiveness, adaptation, and practices-in-use. Organ. Stud. **25**(4), 529–560 (2004)
13. Orlikowski, W.J.: Using technology and constituting structures: a practice lens for studying technology in organizations. Organ. Sci. **11**(4), 404–428 (2000)
14. Thomas, K.W., Tymon, W.G., Jr.: Necessary properties of relevant research: lessons from recent criticisms of the organizational sciences. Acad. Manag. Rev. **7**(3), 345–352 (1982)

15. Schlogl, L., Sumner, A.: The rise of the robot reserve army: automation and the future of economic development, work, and wages in developing countries. Center for Global Development Working Paper (487) (2018)
16. Berman, S.J., et al.: How cloud computing enables process and business model innovation. Strategy Leadersh. **40**(4), 27–35 (2012)
17. Fareri, S., Fantoni, G., Chiarello, F., Coli, E., Binda, A.: Estimating industry 4.0 impact on job profiles and skills using text mining. Comput. Industry **118**, 103222 (2020)
18. Kagermann, H.: Change through digitization—value creation in the age of industry 4.0. In: Management of Permanent Change, pp. 23–45. Springer (2014)
19. United Nations Department of Economic and Social Affairs (UNDESA). Transforming our world: The 2030 agenda for sustainable development. A/RES/70/1 (2015)
20. Bresciani, S., Huarng, K.-H., Malhotra, A., Ferraris, A.: Digital transformation as a springboard for product, process and business model innovation. J. Bus. Res. **128**, 204–210 (2021)
21. Nambisan, S., Lyytinen, K., Majchrzak, A., Song, M.: Digital innovation management. MIS Q. **41**(1), 223–238 (2017)
22. Kiel, D., Arnold, C., Voigt, K.-I.: The influence of the industrial internet of things on business models of established manufacturing companies–a business level perspective. Technovation **68**, 4–19 (2017)
23. Foss, N.J., Saebi, T.: Fifteen years of research on business model innovation: how far have we come, and where should we go? J. Manag. **43**(1), 200–227 (2017)
24. Zott, C., Amit, R., Massa, L.: The business model: recent developments and future research. J. Manag. **37**(4), 1019–1042 (2011)
25. Zott, C., Amit, R.: Business model design: an activity system perspective. Long Range Plan. **43**(2–3), 216–226 (2010)
26. Osterwalder, A., Pigneur, Y.: Business Model Generation: A Handbook for Visionaries, Game Changers, and Challengers, vol. 1. Wiley (2010)
27. Andreassen, T. W., et al.: Business model innovation and value-creation: the triadic way. J. Serv. Manage. (2018)
28. Turetken, O., Grefen, P., Gilsing, R., Adali, O.E.: Service-dominant business model design for digital innovation in smart mobility. Bus. Inf. Syst. Eng. **61**, 9–29 (2019)
29. Paiola, M., Gebauer, H.: Internet of things technologies, digital servitization and business model innovation in BtoB manufacturing firms. Ind. Mark. Manage. **89**, 245–264 (2020)
30. Aagaard, A., Presser, M., Andersen, T.: Applying iot as a leverage for business model innovation and digital transformation. In: 2019 Global IoT Summit (GIoTS), pp. 1–5 (2019)
31. Porter, M.E., Heppelmann, J.E.: How smart, connected products are transforming competition. Harv. Bus. Rev. **92**(11), 64–88 (2014)
32. Van Alstyne, M.W., Parker, G.G., Choudary, S.P.: Pipelines, platforms, and the new rules of strategy. Harv. Bus. Rev. **94**(4), 54–62 (2016)
33. Kenney, M., Zysman, J.: The rise of the platform economy. Issues Sci. Technol. **32**(3), 61–69 (2016)
34. Winter, J.: Digital business model innovation–empirical insights into the drivers and value of artificial intelligence. Int. J. Comput. Technol. **21**(1), 63–75 (2021)
35. Velu, C.: Business model cohesiveness scorecard: implications of digitization for business model innovation. In: Handbook of Digital Innovation, pp. 179–197. Edward Elgar Publishing (2020)
36. Gioia, D.A., Corley, K.G., Hamilton, A.L.: Seeking qualitative rigor in inductive research: notes on the Gioia methodology. Organ. Res. Methods **16**(1), 15–31 (2013)

Research on Dynamic Risk-Avoidance Route Planning for Multi-drone Collaborative Delivery in Complex Urban Environments

Jiang Runxue and Mi Chuanmin(✉)

School of Economics and Management, Nanjing University of Aeronautics and Astronautics, Nanjing 211106, China
cmmi@nuaa.edu.cn

Abstract. To address challenges such as dynamic obstacles, temporary no-fly zones, multi-drone collaborative conflicts, and energy consumption constraints in complex urban environments, a dynamic risk-averse route planning method for multi-drone collaborative delivery is proposed. First, a collaborative delivery model for multi-distribution centers is constructed, integrating a 3D grid-based method to quantify obstacle risks and energy constraints. The urban airspace is stratified into three flight levels, and a dynamic risk update mechanism is designed. Second, a Hybrid A*-Whale Optimization Algorithm (Hybrid A*-WOA) is proposed, combining the improved A* algorithm's local route search capability with the global optimization characteristics of the Whale Optimization Algorithm. This hybrid approach achieves collaborative optimization of task allocation and dynamic obstacle avoidance for multi-drone systems. Furthermore, a flight-level comprehensive evaluation function is introduced to balance the risk and time cost through weight coefficients and dynamically select the optimal flight altitude.

Keywords: multi-drone · collaborative delivery · route planning · hybrid A*-WOA · risk avoidance

1 Introduction

With the rapid development of e-commerce and the surge in consumer demand for instant delivery, drone delivery has emerged as a key technology to address the "last-mile" logistics challenge, owing to its ability to overcome ground traffic constraints (Pons-Prats et al., 2022), reduce response time, and lower carbon emissions. The global drone delivery market is projected to grow from 11.27 billion in 2024 to 34.5 billion by 2032 (CAGR: 15.01%), reflecting not only technological maturity but also its immense potential in urban logistics. Leading companies such as Meituan, JD.com in China, and Google and Amazon (Aurambout et al., 2019) in the U.S. are actively developing low-altitude drone delivery systems. By 2024, Meituan's drones had completed over 450,000 orders across multiple Chinese cities, covering more than 90,000 product categories, demonstrating practical feasibility. However, the complexity of urban environments imposes higher requirements on drone route planning, risk avoidance, and

X. Xie et al. (Eds.): ICSS 2025, LNOR, pp. 39–50, 2026.
https://doi.org/10.1007/978-3-032-13116-4_4

collaborative management, including challenges such as dynamic obstacles, temporary no-fly zones, multi-drone coordination conflicts, and energy constraints.

Existing studies have explored aspects such as route optimization, multi-objective collaboration, and risk control. In terms of route planning, approaches include mixed-integer programming (Dorling et al., 2017), dynamic task adjustment (Song et al., 2018), and multi-objective optimization models (Cheng et al., 2020; Ozkan et al., 2023). For multi-drone collaboration (Li et al., 2023), hierarchical route planning for heterogeneous drones (Wen et al., 2022) and flight-level stratification strategies (Kim et al., 2024) have provided novel insights for 3D route optimization. Risk control research enhances route safety through quantitative environmental modeling (Shao et al., 2022), real-time trajectory adjustments (Lee et al., 2022), and bio-inspired optimization algorithms (Zhu et al., 2022; Wang et al., 2024). In energy management, scholars have analyzed multi-depot energy consumption patterns (Torabbeigi et al., 2020) and the nonlinear impact of parcel weight on flight range (Kyriakakis et al., 2023).

However, significant limitations persist in existing methods: most route planning relies on static environmental assumptions, failing to integrate updates for temporary no-fly zones or dynamic obstacle predictions (Evers et al., 2014); energy optimization and risk avoidance are studied in isolation, lacking coupled analysis (Vural et al., 2021); multi-drone collaborative scheduling focuses predominantly on task allocation without deep integration with real-time route adjustments (Li et al., 2025). Additionally, current risk models primarily address single factors (e.g., weather or static obstacles) and lack comprehensive quantification of multi-dimensional dynamic risks (Tadic et al., 2024). These shortcomings hinder existing approaches from meeting the demands of efficient, safe, and collaborative drone delivery in complex urban environments.

2 Problem Description and Model Establishment

2.1 Problem Description

In urban drone delivery networks, multiple distribution centers (drone takeoff points) and customer-oriented logistics delivery points (stop points for package drop-offs) coexist. During operations, distribution centers are staffed with professionals responsible for drone loading and battery replacement, thereby functioning as charging stations. Drones must ascend from these distribution centers, navigate to customer locations, and descend to stop points for package delivery (Kim et al., 2024), as shown in Fig. 1. In the absence of temporary no-fly zones, dynamic obstacles, or insufficient battery levels, drones return to their distribution centers along the same route used for delivery. Otherwise, they replan their routes to bypass restricted zones or divert to nearby distribution centers for battery replacement before returning. This study introduces a multi-flight-level concept, categorizing flight altitudes into three tiers: 50 m, 80 m, and 110 m, as shown in Fig. 2. The objective is to optimize drone routes and task allocation in complex urban environments considering obstacles at varying heights, temporary no-fly zones, and route conflicts to achieve minimal total delivery time and risk in multi-drone collaborative delivery. The following assumptions are made:

1) Weather conditions and unexpected events are disregarded during delivery, assuming uninterrupted operations.

2) Drones maintain predetermined flight altitudes without mid-flight adjustments.
3) All drones are identical in specifications, including dimensions, payload capacity, and flight range.

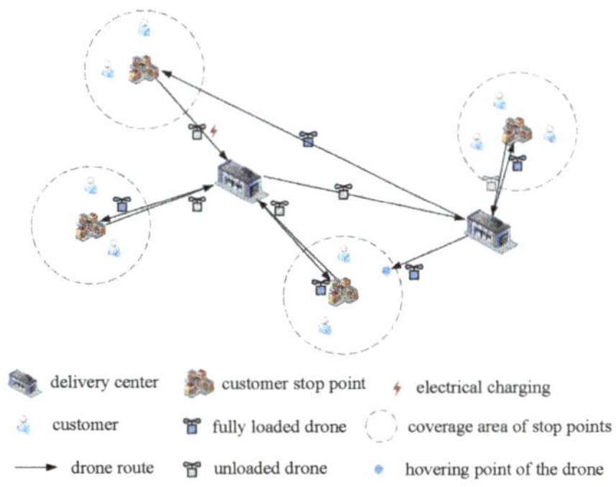

Fig. 1. Schematic diagram of drone delivery under multi-distribution centers

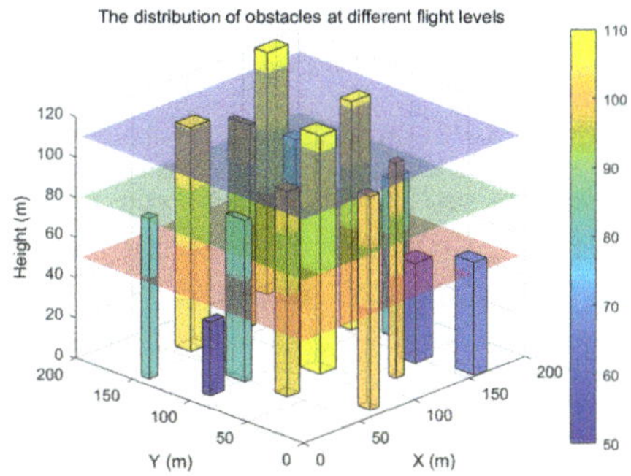

Fig. 2. Obstacle distribution across different flight levels

2.2 Urban Low-Altitude Environmental Modeling

This paper used the grid method for spatial modeling. Fixed obstacles such as buildings are simplified as cuboids occupying continuous grids, marked as non-traversable zones. A grid risk degree is introduced to quantify the danger of traversable grids. For grids containing obstacles, the risk degree is set to 1. For obstacle-free grids, the risk degree is calculated as the ratio of adjacent 26 grids containing obstacles to the total adjacent grids:

$$\phi_k = \frac{N_{obs}(k)}{N_{sur}(k)} \tag{1}$$

Where, $N_{obs}(k)$ is the number of adjacent grids with obstacles, and $N_{sur}(k)$ is the total number of adjacent grids. Higher risk degrees indicate denser obstacle proximity, prioritizing avoidance during route planning.

For mobile obstacles, grid risk degrees are updated using real-time perception. If a grid is repeatedly occupied by dynamic obstacles within a time window Δt, its risk degree increases exponentially. n_{hit} is the obstacle occurrence count within Δt, and λ is the risk decay coefficient.

$$\phi_k(t + \Delta t) = \phi_k(t) \cdot e^{\lambda n_{hit}} \tag{2}$$

2.3 Model Establishment

Objective function:

$$\min\{\max(T^{u_a^o}\}\}, \quad \forall u_a^o \in U, \quad \forall o \in O \tag{3}$$

$$T^{u_a^o} = \sum_{n_m^o \in N^o} \sum_{l \in L} \left(d_{o,n_m^o}^{u_a^o,l} / v_u + d_{n_m^o,o}^{u_a^o,l} / v_u + 2 \cdot at_l + 2 \cdot dt_l + s_{n_m^o}^{u_a^o} \right) \cdot x_{l,n_m^o}^{u_a^o}, \ \forall u_a^o \in U, \ \forall o \in O \tag{4}$$

Objective (3) minimizes the maximum delivery completion time across all drones. Constraint (4) defines the total time for a single drone to complete tasks, including delivery and return durations, ascent/descent times at distribution centers and customer stops, and waiting time due to delivery conflicts.

1) Energy Consumption Constraints

Constraints (5, 6, 7, 8, 9, and 10) are unloaded horizontal energy consumption (Lu et al., 2024), loaded horizontal energy consumption, loaded vertical ascent/descent energy consumption, unloaded vertical ascent/descent energy consumption, hovering energy consumption (Zhang et al., 2021), total energy consumption.

$$E_1 = \frac{(m_t + m_b) g \mu}{3600 \vartheta_v \eta} d_{n_m^o,o}^{u_a^o,l}, \forall l \in L, \ \forall u_a^o \in U^o, \ \forall o \in O, \ \forall n_m^o \in N^o \tag{5}$$

$$E_2 = \frac{\left(m_t + m_b + q_{n_m^o}^{u_a^o} \right) g \mu}{3600 \vartheta_v \eta} d_{o,n_m^o}^{u_a^o,l}, \ \forall l \in L, \ \forall u_a^o \in U^o, \ \forall o \in O, \ \forall n_m^o \in N^o \tag{6}$$

$$E_3 = \frac{\left(m_t + m_b + q_{n_m^o}^{u_a^o} \right) g h_l}{\eta}, \forall l \in L, \ \forall u_a^o \in U^o, \ \forall o \in O, \ \forall n_m^o \in N^o \tag{7}$$

$$E_4 = \frac{(m_t + m_b) g h_l}{\eta}, \ \forall l \in L \tag{8}$$

$$E_5 = \frac{\sqrt[3/2]{\left(m_t + m_b + q_{n_m^o}^{u_a^o} \right) g}}{\eta \sqrt{2 \rho S}} \cdot \frac{t_{hor}}{60}, \ \forall u_a^o \in U^o, \ \forall o \in O, \ \forall n_m^o \in N^o \tag{9}$$

$$E_{total} = E_1 + E_2 + 2 \cdot E_3 + 2 \cdot E_4 + E_5 \tag{10}$$

Battery safety constraints:

$$E_{cur} - E_{total} < E_{\max} - E_{\max} / \mu \tag{11}$$

$$E_{cur} - (E_2 + 2 \cdot E_3 + E_5) < E_{\max} - E_{\max} / \mu \tag{12}$$

$$E_{cur} > (E_1 + 2 \cdot E_4) \tag{13}$$

If unsatisfied battery safety constraints, the drone must recharge at the nearest distribution center:

$$E_6 = \frac{(m_t + m_b) g \mu}{3600 \vartheta_v \eta} d_{n_m^o,j}^{u_a^o,l}, \forall l \in L, \ \forall u_a^o \in U^o, \ \forall j \in O, j \neq \text{o}, \forall n_m^o \in N^o \tag{14}$$

$$E_7 = \frac{(m_t + m_b) g \mu}{3600 \vartheta_v \eta} d_{j,o}^{u_a^o,l}, \forall l \in L, \ \forall u_a^o \in U^o, \ \forall j \in O, j \neq \text{o}, \forall o \in O \tag{15}$$

$$E_{cur} > E_6 + 2 \cdot E_4, \ E_7 + 2 \cdot E_4 < \frac{E_{\max}}{\mu} \tag{16}$$

2) Payload Constraints

$$\sum_{n_m^o \in N^o} \sum_{l \in L} q_{n_m^o}^{u_a^o} x_{l,n_m^o}^{u_a^o} \leq W_{\max}^{u_a^o}, \ \forall o \in O, \ \forall u_a^o \in U \tag{17}$$

3) Altitude Constraints

Safe separation intervals (longitudinal, lateral, vertical):

$$H_{\min} \leq \forall z_i \leq H_{\max} \tag{18}$$

$$\begin{cases} | x_i - x_j | \geq d_{lon} \\ | y_i - y_j | \geq d_{lat} \\ | z_i - z_j | \geq d_{vert} \end{cases}, \ \textit{At least two inequalities hold.} \tag{19}$$

3 Algorithm Design

To address the dynamic risk avoidance problem in multi-drone collaborative delivery within complex urban environments, a hybrid A*-whale optimization algorithm (Hybrid A*-WOA) is designed to solve it. The A* algorithm demonstrates high efficiency and precision in route planning. Its heuristic search mechanism enables rapid identification of the shortest feasible routes for drones in complex urban environments, particularly excelling in static obstacle avoidance and real-time route adjustments. As a result, it has been widely adopted in drone route planning (Yoo et al. 2025). On the other hand, the WOA as a swarm intelligence method, specializes in global search and multi-objective optimization for complex problems (Seyedali et al. 2016). It effectively addresses NP-hard challenges in multi-drone collaborative task allocation (Zhang et al. 2024), achieving optimal order-to-drone matching. The WOA designed in this study is responsible for optimizing order allocation and flight altitude selection, while the improved A* algorithm performs route planning based on the allocation results. This algorithm combines 3D grid-based environmental modeling and multi-objective optimization to achieve collaborative optimization of task allocation, route planning, and dynamic obstacle avoidance for multiple drones.

3.1 Flight-Level Priority in Routes

In complex urban environments, three flight-level routes exist between each distribution center and customer stop. The cumulative risk degree and delivery time vary across these levels, necessitating a comprehensive analysis to prioritize flight-level selection. A comprehensive evaluation function is introduced to integrate cumulative risk and delivery time:

$$P(l) = \alpha \cdot R(l) + (1 - \alpha)T(l) \tag{20}$$

Where, $P(l)$ is comprehensive evaluation value for flight level l. $R(l) = \sum_{k \in path(l)} \phi_k$ is cumulative risk degree of flight level l. $T(l) = d_{o,n_m^o}^{u_a^o,l}/v_u + at_l + dt_l$ is delivery time for flight level l. α is weight coefficient ($0 \leq \alpha \leq 1$) reflecting risk aversion preference. Flight levels are ranked by $P(l)$, with lower values indicating higher priority. The level with the minimum $P(l)$ is selected as the optimal choice. In dynamic scenarios, $R(l)$ and $T(l)$ are updated in real time to ensure continuous optimality. In this paper, the weighting method is used to transform the multi-objective function into a linear objective function. Considering the difference in units, the unified scale requires the time cost and cumulative risk cost, and the data are mapped uniformly on the [0,1] interval using the min-max normalization method.

$$T(l) = \frac{T(l) - \min(T(L))}{\max(T(L)) - \min(T(L))}, R(l) = \frac{R(l) - \min(R(L))}{\max(R(L)) - \min(R(L))} \tag{21}$$

3.2 Hybrid A*-WOA

1) Generate initial solution

Real-number encoding is adopted to represent task sequences. For example, a task sequence "1-3-2-4" corresponds to customer stop codes "1-2-3-1." To enhance optimization, sequences are reordered based on payload constraints. For instance, "1-4-3-2" may be generated, mapping to customer stops "1-1-2-3," allowing a single drone to handle multiple tasks at the same stop within payload limits.

2) Improved A* algorithm for route construction

To overcome the inefficiency of traditional A* in large-scale grids, we integrate line-of-sight (LOS) detection and local obstacle circumvention strategies. The comprehensive cost function for route planning is defined as:

$$f(n) = g(n) + h(n) + \gamma \cdot \Phi(n) \tag{22}$$

$g(n) = \sum_{k \in path} d_k$ is accumulated route length from the start node to node n. $\Phi(n) = \sum_{k \in path} \phi_k$ is cumulative risk along the route. $h(n) = \sqrt{(x_n - x_{goal})^2 + (y_n - y_{goal})^2}$ is euclidean distance from node n to the goal. γ is risk weight coefficient. Obstacle Avoidance Steps:

Step 1: Utilize Bresenham's line algorithm (Kaleem et al. 2021)' and 'Kaleem MK, Verma D, Idrisi MJ, Ieee: Generalization of Line Drawing Algorithm - An effective approach to minimize the error in the existing Bresenham's Line Drawing Algorithm. 2021 International Conference on Emerging Smart Computing and Informatics (ESCI), 2021:516-521.'
Step 2: Upon obstacle detection, perform a localized A* search algorithm (He et al. 2022) within a restricted radius to bypass the obstacle.
Step 3: Resume LOS-based navigation post-circumvention; repeat Steps 1–2 until reaching the goal.

3) Position update process

Phase 1: Encircling Prey.

$$D = \left|C \cdot X^*(t) - X(t)\right|, \ X(t+1) = X^*(t) - A \cdot D \quad (23)$$

$$A = 2a \cdot r - a, \ C = 2 \cdot r \quad (24)$$

Phase 2: Bubble-Net Attack. Select between shrinking encircling or spiral motion with probability p.

$$X(t+1) = \begin{cases} X^*(t) - A \cdot D & if \ p < 0.5 \\ D^{'} \cdot e^{bl} \cdot \cos(2\pi l) + X^*(t) & if \ p \geq 0.5 \end{cases} \quad (25)$$

Phase 3: Random Search.

$$D = |C \cdot X_{rand} - X|, \ X(t+1) = X_{rand} - A \cdot D \quad (26)$$

The objective function (3) is used to evaluate solutions, prioritizing those with shorter completion times and constraint satisfaction to accelerate convergence.

4 Simulation Experiment and Result Analysis

4.1 Data Sources and Experimental Settings

Parameters are configured as shown in Table 1, referencing established literature. Task data from October 24, 2022, in a selected urban area are analyzed. The experiment involves 2 distribution centers and 5 customer stops, with tasks generated within 5 min of merchant order readiness. Each center is equipped with 5 homogeneous drones.

Table 1 Simulate experimental parameters

Experimental Parameter	Symbol	Value	Experimental Parameter	Symbol	Value
Drone horizontal speed	v_u	16 m/s	Drone ascent speed	v_a	5 m/s
Drone descent speed	v_d	4 m/s	Weight coefficient	α	0.5
Maximum battery capacity	$E_{\max}$	154 W•h	Drone maximum payload	$W_{max}^{u_a^o}$	5 kg
Grid cell edge length	l_{grid}	5 m	Drone weight (body + battery)	$m_t + m_b$	10.1 kg
Battery safety factor	μ	1.25	Gravitational acceleration	g	9.8 m/s^2
Standard air density	ρ	1.225 kg/m^3	Rotor area	S	0.12m^2
Drone ascent-to-drag ratio	ϑ_v	3.5	Drone energy conversion efficiency	η	0.66

4.2 Simulation Experiment Results

The proposed Hybrid A*-Whale Optimization Algorithm is implemented to achieve task allocation and route planning for multi-drone systems under multiple distribution centers. The WOA population size is set to 10, the risk weight coefficient is configured as 100, and the algorithm runs for 100 iterations. The final task allocation scheme for drone delivery is presented in Table 2, with corresponding route planning results illustrated in Figs. 3 and 4.

Table 2 Task allocation scheme

centers	Drone	Risk	Order	Flight high	Arrival time at each node(s)
1	1	0	[14;16]-[7;12]	50-50	T1(110.03)-T01(220.05)-T4(348.05)-T01(476.04)
	2	7.38	[2;3]-[9]	50-50	T5(157.65)-T01(315.3)-T5(472.96)-T01(630.61)
	3	6.46	[10;4;11]-[1;5;6]	80-80	T1(146.69)-T01(272.72)-T2(466.66)-T01(660.59)
	4	9.23	13	50	T3(285.76)-T02(404.62)-T01(614.08)
	5	3.07	[15;8]	110	T5(204.32)-T01(394.41)
2	1	16.04	30	50	T2(304.87)-T01(484.53)-T02(693.99)
	2	0	[19;24;26;17]-[28]	50-50	T3(118.86)-T02(237.72)-T3(356.59)-T02(475.45)
	3	6.92	[20;27]-[29]	50-50	T4(136.12)-T02(272.24)-T1(413.46)-T02(554.69)
	4	0	[21;25;18]	80	T3(155.53)-T02(290.39)
	5	0	[22;23]	110	T3(187.53)-T02(338.39)

Based on the task allocation results, drones are assigned distinct mission sequences according to their respective distribution centers, customer stop locations, and risk factors. T01 and T02 denote distribution center nodes, while other labels represent customer stop nodes. After reaching a delivery node, drones with sufficient battery return to their assigned distribution centers to await subsequent tasks. For example, drone 1 from distribution center 1 initially carries orders 14 and 16 to customer stop 1 at a flight altitude of 50 m. After completing delivery, it returns to distribution center 1 and proceeds to customer stop 4 for orders 7 and 12, reassigned to a new flight altitude. If a drone lacks sufficient energy to return, it diverts to the nearest distribution center for recharging. For example, drone 1 from distribution center 2, after delivering order 30, recharges at distribution center 1 before returning. Overall, all drones successfully completed their assigned tasks following planned routes and sequences, demonstrating effective collaborative delivery in complex urban environments.

Furthermore, cumulative risk values and delivery times vary across flight altitudes. For example, drone 5 from distribution center 1 operating at 110 m completed delivery to node T5 with a risk value of 3.07 and an arrival time of 204.32 s. In contrast, drone 5 from distribution center 2, also at 110 m, delivered to node T3 with zero risk and an arrival time of 187.53 s. The comprehensive evaluation function dynamically selects optimal flight altitudes to balance risk and time costs.

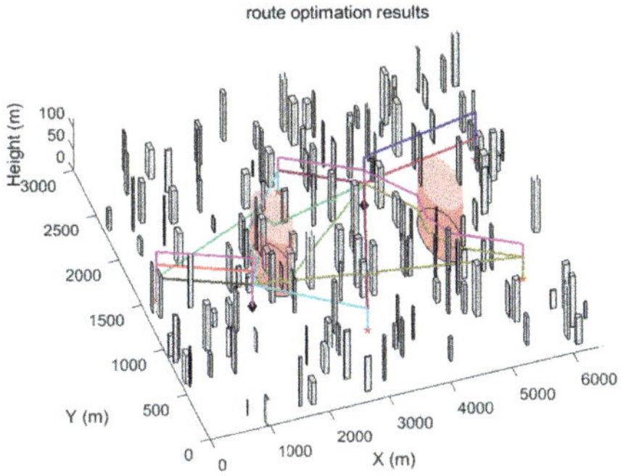

Fig. 3. 3D Route Map

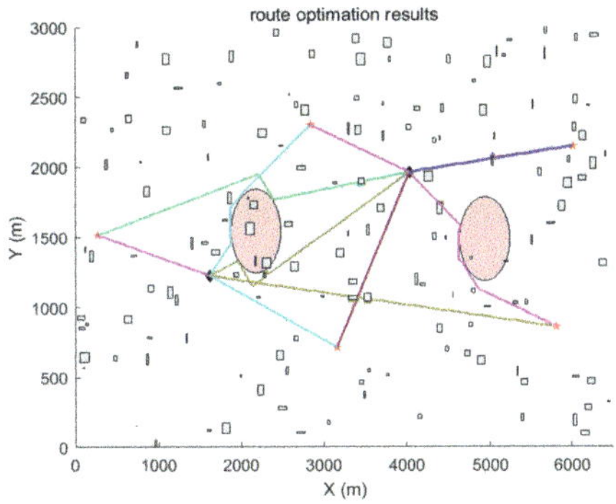

Fig. 4. Overhead Route Map

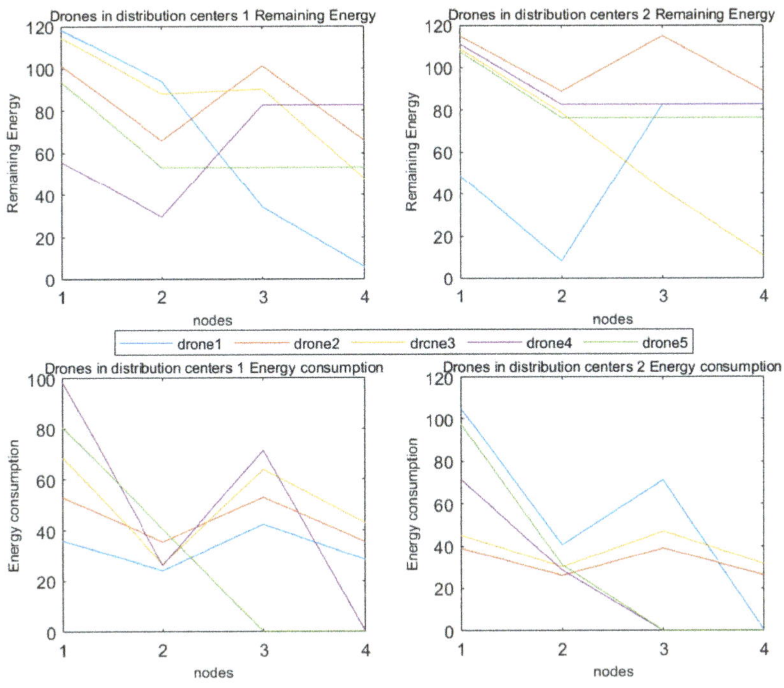

Fig. 5. Energy consumption and remaining battery levels across distribution centers

As shown in Fig. 3, the 3D route map clearly visualizes drone trajectories across urban airspace. Drones navigate through different altitude layers, avoiding obstacles and temporary no-fly zones while selecting safe and efficient routes. The map highlights their tendency to avoid high-risk areas and dynamically adjust altitudes based on real-time risk assessments. Figure 4 demonstrating obstacle avoidance strategies for ground-level infrastructure and customer stop distributions. This approach minimizes conflicts with ground facilities and other drones, enhancing delivery safety and efficiency.

Figure 5 illustrates the dynamic changes in energy consumption and remaining battery levels as drones reach nodes. During flight, energy is continuously depleted due to aerodynamic resistance, payload transport, and altitude adjustments. For example, drone 1 from distribution center 1 initially departs with orders 14 and 16 for customer stop 1 at 50 m, arriving in 110.03 s with high remaining battery. After returning to recharge, it proceeds to customer Stop 4 for orders 7 and 12, showing reduced battery levels. Similarly, drone 1 from distribution center 2 recharges at distribution center 1 after delivering order 30, resulting in a significant recovery of battery capacity.

4.3 Sensitivity Analysis

Scenario: Dynamic Emergence of Temporary No-Fly Zones.

Three temporary no-fly zones (simulating emergencies) are randomly introduced during mission execution, each with a 200-m radius and 30-min duration. Drones in flight continue their original routes to the next node before replanning subsequent routes. As shown in Fig. 6, route adjustments lead to extended flight distances and increased delivery times. For instance, drone 2 at distribution center 2 was originally scheduled to return to node T02 in 272.23 s. However, due to the emergence of a no-fly zone, the actual arrival time increased to 281.92 s.

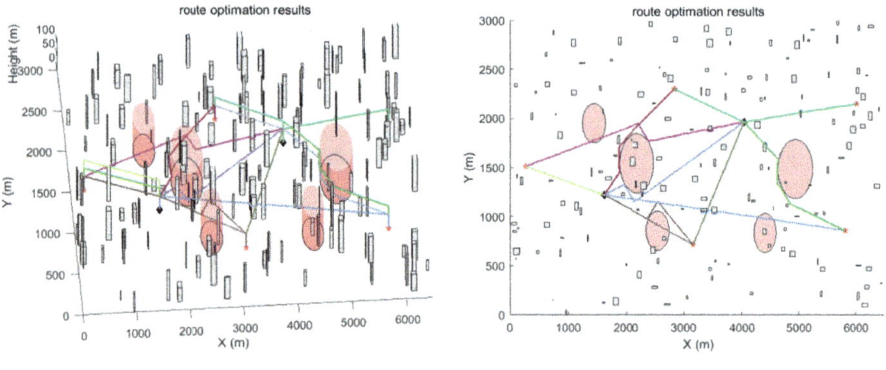

Fig. 6. Route replanning under temporary no-fly zones

5 Conclusion

To address challenges such as obstacles at varying flight altitudes, temporary no-fly zones, and route conflicts in complex urban environments, this study investigates multi-drone collaborative task allocation and route planning. A cooperative delivery model for multi-distribution centers is constructed, aiming to minimize the maximum delivery completion time and cumulative risk degree across all drones. For this model, a Hybrid A*-Whale Optimization Algorithm is designed by integrating the A* algorithm's local search capability with the global optimization strengths of the whale algorithm. However, the current research has limitations requiring further improvement. For instance, the model assumes fixed drone flight altitudes and ignores weather impacts. Future studies could introduce dynamic altitude adjustment strategies and meteorological risk prediction modules to enhance adaptability. Additionally, the analysis focuses on single-task scenarios; integrating dynamic task updates for continuous or multi-phase operations would better reflect real-world logistics demands.

Acknowledgments. This work is supported by the Fundamental Research Funds for the Central Universities, NO.NJ2024025.

References

Pons-Prats, J., Živojinović, T., Kuljanin, J.: On the understanding of the current status of urban air mobility development and its future prospects: commuting in a flying vehicle as a new paradigm. Transp. Res. Part E Logist. Transp. Rev. **166**, 102868 (2022)

Aurambout, J.-P., Gkoumas, K., Ciuffo, B.: Last mile delivery by drones: an estimation of viable market potential and access to citizens across European cities. Eur. Transp. Res. Rev. **11**, 30 (2019)

Dorling, K., Heinrichs, J., Messier, G.G., Magierowski, S.: Vehicle routing problems for drone delivery[J]. IEEE Trans. Syst. Man Cybern. Syst. **47**(1), 70–85 (2017)

Song, B.D., Park, K., Kim, J.: Persistent UAV delivery logistics: MILP formulation and efficient heuristic[J]. Comput. Ind. Eng. **120**, 418–428 (2018)

Cheng, C., Adulyasak, Y., Rousseau, L.M.: Drone routing with energy function: formulation and exact algorithm[J]. Transp. Res. B Methodol. **139**, 364–387 (2020)

Ozkan, O.: Multi-objective optimization of transporting blood products by routing UAVs: the case of Istanbul. Int. Trans. Oper. Res. **30**, 302–327 (2023)

Li, H., Wang, F., Zhan, Z.: Drone routing problem with swarm synchronization[J]. Eur. J. Oper. Res. (2023)

Wen, X., Guohua, W.: Heterogeneous multi-drone routing problem for parcel delivery[J]. Transp. Res. C. **141**(103763), 1–22 (2022)

Kim, Y., Jeong, H.Y., Lee, S.: Drone delivery problem with multi-flight level: machine learning based solution approach. Comput. Ind. Eng. **197**, 110565 (2024)

Shao, Q., Li, J., Li, R., Zhang, J., Gao, X.: Study of urban logistics drone path planning model incorporating service benefit and risk cost. Drones. **6**(12), 418 (2022)

Lee, S., Hong, D., Kim, J., et al.: Congestion-aware multi-drone delivery routing framework[J]. IEEE Trans. Veh. Technol. **71**(9), 9384–9396 (2022)

Zhu, H., Wang, Y., Li, X.: UCAV path planning for avoiding obstacles using cooperative co-evolution spider monkey optimization. Knowl.-Based Syst. **246**, 108713 (2022)

Wang, Q., Xu, M., Hu, Z.: Path planning of unmanned aerial vehicles based on an improved bio-inspired tuna swarm optimization algorithm. Biomimetics. **9**(7), 388 (2024)

Torabbeigi, M., Lim, G.J., Kim, S.J.: Drone delivery scheduling optimization considering payload-induced battery consumption rates[J]. J. Intell. Robot. Syst. **97**(3), 471–487 (2020)

Kyriakakis, N.A., Aronis, S., Marinaki, M., et al.: A GRASP/VND algorithm for the energy minimizing drone routing problem with pickups and deliveries[J]. Comput. Ind. Eng. **109340**, 1–13 (2023)

Evers, L., Dollevoet, T., Barros, A.I., Monsuur, H.: Robust UAV mission planning. Ann. Oper. Res. **222**(1), 293–315 (2014)

Vural, D., Dell, R.F., Kose, E.: Locating unmanned aircraft systems for multiple missions under different weather conditions. Oper. Res. Int. J. **21**, 725–744 (2021)

Li, Z., Li, S., Lu, J., Wang, S.: Air route network planning method of urban low-altitude logistics UAV with double-layer structure. Drones. **9**(3), 193 (2025)

Tadić, S., Krstić, M., Veljović, M., Čokorilo, O., Milovanović, M.: Risk analysis of the use of drones in city logistics. Mathematics. **12**(8), 1250 (2024)

Lu, F., Jiang, R., Bi, H., Gao, Z.: Order distribution and routing optimization for takeout delivery under drone–rider joint delivery mode. J. Theor. Appl. Electron. Commer. Res. **19**(2), 774–796 (2024)

Zhang, J., Campbell, J.F., Sweeney II, D.C., Hupman, A.C.: Energy consumption models for delivery drones: a comparison and assessment. Transp. Res. Part D: Transp. Environ. **90**, 102668 (2021)

Yoo, Y.-D., Moon, J.-H.: Study on A-Star algorithm-based 3D path optimization method considering density of obstacles. Aerospace. **12**(2), 85 (2025)

Mirjalili, S., Lewis, A.: The whale optimization algorithm. Adv. Eng. Softw. **95**, 51–67 (2016)

Zhang, H., You, M., Zhao, Z., Li, J.: Improving multi-aspect QoE for large-scale O2O order allocation and distribution problem using social behavior whale optimization. Asia-Pac. J. Oper. Res. **42**(01), 2440019 (2024)

He, Z., Liu, C., Chu, X., Negenborn, R.R., Wu, Q.: Dynamic anti-collision A-star algorithm for multi-ship encounter situations. Appl. Ocean Res. **118**, 102995 (2022)

Li, X., Shao, X.: Fourth International Conference on Machine Vision (ICMV 2011): Machine Vision, Image Processing, and Pattern Analysis (2012)

Balancing Innovation and Values: How Culture Shapes AI Regulation

Debarya Dutta(✉)

University of Cambridge, Cambridge, UK
dd644@cam.ac.uk

Abstract. This paper examines how regulations in artificial intelligence (AI) are created, with a primary focus on the European Union's AI Act. Employing a comparative qualitative analysis grounded in social contract theory, historical institutionalism, and cultural theory, it explores how cultural, historical, and political contexts shape AI regulation. The EU AI Act is analyzed as a central case study to demonstrate how societal values and historical experiences influence regulatory approaches. To further contextualize these insights, China's AI policy is used as a comparative example, highlighting how differing cultural and historical factors drive contrasting regulatory strategies. The paper makes two key contributions: it illustrates how AI regulation and progress in AI is shaped by cultural and historical factors, and explains why AI policies that align with a society's values are more likely to succeed. By offering a clear understanding of the interplay between culture, regulation, and technological progress, the paper contributes to the ongoing discourse on creating effective AI policies to balance regulation and innovation.

1 Introduction

As China accelerates AI deployment, Europe adopts a cautious stance, prioritizing regulation. This contrast prompts a key question: Why do nations differ in their regulatory approaches to innovation? AI innovation drives economic growth, social progress, and global influence, but also poses risks such as algorithmic bias, privacy breaches, and ethical conflicts in autonomous systems. Regulation sets limits to ensure technology aligns with public trust and ethical values. Europe's focus on privacy and human dignity reflects its historical experiences with political oppression and technological misuse, whereas China's state-driven strategy prioritizes technological dominance and sovereignty. Analyzing these divergent strategies reveals how cultural and historical contexts shape regulatory priorities and, consequently, the course of technological advancement.

This paper explores how regulation is shaped by cultural and historical influences that are key to fostering ethical and sustainable technological development. Drawing on social contract theory, it contends that regulation is crucial for upholding societal trust and addressing the ethical challenges posed by AI. Historical institutionalism shows how past experiences influence the structure of current institutions, while cultural theory highlights the role of societal values in shaping regulatory priorities. By combining these perspectives, the paper demonstrates how cultural and historical contexts guide the interplay between regulation and innovation.

X. Xie et al. (Eds.): ICSS 2025, LNOR, pp. 51–65, 2026.
https://doi.org/10.1007/978-3-032-13116-4_5

Building on these concepts, the EU AI Act serves as a compelling case study for illustrating how ingrained cultural and historical factors inform regulatory choices. The EU AI Act embodies Europe's commitment to human rights, standing as the world's first comprehensive regulatory framework for artificial intelligence. The rationale behind the EU AI Act's design reveals the influence of historical experiences and societal values on regulatory strategies. To further illustrate this point, the paper examines China's AI policy, demonstrating how its regulatory choices have fostered a unique model of innovation driven by centralized governance and a focus on technological sovereignty. By juxtaposing the EU AI Act with China's AI strategy, this paper highlights how it affects successful regulation.

In Sect. 2, we will debate on whether emerging technologies posit regulation as either a stifle to progress or a necessary enabler. Section 3 explores the literature explaining AI regulation. Section 4 focuses on why regulation is necessary. Section 5 explores how cultural, social, and historical contexts shape the regulation versus innovation debate. Section 6.1 introduces the EU AI Act, followed by Sect. 6.2, which addresses the challenges it poses. Sections 6.3 and 6.4. delve into the cultural underpinnings of the AI regulations in the EU and China, respectively. Section 7 considers an opposing view around universal regulations. Finally, Sect. 8 concludes.

2 The Regulatory Paradox: Hindrance and Catalyst for AI

The governance of emerging technologies like artificial intelligence is defined by a central debate: does regulation act primarily as a brake on progress or as a necessary guardrail that enables sustainable innovation? One side of this debate argues that overbearing regulation can stifle advancement and prompt innovators to relocate their efforts to jurisdictions with less stringent frameworks. This concern is not merely theoretical. In the semiconductor industry, for example, the Dutch firm ASML is reconsidering its location due to restrictive export controls, while studies on French regulations show they can increase costs for firms, reducing innovation intensity by 5.4% and overall welfare by 2.2% [1]. Such policies often impose a disproportionate burden on small and medium-sized enterprises (SMEs), limiting their capacity for growth and competition.

Although federal regulations since 2012 have imposed an estimated $3.079 trillion in costs, studies reveal they have also encouraged cleaner technologies, improved operational efficiency, and supported long-term sustainable development [2]. U.S. policies like the Clean Air Act demonstrate how well-designed frameworks foster cleaner technologies, operational efficiency, and sustainable development. Similarly, laws like the Endangered Species Act balance high standards with societal benefits such as biodiversity protection. Effective regulations build trust by addressing safety, privacy, and public health, as seen with U.S. seatbelt laws saving over 226,000 lives between 1975 and 2006 [3]. These outcomes show how regulations, when effectively implemented, create conditions for both technological advancement and public welfare. While stringent regulations can sometimes hinder innovation and prompt firms to relocate, well-crafted regulatory frameworks can drive technological progress by fostering trust, enhancing efficiency, and ensuring long-term sustainability.

3 Literature Review

The literature on AI regulation highlights diverse theoretical and practical approaches but reveals a clear gap in addressing how cultural and historical contexts shape effective regulatory frameworks. Few scholars advocate for a global governance model, such as an International Artificial Intelligence Organization (IAIO), to address AI's borderless nature and regulatory inertia [4]. Their focus on the principal-agent problem underscores tensions between societal welfare and technological progress but overlooks how cultural values influence regulatory priorities across nations. Similarly, Scherer emphasizes AI's autonomy as a regulatory challenge, advocating for flexible architectures prioritizing transparency and human oversight [5]. While they align with social contract theory, their analyses remain technical, neglecting the socio-cultural dynamics that determine regulatory legitimacy.

Cultural and historical influences on AI regulation are partially addressed in the literature. Golpayegani links the EU AI Act's risk-based framework to Europe's historical sensitivity to privacy violations, such as data misuse during the Nazi regime, and high digital literacy [6]. In contrast, China's AI policies, like the Next Generation Artificial Intelligence Development Plan, within Confucian values and historical imperatives like the Century of Humiliation, prioritize technological sovereignty. These studies confirm that cultural alignment enhances regulatory success but focus narrowly on specific jurisdictions without integrating broader theoretical frameworks. Bresnahan and Trajtenberg and Eloundou et al. frame AI as a general-purpose technology requiring tailored regulations [7, 8]. However, their analyses remain descriptive, lacking a cohesive framework to explain why cultural and historical factors drive regulatory outcomes. This gap is evident in the absence of multidisciplinary approaches combining social contract theory, historical institutionalism, and cultural theory to elucidate the interplay of culture, history, and AI governance.

This paper addresses this obvious gap by offering an integrative framework that explains how and why cultural and historical contexts shape AI regulation. By synthesizing social contract theory, historical institutionalism, and cultural theory, it provides a novel lens to understand why AI policies succeed when aligned with societal values, extending beyond the technical and legal focus of prior studies.

4 The Necessity of Regulation

Social contract theory provides a framework for justifying the need of state regulation. The theory proposes that individuals agree, either explicitly or implicitly, to form a society and establish a governing body to maintain order, protect rights, and promote the common good [9]. The legitimacy of such governance arises from the consent of the governed, ensuring that power is exercised justly and aligns with societal needs. The government governs through regulations enforced by legal mechanisms, guiding both individual and organizational behavior [10]. While the need to maintain a social contract is evident in democratic systems, maintaining a social contract is vital, even in autocracies. Although autocratic regimes lack democratic freedoms, they must still fulfill citizens' basic needs such as security, economic growth, and essential services to ensure legitimacy and stability. When they fail to meet these expectations, as evidenced during the Arab Spring, unrest can erupt.

Regulating AI is essential to maintaining the social contract, which guarantees equal fundamental rights, preserves lives, and ensures societal order. While technological advancements in AI offer significant benefits, they also pose substantial risks, such as privacy violations and the potential for biased systems that can cause discrimination by reinforcing societal prejudices [11]. These biased AI systems often worsen disparities, leading to unfair treatment of minority and marginalized groups. Such outcomes violate the social contract, which emphasizes fairness, equality, and the protection of fundamental rights. This breach of the social contract becomes even more concerning when we think of AI-driven weapon systems. Governments have a responsibility to protect civilian lives, yet without strong regulatory frameworks, autonomous AI-powered weapons could operate independently, making life-or-death decisions with devastating consequences [12]. This lack of control increases the risk of uncontrollable conflicts and civilian casualties, further highlighting the urgent need for comprehensive regulation.

The cryptocurrency ecosystem highlights how inadequate regulation can expose individuals to significant financial and social risks, undermining trust between citizens and the government. Only one-third of countries have consumer protection rules for crypto, despite the market's rapid growth and its parallels to the 2007–08 financial crisis [13]. This lack of oversight has led to widespread fraud, with Americans losing $679 million to crypto scams in the first half of 2024 [14]. The Silk Road marketplace further demonstrates how unregulated technology use can facilitate illegal activities, eroding financial security and public trust [15]. The platform exclusively utilized Bitcoin for anonymous transactions, allowing users to buy and sell illegal goods, including drugs, firearms, and stolen data, without detection.

While in theory a social contract can be maintained without formal regulation through mechanisms like social norms, self-regulation, and market dynamics, these approaches often face significant limitations in practice. Communities may rely on shared values, reputation systems, or voluntary industry standards to guide behavior, but social norms often fail to constrain powerful corporations or influential entities whose actions have widespread societal impacts, as these actors frequently prioritize profits over ethical considerations [16]. Similarly, self-regulation can create conflicts of interest, with industries favoring their own objectives over public welfare, as evidenced by environmental damage caused by weak corporate oversight. Market-based solutions, while effective in driving innovation and rewarding ethical practices, tend to disproportionately benefit consumers with purchasing power, leaving marginalized groups vulnerable to exploitation or exclusion [17]. Decentralized systems, such as cryptocurrencies, also face significant challenges. The infamous Silk Road marketplace shows how the lack of centralized oversight can facilitate illegal activities, undermining societal trust and security. Without a central authority to enforce rules or ensure accountability, these approaches lack the uniformity, scalability, and legal backing needed to address systemic or transnational problems effectively.

5 How Historical Events and Societal Attitudes Influence Regulation

The evolution of social contracts is best understood as a relationship shaped by historical events, and cultural values. At its core, the social contract represents an ongoing agreement between the state and its citizens about power and responsibilities, rooted in both historical experiences and shared societal belief. Social contract is a dynamic entity influenced by historical processes and social norms. Here, historical institutionalism provides insight by showing how key events, such as colonial rule or revolutions, establish systems and expectations that continue to shape societies over time [18]. Building on this, cultural theory adds a crucial perspective by explaining how values and identities shape what people expect from their governments and what they are willing to contribute in return [19]. Together, these frameworks create a more complete picture of how social contracts evolve and adapt.

This perspective highlights both the differences and similarities in social contracts. Post-colonial states often retain centralized authority, reflecting the legacy of colonial rule and Hobbes's idea of a strong state [20]. Over time, these systems evolve, shaped by cultural traditions and increasing demands for democracy. On the other hand Scandinavian welfare states embody Rousseau's concept of the collective will. Their foundations lie in early cooperative societies and labor movements that advocated for social protections [21]. Additionally, relatively homogeneous populations and less rigid social hierarchies helped build trust among citizens, making policies like universal healthcare and progressive taxation widely accepted.

This framework also helps explain how societies respond to crises. Events like the COVID-19 pandemic force governments to adapt, focusing on public health, economic support, and digital governance. Responses to COVID-19 are shaped by historical legacies and cultural expectations. Collectivist societies in East Asia embraced stringent measures, such as lockdowns and digital tracking, as part of their shared responsibility for public welfare [22]. By contrast, individualistic societies viewed similar measures as threats to personal freedom, leading to resistance. The pandemic also emphasized the growing importance of digital governance. Governments relied heavily on technology for tasks like contact tracing, vaccine distribution, and managing remote work.

The social contract is neither static nor purely theoretical. It is a dynamic framework that evolves with shifts in history, culture, and governance, enabling societies to address pressing modern issues such as climate change, digital privacy, and economic inequality. By adapting to these challenges, societies can reshape their social contracts while remaining anchored in their unique histories and values. Far from being an abstract idea, the social contract serves as a practical tool for balancing power, fairness, and goals in a changing world. This perspective allows us to explore why some countries prioritize strict AI regulations, even at the expense of innovation, while others emphasize fostering innovation, often compromising ethical safeguards. Despite their contrasting approaches, both strategies reflect efforts to preserve the social contract within their specific historical and cultural contexts.

6 EU AI Act

We will explore the regulation-innovation debate through the lens of the EU AI Act.

6.1 Overview of the EU AI Act and Key Provisions

The EU AI Act establishes a regulatory framework that focuses on ensuring that AI systems used in the European Union are safe and respect fundamental rights. Drafted within the context of the EU's Digital Strategy, it represents the first legal framework globally designed to regulate AI systems comprehensively [23].

The EU AI Act categorizes AI systems into four risk levels: minimal, limited, high, and unacceptable risk. High-risk AI systems, which pose significant threats to safety, health, or fundamental rights, are subject to the most stringent regulatory requirements. These systems must adhere to rigorous standards [23]. They must have robust risk management systems to identify, assess, and mitigate potential risks. High-quality, unbiased datasets must be used for training, and strict data governance practices must be followed. Detailed logs and documentation must be maintained to enable traceability and accountability [23] Effective human oversight mechanisms must be implemented to monitor AI systems and intervene when necessary. AI systems must also be transparent and explainable, allowing users to understand how decisions are made.

Both providers and deployers of AI systems are held accountable under the EU AI Act [24]. Providers are responsible for developing and placing AI systems on the market, ensuring compliance with regulatory requirements. Deployers, on the other hand, are responsible for putting AI systems into service and ensuring their appropriate use. To enforce compliance, the EU empowers regulatory authorities to conduct audits and inspections. Non-compliance with the Act can result in significant penalties, including fines and restrictions on AI deployment [25]. By establishing a clear regulatory framework and enforcing strict compliance measures, the EU aims to promote responsible AI development and usage.

To support innovation, the Act introduces regulatory sandboxes, allowing businesses to test AI technologies in controlled environments without facing the full compliance burden [26]. Additionally, it emphasizes collaboration among stakeholders industry, academia, and policymakers to encourage experimentation and knowledge sharing. These mechanisms aim to position the EU as a global leader in ethical and responsible AI development while fostering safe technological advancement.

Grounded in principles of precaution, transparency, and accountability, the EU AI Act reflects a regulatory approach that prioritizes human dignity and fundamental rights. The risk-based categorization system demonstrates how the precautionary principle is operationalized, tailoring regulatory requirements to the potential for harm. Transparency measures, such as detailed documentation and explainability, ensure public trust and meaningful oversight. Together, these elements create a framework that aligns technological innovation with societal values, addressing both opportunities and risks in the digital age.

6.2 Challenges Posed by the EU AI Act

While the EU AI Act aims to ensure ethical use and safety of AI technologies, its stringent requirements and comprehensive framework may inadvertently slow the velocity of AI innovation. Compliance involves extensive documentation and transparency, testing, risk management procedures, and ensuring alignment with data governance standards, all of which collectively extend development timelines and resources. Experts estimate the range for compliance costs for large organizations varies from €2 million to €10 million per year, while smaller firms may face costs of approximately €250,000 annually [27]. These expenses divert significant resources from R&D, limiting the ability of companies to innovate effectively. A survey indicated that 30% of firms might reduce their R&D budgets by about 15% to allocate funds toward compliance, suggesting a potential loss of innovation funding on the order of €750,000 for medium-sized enterprises. Such financial constraints are likely to stifle innovation within the EU. Approximately 25% of surveyed small to medium-sized enterprises (SMEs) indicated they would consider relocating operations outside the EU due to stringent regulatory requirements [28]. The cumulative effect of these financial and operational challenges may not only suppress innovation but also weaken the EU's competitive edge in emerging technologies, with the AI Act posing a particularly significant obstacle.

The EU AI Act requires AI systems to undergo risk categorization and thorough conformity assessments significantly delays market entry, as procedural complexity and ongoing compliance checks extend launch timelines. Such delays give a competitive edge to firms outside the EU developing under lighter regulatory frameworks, enabling them to innovate and deploy solutions more rapidly [29]. The EU's AI lag could compromise national security. Non-EU countries are advancing AI military technologies, including autonomous systems and surveillance. This risks leaving the EU behind in critical defense capabilities and innovations [30]. Countries leading in AI innovation will shape global standards and ethics, potentially diminishing the EU's influence. While the EU AI Act aims to create a principled framework, its stringent regulations risk slowing innovation and causing innovation flight, especially for SMEs. To maintain global leadership, the EU must balance regulation with flexibility, ensuring its measures support rather than hinder progress.

6.3 Cultural Foundations of the EU's Approach to Regulation

The EU AI Act mirrors European society's historical and cultural relationship with technological innovation, shaped by centuries of intellectual experience and lessons from social and political upheavals. Rooted in traditions of social democracy and a proactive stance toward technological risks, Europe's regulatory framework prioritizes ethical safeguards and privacy.

Historically, Europe's response to emerging technologies has a precedent for immediate regulatory intervention. France's establishment of image rights in 1858 and Germany's development of "personality rights" to address unauthorized photographs show how Europe seeks to preempt harm rather than respond reactively [31]. These historical experiences have influenced the EU AI Act, which builds on this tradition by incorporating mechanisms such as risk-based categorization and transparency requirements. For instance, the classification of high-risk AI applications, such as facial recognition for policing, reinforces Europe's commitment to ethical oversight, ensuring that innovation aligns with privacy and human rights laws. This commitment is reinforced in Article 2(7) and Recital 69, which mandate that existing EU laws on privacy and data protection must be adhered to when deploying AI technologies [6].

These historical and cultural drivers are further reinforced by Europe's high levels of digital literacy. With nine of the top ten most digitally literate countries located in the European Union, public demand for privacy protection is not only strong but also informed. This creates a regulatory environment where trust in digital systems becomes both a societal expectation and a policy imperative. The EU AI Act leverages this dynamic to align governance with public values, fostering a regulatory framework that resonates with European citizens while ensuring accountability from AI developers.

However, Europe's emphasis on precaution is not just rooted in intellectual traditions but in the collective trauma of its past. The misuse of data during the Nazi regime, where tools like census records and IBM punch cards were repurposed for genocide, left an indelible mark on European governance. This historical lesson underscores the dangers of unregulated technological advancement, driving the EU to embed principles of accountability and harm prevention into its regulatory approach. Transparency mandates in the EU AI Act, requiring developers to disclose decision-making processes, directly reflect these historical lessons, ensuring that innovation serves societal well-being rather than risking misuse.

Europe's deep concern with existential risks stems from a history punctuated by devastating events that have shaped its collective consciousness. Two World Wars, the Holocaust, the Cold War, and repeated periods of political turmoil have instilled a deep awareness of fragility and a powerful desire to preserve hard-won peace and stability. Indeed, the very creation of the European Union was driven by the imperative to prevent future continental catastrophes [32]. Consequently, this historical trauma has instilled a heightened sensitivity to existential threats, driving issues such as AI weaponization and the potential for humanity-ending AI to the center of public and political discourse in Europe. This focus is further amplified by Europe's comparatively free press and robust civil society. In contrast to more authoritarian states like China—where, despite experiences of war and pandemics, state-controlled media often emphasize narratives of national strength and resilience—European media operate with greater independence [33]. This freedom enables open, often uncomfortable discussions about worst-case scenarios, fostering a more critical and transparent exploration of existential threats. This historical sensitivity to existential risks is reflected in the EU AI Act, which prioritizes a precautionary and ethical approach to regulating AI. The Act explicitly addresses high-risk AI applications, such as AI weaponization, facial recognition, and systems that could undermine human rights or safety. By classifying certain AI uses like autonomous weapons or surveillance systems as "high-risk" or even "unacceptable" the EU AI Act seeks to mitigate existential harms [34]. The EU AI Act reflects lessons from historical traumas and public expectations, aligning with Europe's cultural values. This alignment ensures that innovation progresses responsibly, reinforcing Europe's position as a leader in ethical and precautionary regulation.

6.4 Cultural Heritage and Historical Legacy in China's AI Regulation and Innovation

China's journey to AI has been shaped by historical events like the Opium Wars and the 'Century of Humiliation', which left a legacy of vulnerability and a drive for self-reliance [35]. These events, including "unequal treaties," territorial losses, and Western economic domination, eroded China's sovereignty and technological resources, while the 1937 Nanjing Massacre under Japanese aggression deepened these scars [35]. These events fostered a collective memory of vulnerability and eroded sovereignty, driving the state to reposition its legitimacy by offering technological sovereignty and economic security as essential components of its modern social contract with citizens. Resistance to foreign interference and the pursuit of technological and economic progress have become central to China's efforts to avoid further "humiliations" while strengthening its global standing and gaining recognition as a major power. Leaders like President Xi Jinping have emphasized this vision through the "China Dream," connecting the nation's resurgence to its historical achievements and its 5,000-year-old civilization [36]. The NGAIDP represents a continuation of this trajectory, serving as a modern social contract that ensures technological self-reliance as a safeguard against external threats [37].

The NGAIDP outlines ambitious milestones, including establishing China as the global AI leader by 2030 and achieving 70% domestic production of critical AI-enabling technologies, such as semiconductors, by 2025 [37]. Policies under the NGAIDP include fostering state-funded research and industrial partnerships, as seen in the development of AI hubs in cities like Beijing and Shanghai. Shanghai has cultivated 250,000 AI professionals (up from 100,000 in 2018) [38]. In recent history U.S. sanctions on Huawei and restrictions on semiconductor exports highlighted the risks of foreign dependency, accelerating China's focus on R&D in 5G, HarmonyOS, and domestic chip production. Through these initiatives, the Chinese government upholds its social contract by addressing vulnerabilities and ensuring the nation's technological and economic resilience.

China's regulatory environment has also adapted to promote innovation. Policies like the Cybersecurity Law (2017) and Data Security Law (2021) emphasize the protection of critical technologies and data sovereignty, creating a secure ecosystem for domestic firms. These regulations limit foreign access to sensitive resources, encouraging local companies to innovate independently while fostering collaboration between state-owned enterprises, private firms, and academic institutions. Modern Chinese society embodies many Confucian principles, including a strong emphasis on education, and a focus on social harmony [39]. These values are reflected in China's AI policy, which emphasizes centralized authority and collective welfare.

The modern social contract reflects Confucian ideals, prioritizing collective welfare and societal harmony over individual autonomy. While not explicitly cited in policy documents, these principles shape decision-making, balancing technological advancement with social cohesion. President Xi Jinping's frequent references to Confucian teachings underscore their enduring relevance, ensuring AI policies resonate with cultural norms and reinforce the social contract [40]. Policies like Shenzhen's AI-driven surveillance systems illustrate the state's trade-off with citizens: reduced privacy is exchanged for enhanced public security and stability. From a social contract perspective, this arrangement reaffirms the state's role as the guarantor of societal welfare, aligning with Confucian principles that emphasize the collective good. Reflecting this perspective, 94% of Chinese consumers were willing to share personal data for convenience, more so than consumers in other countries [41]. Laws like the Personal Information Protection Law (PIPL) further treat data as a strategic national asset to serve state-defined priorities [21]. However, these trade-offs raise questions about the ethical dimensions of such policies and their alignment with broader global norms. For instance, while Confucian ideals might justify these measures domestically, they challenge the universal ethical standards that prioritize individual rights, exposing tensions in fulfilling the terms of this contract in an interconnected world.

China's reliance on exports to sustain employment for its large population reflects a critical aspect of its social contract: the government's role in ensuring economic stability and opportunity. AI sovereignty plays a key role in this strategy, positioning China as a global leader in shaping AI standards while addressing domestic priorities. By embedding "Chinese characteristics" into its policies, China promotes its governance model as a blueprint for nations seeking rapid modernization. For instance, countries like Zimbabwe and Uganda have adopted AI-driven surveillance systems inspired by China's approach, emphasizing state control and public safety [42]. These systems, often bundled with financial loans and technical support, make China's model particularly appealing to developing nations. This approach allows China to expand its geopolitical influence while fostering a network of countries aligned with its technological and political values.

On the other hand China's centralized governance model often clashes with the collaborative nature of global AI development. Strict data localization requirements and a state-driven approach hinder international research partnerships, contrasting with initiatives like the Global Partnership on Artificial Intelligence (GPAI), which champions multilateral cooperation [43]. This divide is reflected in China's limited role in shaping widely accepted ethical AI standards, such as those promoted by the EU and U.S., which prioritize transparency, accountability, and individual privacy. While this limitation is partly self-imposed stemming from China's prioritization of state sovereignty and reluctance to align with globally accepted norms it is also influenced by resistance from other nations. Western democracies, in particular, view China's policies, including AI-driven surveillance and strict data localization laws, as incompatible with their values. As a result, China has been excluded from key ethical AI standard-setting initiatives, such as GPAI. China's AI governance model struggles to appeal beyond its sphere of influence. These policies also pose challenges for domestic firms looking to expand internationally. Regulations like the Cybersecurity Law and PIPL create friction by increasing costs and limiting flexibility for companies such as Alibaba and Tencent [44]. Government oversight of AI algorithms stifles private sector autonomy, as seen with SenseTime and Megvii, which face regulatory hurdles in AI development [45]. The alignment of AI with national security goals further restricts international collaborations, frustrating researchers [46].

China's AI strategy reveals the inherent tensions of embedding historical imperatives into a modern governance framework. The invocation of the "Century of Humiliation" as a legitimizing narrative for technological sovereignty underscores a state-driven social contract that prioritizes collective security and national resilience. However, this framework relies heavily on centralized control, which, while effective in addressing immediate vulnerabilities, poses long-term challenges for fostering innovation and aligning with global ethical norms. The reliance on Confucian ideals, particularly the prioritization of societal harmony over individual rights, positions China's governance as culturally resonant domestically but increasingly contentious internationally. This cultural and political asymmetry creates friction in China's attempts to shape global AI standards, as its state-centric model contrasts sharply with liberal democratic values of transparency and individual accountability. Thus, China's AI strategy, while ambitious and historically informed, must grapple with the dual challenge of sustaining domestic legitimacy while negotiating a position in a global AI landscape that often resists its governance ideals.

7 Universal Regulations

Universal regulations often falter because they prioritize the interests of dominant nations while disregarding the cultural, political, and economic diversity of the global community. Although these frameworks are ostensibly designed to promote equality, their implementation often reveals deep-seated imbalances. The regulatory design process is frequently spearheaded by powerful countries or regions, embedding their values and objectives into global structures. This dynamic marginalizes alternative perspectives, transforming what should be platforms for equal governance into instruments of geopolitical influence. The evolving economic and geopolitical landscape further complicates the creation of new regulations. The era when the West could unilaterally craft rules and export them to the rest of the world has given way to a multipolar reality [47]. An example is the Budapest Convention on Cybercrime, which illustrates how treaties shaped by Western priorities can alienate major global powers. Nations like China and Russia view such agreements as threats to their sovereignty, highlighting the pitfalls of exclusionary frameworks [48]. This shows limited effectiveness of universal regulations.

8 Conclusion

Regulation and innovation in AI are intertwined forces, each shaped by a nation's social, historical, and cultural foundations. As this paper illustrates, frameworks like the EU AI Act and China's sovereignty-driven policies go beyond reactive measures, reflecting distinct societal priorities and ambitions. The EU demonstrates how principled oversight can embed human dignity and public trust in technological progress, while China underscores the potency of centralized governance for addressing urgent national imperatives. Theoretical perspectives help explain these dynamics. Social contract theory highlights the importance of regulation in maintaining order and addressing citizens' needs. Historical institutionalism shows how past events shape governance and social contracts, while cultural theory emphasizes the role of society in setting societal priorities. The future of AI regulation thus hinges on recognizing how social contracts adapt in the wake of major historical changes and in response to evolving cultural norms.

Building on this understanding, emerging economies offer a promising avenue for developing innovative AI governance models tailored to their specific needs and challenges. For example, participatory councils in South Asia could prioritize human development by tailoring AI policies to address education, healthcare, and job creation needs for vast and diverse populations. In Africa, dynamic benchmarks could measure AI's impact on community well-being and sustainability, using locally relevant metrics such as access to clean energy, public safety, and equitable economic growth. Achieving culturally informed governance requires a multidisciplinary approach. Social scientists and historians, working alongside technologists and policymakers, can design frameworks that integrate societal complexities and historical contexts into actionable policies. Academia plays a critical role by fostering collaboration across disciplines and borders, enabling research that addresses real-world challenges and produces solutions tailored to diverse populations. For instance, academic institutions could host international forums to explore how cultural values influence AI adoption and regulation, generating insights that guide policymakers in crafting adaptive frameworks. This approach helps create AI regulations that are not only innovative but also culturally resonant.

AI regulation must be a dynamic and collaborative process that balances fairness with innovation. Policymakers should actively engage communities to ensure governance reflects cultural values, ethical priorities, and societal needs. This involves fostering open public discussions, improving digital literacy, and creating platforms where diverse voices can shape the evolving social contract. For instance, privacy-focused regulations, even if they limit convenience, must be grounded in cultural understanding to gain acceptance and build trust. This inclusive and forward-thinking approach ensures AI regulation not only meets current challenges but also paves the way for equitable and sustainable progress.

References

1. Leary, S.O.: ASML is hinting at leaving the Netherlands: here's why, https://dutchreview.com/. Accessed: 17 Nov 2024. [Online]. Available: https://dutchreview.com/news/why-does-asml-hint-at-leaving-netherlands/
2. Wilkinson, L.: Share, Post, Print, Email, and License, Executives expect complying with AI regulations will increase tech costs, CIO Dive. Accessed: 16 Nov 2024. [Online]. Available: https://www.ciodive.com/news/enterprise-cost-increase-ai-regulation-security-data/724345/
3. Government Regulation: Costs Lower, Benefits Greater Than Industry Estimates. Accessed: 16 Nov 2024. [Online]. Available: http://bit.ly/1LM7ayo
4. Erdélyi, O.J., Goldsmith, J.: Regulating artificial intelligence: proposal for a global solution. Gov. Inf. Q. **39**(4), 101748 (2022). https://doi.org/10.1016/j.giq.2022.101748
5. Scherer, M.U.: Regulating Artificial Intelligence Systems: Risks, Challenges, Competencies, and Strategies, p. 2609777. Social Science Research Network, Rochester (2015). https://doi.org/10.2139/ssrn.2609777
6. Cabrera, L.L.: EU AI Act Brief – Pt. 2, Privacy & Surveillance. Center for Democracy and Technology. Accessed: 16 Dec 2024. [Online]. Available: https://cdt.org/insights/eu-ai-act-brief-pt-2-privacy-surveillance/
7. Bresnahan, T.F., Trajtenberg, M.: General purpose technologies 'Engines of growth'? J. Econ. **65**(1), 83–108 (1995). https://doi.org/10.1016/0304-4076(94)01598-T

8. Eloundou, T., Manning, S., Mishkin, P., Rock, D.: GPTs are GPTs: an early look at the labor market impact potential of large language models. arXiv: arXiv:2303.10130, 21 Aug 2023. https://doi.org/10.48550/arXiv.2303.10130
9. Aman, S.E.: Article review: an overview of legitimacy theory as applied within the social and environmental accounting literature. Int. J. Adv. Res. Econ. Financ. **5**(4) (2023)
10. Braunerhjelm, P., Andersson, M., Blind, K., Eklund, J.E.: Handbook of innovation and regulation: introductory chapter. In: Braunerhjelm, P., Andersson, M., Blind, K., Eklund, J.E. (eds.) Handbook of Innovation and Regulation, pp. 1–20. Edward Elgar Publishing (2023). https://doi.org/10.4337/9781800884472.00005
11. Chen, Z.: Ethics and discrimination in artificial intelligence-enabled recruitment practices. Humanit. Soc. Sci. Commun. **10**(1), 1–12 (2023). https://doi.org/10.1057/s41599-023-02079-x
12. Etzioni, A., Etzioni, O.: Pros and Cons of Autonomous Weapons Systems. Army University Press. Accessed: 7 Dec 2024. [Online]. Available: https://www.armyupress.army.mil/Journals/Military-Review/English-Edition-Archives/May-June-2017/Pros-and-Cons-of-Autonomous-Weapons-Systems/
13. Brownstein, G.: Three Challenges in Cryptocurrency Regulation. Atlantic Council. Accessed: 7 Dec 2024. [Online]. Available: https://www.atlanticcouncil.org/blogs/econographics/three-challenges-in-cryptocurrency-regulation/
14. Caporal, J.: Crypto and Investment Scam Statistics for 2024. The Motley Fool. Accessed: 7 Dec 2024. [Online]. Available: https://www.fool.com/research/crypto-investment-scams/
15. Thomaz, F., Salge, C., Karahanna, E., Hulland, J.: Learning from the Dark Web: leveraging conversational agents in the era of hyper-privacy to enhance marketing. J. Acad. Mark. Sci. **48**(1), 43–63 (2020). https://doi.org/10.1007/s11747-019-00704-3
16. Kramer, M., Kania, J.: Changing the Game (SSIR). Accessed: 15 Dec 2024. [Online]. Available: https://ssir.org/articles/entry/changing_the_game
17. Friedline, T., Oh, S., Klemm, T., Kugiya, J.: Exclusion and Marginalization in Financial Services: Frontline Employees as Street-Level Bureaucrats (2020)
18. Thoenig, J.-C.: Institutional theories and public institutions: new agendas and appropriateness. In: The SAGE Handbook of Public Administration, pp. 169–179. Sage, London (2012). https://doi.org/10.4135/9781446200506.n11
19. Kao, Y.-H., Sapp, S.G.: The effect of cultural values and institutional trust on public perceptions of government use of network surveillance. Technol. Soc. **70**, 102047 (2022). https://doi.org/10.1016/j.techsoc.2022.102047
20. Parashar, S., Schulz, M.: Colonial legacies, postcolonial 'selfhood' and the (un)doing of Africa. Third World Q. **42**(5), 867–881 (2021). https://doi.org/10.1080/01436597.2021.1903313
21. Yao, Y., Yang, F.: Overcoming personal information protection challenges involving real-world data to support public health efforts in China – Consensus. Accessed: 11 Dec 2024. [Online]. Available: https://consensus.app/papers/overcoming-personal-information-protection-challenges-yao-yang/dfdff6043c145df6aa8dab87ab1daa7c/?extracted-answer=China%27s+Personal+Information+Protection+Law+%28PIPL%29+provides+a+consistent+legal+framework+for+protecting+personal+information%2C+including+sensitive+medical+health+data%2C+but+further+clarification+is+needed+regarding+specific+issues+like+%22separate+consent%22+and+cross-border+data+transfer+requirements.&q=China%27s+Personal+Information+Protection+Law+%28PIPL%29+imposes+stringent+data+privacy+regulations%2C+particularly+for+cross-border+data+transfers&pro=on
22. Wu, P., Sun, Z.: Culture and political regimes: how have they influenced the effectiveness of COVID-19 response policy? Heliyon. **10**(14) (2024). https://doi.org/10.1016/j.heliyon.2024.e34271

23. Iks, F.: The European Artificial Intelligence act overview and recommendations for compliance. May 2024. Accessed: 29 Oct 2024. [Online]. Available: https://www.semanticscholar.org/paper/The-European-Artificial-Intelligence-Act-Overview-Iks/024e481a2b4a847559b74e1cb5c1e36255a08d80
24. Article 16: Obligations of Providers of High-Risk AI Systems | EU Artificial Intelligence Act. Accessed: 12 Dec 2024. [Online]. Available: https://artificialintelligenceact.eu/article/16/
25. Söderlund, K., Larsson, S.: Enforcement design patterns in EU Law: an analysis of the AI Act. Digit. Soc. **3**(2), 41 (2024). https://doi.org/10.1007/s44206-024-00129-8
26. Buocz, T., Pfotenhauer, S., Eisenberger, I.: Regulatory sandboxes in the AI Act: reconciling innovation and safety? Law Innov. Technol. **15**(2), 357–389 (2023). https://doi.org/10.1080/17579961.2023.2245678
27. Hauer, M.P., Krafft, T.D., Sesing-Wagenpfeil, A., Zweig, K.: Quantitative study about the estimated impact of the AI Act. arXiv: arXiv:2304.06503. 29 Mar 2023. https://doi.org/10.48550/arXiv.2304.06503
28. Doekes, D., Ramackers, D.G., van Stein, D.N.: Computer Science & Economics AI Legislation in the EU Financial Sector: Challenges and Compliance Strategies. Accessed: 25 Oct 2024. [Online]. Available: https://www.semanticscholar.org/paper/Computer-Science-%26-Economics-AI-Legislation-in-the-Doekes-Ramackers/952c143d59cf351aedcf89a3779cb44b9be8a52c
29. Schneider, H.: Europe's AI regulation will stifle innovation GIS Reports. Accessed: 17 Nov 2024. [Online]. Available: https://www.gisreportsonline.com/r/ai-act-eu-regulation-innovation/
30. Scharre, P.: Debunking the AI Arms Race Theory. Texas National Security Review. Accessed: 17 Nov 2024. [Online]. Available: https://tnsr.org/2021/06/debunking-the-ai-arms-race-theory/
31. Synodinou, T.: Image right and copyright law in Europe: divergences and convergences. Laws. **3**(2) (2014). https://doi.org/10.3390/laws3020181
32. Pylypenko, D., Spahija, D.: Historical significance of Germany, France and Belgium in the creation of the European Union. Foreign Aff. **33**(1) (1970). https://doi.org/10.46493/2663-2675.33(1).2023.8-15
33. Skrzypczak, J.: The commercialisation of propaganda and the independence of the fourth estate – a contribution to the discussion on contemporary threats to media freedom. Przegląd Politol. **2**, 101–117 (2023). https://doi.org/10.14746/pp.2023.28.2.7
34. Golpayegani, D., Pandit, H.J., Lewis, D.: AIRO: an ontology for representing AI risks based on the proposed EU AI Act and ISO risk management standards. In: Dimou, A., Neumaier, S., Pellegrini, T., Vahdati, S. (eds.) Studies on the Semantic Web. IOS Press (2022). https://doi.org/10.3233/SSW220008
35. Tijerina, L.L., Baig, M.A.: China's People's Liberation Army: restructuring and modernization. Open Mil. Stud. **2**(1), 84–95 (2022). https://doi.org/10.1515/openms-2022-0127
36. Wang, H.H.: 'Century of Humiliation' Complicates U.S.-China Relationship. Forbes. Accessed: 14 Dec 2024. [Online]. Available: https://www.forbes.com/sites/helenwang/2015/09/17/century-of-humiliation-complicates-us-china-relationship/
37. Webster, G., Creemers, R., Kania, E., Triolo, P.: Full Translation: China's 'New Generation Artificial Intelligence Development Plan' (2017). DigiChina. Accessed: 9 Dec 2024. [Online]. Available: https://digichina.stanford.edu/work/full-translation-chinas-new-generation-artificial-intelligence-development-plan-2017/
38. Shenggao, Y.: Great potential for AI companies in Shanghai – Chinadaily.com.cn. Accessed: 14 Dec 2024. [Online]. Available: https://epaper.chinadaily.com.cn/a/202412/13/WS675b8966a3105c25b38ee755.html

39. de Vigo, P.: 2014 Topic: Confucianism Tags: Confucianism, R. I. C. Quarterly, and Spring 2014, Confucianism in Modern Chinese Society – ChinaSource. https://www.chinasource.org/. Accessed: 14 Dec 2024. [Online]. Available: https://www.chinasource.org/resource-library/articles/confucianism-in-modern-chinese-society/
40. Zhang, F.: Confucian foreign policy traditions in Chinese history. Chin. J. Int. Polit. **8**(2), 197–218 (2015). https://doi.org/10.1093/cjip/pov004
41. Zhecheng, Q.: Chinese Consumers Most Willing to Trade Privacy for Convenience (2018). Accessed: 14 Dec 2024. [Online]. Available: https://www.sixthtone.com/news/1002467
42. Feldstein, S.: The Global Expansion of AI Surveillance. Carnegie Endowment for International Peace. Accessed: 26 Dec 2024. [Online]. Available: https://carnegieendowment.org/research/2019/09/the-global-expansion-of-ai-surveillance?lang=en
43. About – GPAI. Accessed: 14 Dec 2024. [Online]. Available: https://gpai.ai/about/
44. He, A.: Trade Deals Might Induce Beijing to Bend on Data Restrictions. Centre for International Governance Innovation. Accessed: 14 Dec 2024. [Online]. Available: https://www.cigionline.org/articles/trade-deals-might-induce-beijing-to-bend-on-data-restrictions/
45. Toh, M.: SenseTime's Stock Market Debut May Be Spoiled by US-China tensions | CNN Business. CNN. Accessed: 14 Dec 2024. [Online]. Available: https://www.cnn.com/2021/12/10/investing/sensetime-ipo-hong-kong-intl-hnk/index.html
46. Gilbert, N., Mallapaty, S.: US and China sign new science pact – but with severe restrictions. Nature. (2024). https://doi.org/10.1038/d41586-024-04175-7
47. Roberts, A., Choer Moraes, H., Ferguson, V.: Toward a geoeconomic order in international trade and investment. J. Int. Econ. Law. **22**(4), 655–676 (2019). https://doi.org/10.1093/jiel/jgz036
48. Nguyen, C.L., Golman, W.: Diffusion of the Budapest Convention on cybercrime and the development of cybercrime legislation in Pacific Island countries: 'Law on the books' vs 'law in action'. Comput. Law Secur. Rev. **40**, 105521 (2021). https://doi.org/10.1016/j.clsr.2020.105521

From Proposals to Outcomes: Concept-Aligned Chunking for Cross-Document Relevance Assessment in Research Funding Review

Fengchi Yuan, Keqin Guan, Siyu Chen, Bokui Chen(✉), and Wai Kin Victor Chan(✉)

Institute of Data and Information, Tsinghua Shenzhen International Graduate School, Tsinghua University, Shenzhen 518000, China
{yfc22,gkq23,chen-sy23}@mails.tsinghua.edu.cn, chenbk@tsinghua.edu.cn, chanw@sz.tsinghua.edu.cn

Abstract. Government-funded science and technology innovation projects are vital for driving industrial development and supporting talent cultivation. However, evaluating their outcomes remains a significant challenge, especially when some researchers misattribute unrelated publications to funding projects, raising concerns about research integrity and transparency. This paper focuses on the challenging task of assessing the relevance between project proposals and research outputs, formulated as a long-text matching problem. Due to the fact that even valid research outputs often address only subtopics of the original project objectives, traditional methods, which typically compare entire documents, often fail to provide accurate relevance assessments. To address this, we propose ConceptSplitter, a concept-based chunking method inspired by long-text structuring strategies. As part of a retrieval-augmented generation (RAG) pipeline, ConceptSplitter serves as the chunking module that improves retrieval precision and contextual relevance in large language model inference. To support robust evaluation, we also construct a domain-diverse dataset that mirrors real-world funding scenarios. Experiments on this dataset show that ConceptSplitter outperforms traditional methods by enhancing chunking quality, improving the accuracy of relevance classification, and providing more reliable confidence estimation in large language model outputs.

Keywords: Research funding evaluation · Long-text matching · Retrieval-augmented generation · Pretrained language models

1 Introduction

Government-funded science and technology innovation programs, such as the National Natural Science Foundation of China (NSFC) and the National Key R&D Program, play a key role in advancing national research agendas and fostering industrial development [1, 2]. Funding decisions are typically based on criteria including research potential, industrial applicability, and technical feasibility. However, in practical settings, instances have emerged in which researchers attribute unrelated academic outputs, such as papers

X. Xie et al. (Eds.): ICSS 2025, LNOR, pp. 66–77, 2026.
https://doi.org/10.1007/978-3-032-13116-4_6

or patents, to funded projects to meet reporting requirements [3]. Such misattributions compromise the integrity of the evaluation process and hinder the equitable and efficient allocation of public research resources. The prevalent reliance on subjective expert assessment in these evaluations underscores the urgent need for a more systematic, objective, and data-driven framework for measuring the consistency between project objectives and reported outcomes.

Determining whether a research output substantively aligns with its corresponding project proposal presents a long-text semantic matching challenge. Conventional methods based on lexical similarity or plagiarism detection often fall short, particularly when alignment occurs at a thematic or sub-topic level [4]. Large Language Models (LLMs) offer a promising avenue for capturing these deeper semantic relations [5, 6]; however, their effectiveness in long-form document matching is constrained by token limits and difficulties in preserving contextual continuity across extended texts. Even advanced models such as ChatGLM4 and Qwen2.5-32B are typically restricted to processing between 8 K and 32 K tokens, limiting their capacity for full-document comparison. Retrieval-Augmented Generation (RAG) architectures partially mitigate these challenges by incorporating external knowledge through retrieval mechanisms [7, 8], yet their performance remains heavily contingent on the quality of document chunking and the precision of retrieval [9, 10].

To address these methodological limitations, this paper proposes ConceptSplitter, a novel segmentation approach inspired by the hierarchical structuring paradigm of the CIG framework [11–13]. ConceptSplitter is integrated into the TextSplitter stage of the LLM inference pipeline and is specifically designed to process academic long-form content, including research proposals and their associated outputs. By identifying domain-specific concepts and organizing textual segments around coherent topical clusters, ConceptSplitter produces semantically consistent chunks that enhance the fidelity of downstream semantic matching. To empirically validate this approach, we introduce a real-world benchmark dataset and conduct a series of evaluations demonstrating that ConceptSplitter improves both alignment accuracy and interpretability, contributing to a more rigorous framework for assessing the outcomes of innovation-driven funding initiatives.

The principal contributions of this work are threefold. First, we construct SciFund-Match, a high-quality dataset that enables long-text matching between project proposals and research deliverables, thereby establishing a benchmark for relevance evaluation in academic funding contexts. Second, we propose a concept-aligned chunk construction strategy that supports fine-grained semantic analysis by structuring long-form text around meaningful concept groupings. Third, we develop a comprehensive evaluation framework that rigorously assesses both chunking quality and classification performance, providing a foundation for future advancements in long-text semantic alignment and research impact assessment.

2 Related Work

Recent advancements in long-text matching have significantly enhanced the ability to align scientific project proposals with their corresponding research outcomes. Key developments include semantic similarity modeling, the use of large language models (LLMs) for extended-context reasoning, and retrieval-augmented generation (RAG) techniques to address token limitations and improve matching scalability. Together, these approaches form the foundation for more robust and scalable systems capable of assessing the thematic alignment between project objectives and scientific deliverables (Fig. 1).

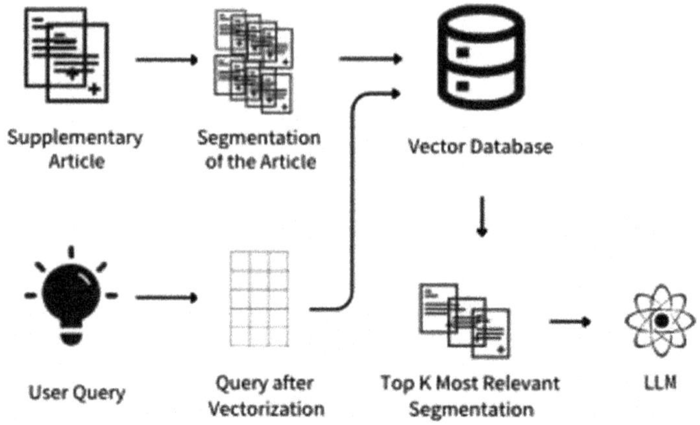

Fig. 1 Workflow of traditional RAG

Traditional methods, such as TF-IDF and word-to-word similarity, struggle with sparsity and fail to capture deep semantics. To address these limitations, Zhang et al. [14] propose a multi-layer coding approach to enhance structure-aware representation. D'Cruz et al. [15] employ ensemble BERT models for domain-specific long-text classification, improving semantic accuracy in sparse texts. Peng et al. [16] integrate multi-granularity embeddings and external knowledge for improved similarity measurement. Jiang et al. [17] introduce SMASH RNN, a hierarchical model lever-aging document structure for enhanced matching, while Askari et al. [18] develop RPRS, a transformer-based re-ranker for efficient long-document retrieval. Despite these advances, challenges remain in computational efficiency and data sparsity. Moreover, conventional similarity metrics often overlook topic-level alignment, limiting their applicability to evaluating the thematic consistency of research funding outcomes [3].

Recent work enhances LLMs' ability to process long contexts through the development of novel benchmarks. Wang et al. [19] propose Ada-LEval, a length-adaptive evaluation framework that supports texts up to 128 k tokens. Wijesiriwardene et al. [20] introduce ANALOGICAL to assess analogy recognition across extended contexts. Bai et al. [21] release LongBench, a bilingual and multitask benchmark spanning 21 datasets, offering a comprehensive platform for evaluating summarization, code completion, and multi-document QA. Additionally, Li et al. [22] introduce GLBench, the first comprehensive benchmark for GraphLLM evaluation, offering valuable insights into the integration of structure and semantics in both supervised and zero-shot scenarios. These efforts reveal limitations in context length, scaling laws, and evaluation diversity, which are equally applicable to textual domains requiring proposal-outcome alignment.

RAG-based architectures address some of these challenges by incorporating external knowledge into LLM pipelines. Chen et al. [23] examine attribution failures in long-form QA and highlight the need for more accurate retrieval mechanisms. Yang et al. [24] propose SMITH, a hierarchical transformer that overcomes self-attention bottlenecks for long documents. Jaiswal and Milios [25] introduce ChunkBERT, which partitions text and applies convolutional layers to enhance memory efficiency while maintaining performance. These approaches significantly improve the effectiveness of LLMs in processing long scientific texts. However, a persistent bottleneck lies in document segmentation: techniques like hierarchical chunking, semantic-aware splitting, and graph-based content organization—such as the proposed ConceptSplitter—seek to increase retrieval precision and reduce redundancy, thereby improving alignment between funding proposals and scientific outcomes [7, 8].

3 Task and Construction

3.1 Problem Definition

We formulate the task of matching a project proposal P with a set of research outcomes R = {r1, r2, …, rn} as a long-text relevance assessment problem. The goal is to compute a relevance score S = (y, Y), where y ∈ {0, 1} indicates whether the outcome is relevant to the proposal, and Y ∈ [0, 1] denotes the confidence. To facilitate comparison, both P and each ri ∈ R are segmented into semantically meaningful chunks, denoted as Cp = {Cp1, …, C pm} and CR = {CR1, …, CRn}. The top-k most similar chunks from both sides are retained, i.e., Cp k ⊆ Cp, CR k ⊆ CR, to focus on the most informative semantic units.

The main challenge lies in designing a chunking strategy that captures topic-level coherence while preserving semantic integrity, which is crucial for improving the accuracy and reliability of the relevance assessment.

3.2 SciFundMatch: A Dataset for Matching Sci-Tech Proposals and Research Outcomes

Accurately linking research outputs to their funding proposals is vital for transparent and accountable grant evaluation. Yet, issues such as the misattribution of unrelated publications continue to undermine existing systems [1, 3, 26]. Expert-led reviews, while prevalent, are subjective and difficult to scale. Despite increasing attention to this problem, there remains a lack of dedicated datasets for long-text matching between proposals and research outcomes—an essential foundation for developing data-driven evaluation methods.

To address this gap, we introduce SciFundMatch, a domain-specific dataset for long-text relevance matching between proposals and research papers. Existing NLP datasets typically focus on short-text similarity and lack the structural complexity needed for modeling academic alignment [27, 28]. Built on the fund relationship table from the Intelligent Innovation Dataset (IIDS) [29], SciFundMatch links 500 high-yield NSFC grants to 2,978 papers across 27 disciplines. Each instance includes one proposal and multiple publications labeled as relevant (1) or irrelevant (0), with negatives sampled from distinct subfields. To improve semantic precision, ChatGLM4-9B is used to regenerate proposals from positively linked papers, reducing label noise and enhancing thematic consistency. The dataset is formatted in JSON structure, where each entry includes the proposal's objectives, innovations, and scientific challenges, along with a list of related papers and their relevance labels.

To assess dataset quality, we conducted classification experiments using LLMs of different scales. The models exhibited balanced performance on positive and negative samples, and the confusion matrix of the Qwen2.5-14B model confirmed a uniform prediction distribution. Additionally, the dataset's variability in text length (Fig. 2) emphasizes the importance of effective chunking strategies for long-document encoding.

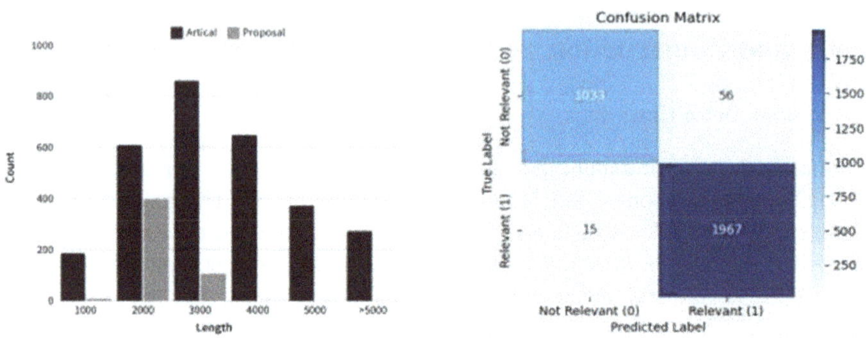

(a) Text Length Distribution in the SciFundMatch Dataset (b) Qwen2.5-14B Classification Task: Confusion Matrix

Fig. 2 Dataset characteristics: length variability and model prediction distribution.

In summary, SciFundMatch serves as a high-quality resource for evaluating proposal-document relevance in real-world funding scenarios. To encourage further research and reproducibility, the dataset is open-source and publicly available at https://openxlab.org.cn/datasets/Lawrence/AcademicTextMatching. It supports downstream tasks such as academic text matching, document chunking, and confidence calibration, providing a realistic testbed for evaluating the relationship between research proposals and academic publications. Researchers and practitioners can leverage the dataset for advancing methods in long-text matching, funding impact analysis, and related NLP applications.

4 Conceptsplitter: Semantic Chunking For LLM-Driven Long-Text Matching

Long academic texts pose challenges for retrieval models due to dense information, technical terminology, and multiple topic transitions, which standard chunking methods fail to handle effectively. To address this, we propose ConceptSplitter, a segmentation approach that preserves semantic integrity by leveraging domain-specific keyword extraction, conceptual clustering, and community detection, inspired by the CIG approach. Unlike length-based chunking, ConceptSplitter generates semantically coherent chunks, enhancing vectorization and ranking for long-text matching.

ConceptSplitter extracts key domain concepts using a hybrid approach that integrates TextRank, TF-IDF, and SciBERT-based Attention Weighting. TextRank [30] ranks words based on co-occurrence relationships, TF-IDF identifies terms with high specificity, and SciBERT assigns contextual importance scores. This combination ensures the extraction of domain-relevant keywords that effectively capture the semantic structure of academic texts.

After extracting key domain concepts, we construct a concept graph G = (v, E), where nodes represent keywords and edges indicate co-occurrence within sentences, weighted by frequency. The Louvain algorithm [31] is applied to identify semantically coherent concept clusters. Sentences containing relevant keywords are then grouped into chunks, ensuring conceptual coherence, maintaining a predefined token limit for efficiency, and minimizing redundancy by reducing keyword overlap across chunks. This approach preserves semantic integrity, facilitating more effective long-text representation and retrieval.

Algorithm 1

Algorithm 1 ConceptSplitter

```
Require: Input text T
Ensure: Chunks C
function CONCEPTSPLITTER(T)
   Step 1: Preprocess Text
   Clean the input text T by removing unnecessary whitespaces and newline characters.
   Step 2: Extract and Fuse Keywords
   Initialize an empty set keywords.
   for each method in {TextRank, TF-IDF, SciBERT} do
      Extract keywords using the current method from text T.
      Add the extracted keywords to keywords.
   end for
   keywords ← TextRank_keywords ∪ TF - IDF_keywords ∪ sciBERT_keywords
   Step 3: Perform Community Detection
   Build a graph G from the keyword set keywords.
   Apply the Louvain algorithm to detect communities in G.
   Step 4: Semantic Chunking
   for each community comm in communities do
      Identify sentences in T that contain terms from comm.
      Group these sentences together into a chunk.
   end for
   return C
end function
```

The final output of ConceptSplitter is a set of structured chunks, each focusing on a specific domain concept.

Compared to fixed-length segmentation methods, our approach ensures that each chunk forms a self-contained, semantically meaningful unit. This improves the performance of downstream tasks, such as document retrieval, similarity computation, and reranking, by providing cleaner and more interpretable representations.

5 Experiments

5.1 Evaluation of Chunk Construction Quality

To evaluate the effectiveness of ConceptSplitter in structuring academic long-text data, we compare it against three baseline segmentation methods: TokenTextSplitter, which divides text into fixed-length windows with token-level overlap; RecursiveCharacterSplitter, which segments based on hierarchical textual boundaries (e.g., paragraphs and sentences); and Small2BigSplitter, which merges semantically similar sub-chunks through an iterative similarity-based process. All methods are configured with a 512-token chunk length, and ConceptSplitter additionally applies TF-IDF keyword extraction (top 50 terms) combined with Louvain community detection to guide topic-based chunk boundaries.

The quality of the resulting chunks is assessed using three complementary metrics. Intra-chunk Semantic Coherence measures the average pairwise cosine similarity of Sentence-BERT embeddings (all-MiniLM-L6-v2) between sentences within each chunk, evaluating semantic coherence. Inter-chunk Lexical Redundancy quantifies vocabulary overlap between chunk pairs using normalized Jaccard similarity, assessing redundancy in terms of shared vocabulary. Finally, Linguistic Fluency is evaluated using GPT-2-based perplexity (PPL), which captures grammatical fluency, with lower perplexity indicating more fluent output (Fig. 3).

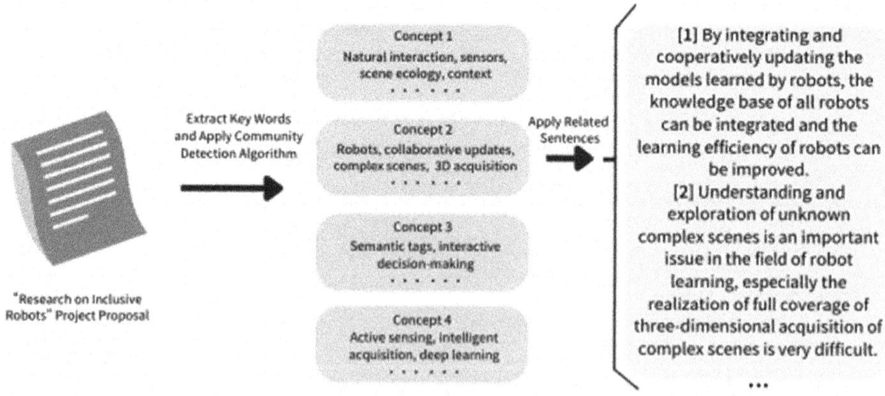

Fig. 3 Overview of the ConceptSplitter Process.

Table 1 Performance Comparison of Chunking Methods

Method	Integrity	Redundancy	Coherence	Runtime
RecursiveCharacterText	0.3067	0.0506	2216	0.0019
Small2Big	0.2934	0.0324	2509	0.2152
TokenText	0.4462	0.0901	55.48	0.0052
ConceptSpitter	0.5540	0.1690	73.86	0.0310
Avg.	0.4348	0.0866	980.3	0.0507

Table 1 presents the results of the evaluation. ConceptSplitter achieves the highest chunk integrity score and semantic coherence, suggesting that it better preserves topic continuity within chunks. Although it introduces slightly more redundancy than other methods due to flexible sentence reuse across chunks, this redundancy remains within an acceptable range. Compared with fixed-length segmentation, ConceptSplitter allows for variable-length, concept-aligned chunking, thereby improving interpretability without significantly increasing computational overhead.

To further understand the robustness of each method under different document lengths, we categorize source texts based on length and re-evaluate chunking behavior. ConceptSplitter maintains a consistent advantage in semantic coherence across varying document sizes, effectively reducing redundancy in longer texts—a challenge often encountered with fixed-length methods. Figure 4 visualizes the comparative performance of ConceptSplitter, highlighting its strength in constructing high-quality, semantically cohesive chunks.

5.2 Funding Consistency Evaluation

We evaluate ConceptSplitter on the Full-text Matching task, which assesses the relevance between a research paper's full text and its project proposal. For comparison, we also evaluate SimNet, ChatGLM4-9B (ZhipuAI/chatglm4- 9b), and Qwen2.5-14B on both Abstract Matching and Full-text Matching.

Experiments are conducted on the SciFundMatch dataset, where each input pairs a proposal with an abstract or full text, and outputs a binary relevance label and a confidence score (0–100). Inference uses greedy decoding within a 1024-token limit for consistency across models (Table 2).

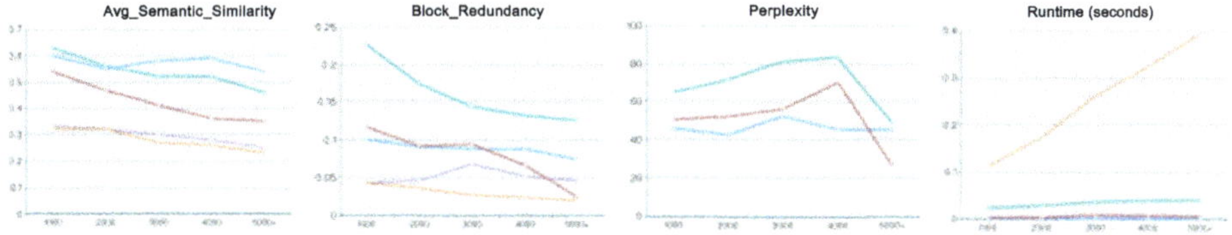

Fig. 4 ConceptSplitter Remains Robust Across Varying Text Lengths

Table 2 Article Matching Evaluation Results. The best values for each metric are highlighted in **bold**.

Model	Acc.	Prec.	Rec.	F1
SimNet (Abstract)	0.8987	0.8646	0.9375	0.8996
SimNet (Fulltext)	0.8991	0.8912	0.9414	0.9156
ChatGLM4-9B (Abstract)	0.7823	0.9720	0.7285	0.8328
ChatGLM4-9B (Fulltext)	0.7579	**0.9769**	0.6391	0.7727
Qwen2.5-14B (Abstract)	0.9737	0.9617	**0.9992**	0.9800
Qwen2.5-14B (Fulltext)	0.9742	0.9636	0.9980	0.9805
ChatGLM4-9B + ConceptSplitter	**0.9754**	0.9720	0.9905	**0.9812**

Our findings highlight the trade-offs between abstract and full-text matching. While full-text matching improves recall by capturing more relevant information, abstract matching tends to inflate precision due to its high-level summaries. However, long-text processing remains challenging due to token constraints and encoding inefficiencies. To address these issues, ConceptSplitter enhances relevance classification by segmenting text into meaningful chunks, thereby preserving context and improving accuracy. Although ChatGLM4-9B prioritizes precision over recall, integrating ConceptSplitter helps balance this trade-off, significantly enhancing recall while maintaining high accuracy.

5.3 Confidence Calibration and Uncertainty Estimation

Confidence calibration ensures a model's predicted confidence aligns with its actual correctness, crucial for reliable scientific relevance assessments and decision-making. We evaluate it using three metrics: Brier Score (mean squared error between confidence and correctness), Expected Calibration Error (ECE, discrepancy between confidence and accuracy across bins), and AUC-ROC (ranking quality of predictions by confidence).

Our analysis highlights the interplay between structured chunking and model scale in confidence calibration and ranking performance. ConceptSplitter improves confidence calibration by enhancing both AUC and interpretability. Specifically, the confidence distribution becomes more concentrated at both extremes, rather than being uniformly spread. This indicates a clearer distinction between relevant and irrelevant matches. Moreover, the threshold for identifying confidently predicted outcomes is significantly lower, which results in better performance when determining the relevance of research outputs to project proposals. Consequently, AUC is improved, as ConceptSplitter allows for more accurate confidence assignment, effectively distinguishing between high-confidence and low-confidence predictions (Fig. 5).

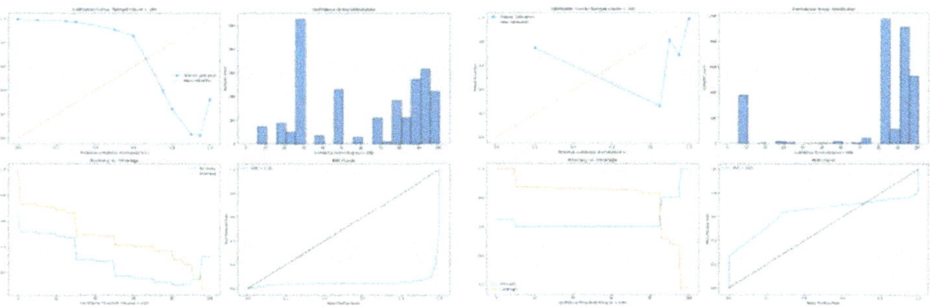

(a) Original ChatGLM4-9B (b) ChatGLM4-9B (ConceptSplitter)

Fig. 5 Comparison of confidence calibration before and after ConceptSplitter

6 Conclusion and Limitations

This study explores the challenge of aligning project proposal with research outcomes in government-funded innovation projects. ConceptSplitter contributes to semantic matching by integrating keyword extraction, concept community detection, and context-aware chunk formation, thereby improving long-text processing for large language models. The SciFundMatch dataset provides a valuable benchmark for academic long-text matching, helping to address a critical evaluation gap (Fig. 6).

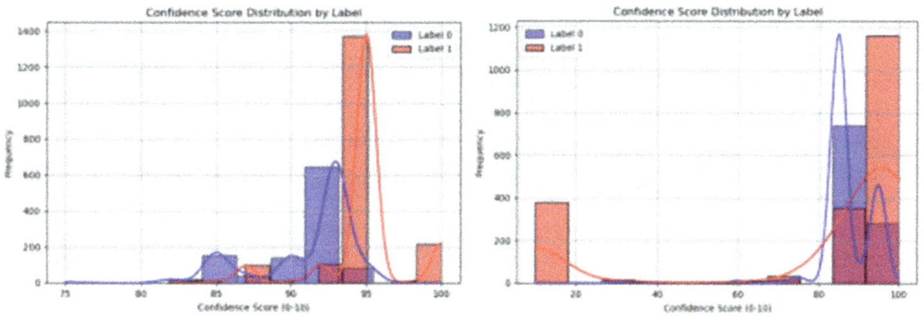

(a) Match based on the Abstract (b) Match based on the Fulltext

Fig. 6 Confidence distribution of the classification task

Despite its strong performance, ConceptSplitter has certain limitations. Discipline-based negative sampling may cause sample imbalance, with the model showing more confident predictions for positive samples (Fig. 6). This skews the confidence distribution toward the positive class. Moreover, reliance on pre-trained models introduces domain bias and limits generalization, underscoring the need for broader evaluation and real-world validation.

Acknowledgements. This research was funded by the Science and Technology Innovation Committee of Shenzhen-Platform and Carrier (International Science and Technology Information Center), the Science and Technology Innovation Commission of Shenzhen

(JCYJ20210324135011030), the Guangdong Pearl River Plan (2019QN01X890), and the National Natural Science Foundation of China (Grant No. 71971127).

References

1. Fu, Y.: The impact of government funding on research innovation: an empirical analysis of Chinese universities. Manag. Decis. Econ. **44**, 285–296 (2022)
2. Wang, S., Li, M., Li, J., Wu, D.: Do researchers from prestigious universities deserve advantages in research funding? Evidence from the national natural science foundation of china. Scientometrics. (2025)
3. Nannan, Y., Dong, Y., de Jong, M.: A helping hand from the government? How public research funding affects academic output in less-prestigious universities in china. Res. Policy. **51**, 104591 (2022)
4. Laban, P., Fabbri, A.R., Xiong, C., Wu, C.-S..: Summary of a haystack: a challenge to long-context llms and rag systems. arXiv preprint (2024)
5. Brown, T., et al.: Language models are few-shot learners. Adv. Neural Inf. Proces. Syst. **33**, 1877–1901 (2020)
6. Chowdhery, A., et al.: Palm: Scaling language modeling with pathways. arXiv preprint arXiv:2204.02311 (2022)
7. Cuconasu, F., et al.: The power of noise: redefining retrieval for rag systems. In: Proceedings of the 47th International ACM SIGIR Conference, p. 11 (2024)
8. Jin, B., Yoon, J., Han, J., Arık, S.Ö.: Long-context llms meet rag: overcoming challenges for long inputs in rag. In: International Conference on Learning Representations (ICLR) (2025)
9. Zhao, S., Yang, Y., Wang, Z., He, Z., Qiu, L.K., Qiu, L.: Retrieval-augmented generation (rag) and beyond: a comprehensive survey on how to make your llms use external data more wisely. arXiv preprint (2024)
10. Jeong, S., Baek, J., Cho, S., Hwang, S.J., Park, J.C.. Adaptive-rag: learning to adapt retrieval-augmented large language models through question complexity. arXiv preprint (2024)
11. Liu, B., Zhang, T., Niu, D., et al. Matching long text documents via graph convolutional networks (2018)
12. Li, W., Xu, J., He, Y., et al.: Coherent comments generation for Chinese articles with a graph-to-sequence model. In: Proceedings of the 57th Annual Meeting of the Association for Computational Linguistics, Florence, Italy (2019)
13. Wagh, R.S., Anand, D.: Legal document similarity: a multi-criteria decision-making perspective. PeerJ Comput. Sci., e262 (2020)
14. Zhang, Y., Zhang, J., Song, G., Lu, Q.: A novel text matching model based on multilayer coding and feature enhancement. In: Proceedings of the 2024 27th International Conferences on Computer Supported Cooperative Work in Design (CSCWD), pp. 974–979. IEEE (2024)
15. D'Cruz, C., Bereder, J.-M., Precioso, F., Riveill, M.: Domain-specific long text classification from sparse relevant information. arXiv preprint arXiv:2408.13253 (2024)
16. Peng, D., Hao, B., Tang, X., Chen, Y., Sun, J., Wang, R.: Learning long-text semantic similarity with multi-granularity semantic embedding based on knowledge enhancement. In: Proceedings of the 2020 1st International Conferences on Control, Robotics and Intelligent System, pp. 19–25 (2020)
17. Jiang, J., Zhang, M., Li, C., Bendersky, M., Golbandi, N., Najork, N.: Semantic text matching for long-form documents. In: Proceedings of the World Wide Web Conferences, pp. 795–806 (2019)

18. Askari, A., Verberne, S., Abolghasemi, A., Kraaij, W., Pasi, G.: Retrieval for extremely long queries and documents with RPRS: a highly efficient and effective transformer-based re-ranker. ACM Trans. Inf. Syst. **42**(5), 1–32 (2024)
19. Wang, C., Duan, H., Zhang, S., Lin, D., Chen, K.: Ada-leval: evaluating long-context llms with length-adaptable benchmarks. arXiv preprint arXiv:2404.06480 (2024)
20. Wijesiriwardene, T., et al.: Analogical–a novel benchmark for long text analogy evaluation in large language models. arXiv preprint arXiv:2305.05050 (2023)
21. Bai, Y., et al.: Longbench: a bilingual, multitask benchmark for long context understanding. arXiv preprint arXiv:2308.14508 (2023)
22. Li, Y., et al.: Glbench: a comprehensive benchmark for graph with large language models. In: Advances in Neural Information Processing Systems 37 (NeurIPS 2024), Datasets and Benchmarks Track, Dataset and Benchmarks Track (2024)
23. Chen, H.-T., Xu, F., Arora, S., Choi, E.: Understanding retrieval augmentation for long-form question answering. arXiv preprint arXiv:2310.12150 (2023)
24. Yang, L., Zhang, M., Li, C., Bendersky, M., Najork, M.: Beyond 512 tokens: Siamese multi-depth transformer-based hierarchical encoder for long-form document matching. In: Proceedings of the 29th ACM International Conferences on Information & Knowledge Management, pp. 1725–1734 (2020)
25. Jaiswal, A., Milios, E.: Breaking the token barrier: chunking and convolution for efficient long text classification with Bert. arXiv preprint arXiv:2310.20558 (2023)
26. Wang, F.F., Guo, W.H., Xue, R., Baron, C., Jia, C.R.: Exploring the subject heterogeneity of scientific research projects funding-example of the Chinese natural science foundation. Inf. Process. Manag. **62**, 104098 (2025)
27. Zhang, Y., Zhang, J., Song, G., Qiang, L.: A novel text matching model based on multilayer coding and feature enhancement. In: Proceedings of the 27th International Conference on Computer Supported Cooperative Work in Design (CSCWD), pp. 974–979. IEEE (2024)
28. Dong, P., Hao, B., Tang, X., Chen, Y., Sun, J., Wang, R.: Learning long-text semantic similarity with multi-granularity semantic embedding based on knowledge enhancement. In: Proceedings of the 1st International Conference on Control, Robotics and Intelligent System, pp. 19–25 (2020)
29. Wu, X., et al.: An intelligent innovation dataset on scientific research outcomes. arXiv preprint, arXiv:2409.06936 (2024)
30. Mihalcea, R., Tarau, P.: Textrank: bringing order into text. In: Proceedings of the 2004 Conference on Empirical Methods in Natural Language Processing (EMNLP 2004), pp. 404–411 (2004)
31. Blondel, V.D., Guillaume, J.L., Lambiotte, R., Lefebvre, E.: Fast unfolding of communities in large networks. J. Stat. Mech. Theory Exp. **2008**(10), P10008 (2008)

Design of Intelligent Traffic Signal Control Using Reinforcement Learning

Chunchen Lin, Ayca Erdogan, and Hongrui Liu(✉)

Industrial and Systems Engineering, San Jose State University, San Jose, CA 95008, USA
hongrui.liu@sjsu.edu

Abstract. An efficient transportation system is fundamental to economic vitality, yet traffic congestion remains a persistent challenge with significant social, environmental, and economic costs. With the growing availability of computational resources and the advancement of machine learning algorithms, adaptive traffic signal control systems based on reinforcement learning (RL) have become a promising solution. This study presents the design and evaluation of intelligent traffic signal control using Deep Q-Networks (DQN) and Double DQN (DDQN). We develop a simulated 6x6 intersection environment to compare the performance of RL-based systems against conventional fixed-cycle traffic control. The results demonstrate that RL-based systems, particularly those using DDQN, substantially outperform the baseline by improving vehicle flow and average velocity, thereby reducing congestion. This work highlights the potential of RL to modernize traffic management and emphasizes DDQN's robustness in mitigating overestimation and enhancing generalization.

Keywords: traffic congestion · traffic signal control · reinforcement learning · deep Q-networks

1 Introduction

Transportation is critical for the economy of a country. Traffic congestion causes severe problems such as environmental pollution, fuel consumption, and long travel times. It has been reported that Americans waste one week a year sitting in congestion in 2024, costing $771 in lost time and productivity according to a report released from INRIX, Inc. [1]. Furthermore, congestion may cause health issues: diesel-powered vehicles such as buses, trucks, and other heavy transport emit nitrogen oxides and particulate matter harmful to human health [2]. Additionally, traffic jams contribute to mental stress due to prolonged confinement in limited space [3]. Overall, transportation systems have wide-ranging negative impacts on health, the environment, and the economy. As economic activity continues to grow, so will the demand for transportation infrastructure. For example, the Bay Area and Los Angeles suffer from chronic traffic congestion due to the rapid expansion of tech and entertainment industries. While large-scale infrastructure solutions such as Elon Musk's Boring Company tunnels aim to ease traffic through underground tunnels, these projects are expensive and time-consuming. In contrast, computational and algorithmic solutions offer faster, more scalable improvements to traffic control.

X. Xie et al. (Eds.): ICSS 2025, LNOR, pp. 78–89, 2026.
https://doi.org/10.1007/978-3-032-13116-4_7

Traditional Traffic Signal Control (TSC) systems rely on preprogrammed cycles and are typically optimized offline using historical data. As traffic patterns are dynamic and non-stationary, these fixed-cycle systems are inherently limited in adapting to real-time conditions. There is an urgent need for adaptive traffic control systems that can respond to real-time conditions to reduce congestion and improve efficiency. Recent advances in artificial intelligence and machine learning offer promising avenues for modernizing traffic control. Adaptive Traffic Signal Control (ATSC) systems have been proposed to adjust control parameters dynamically based on real-time traffic data captured by road sensors and cameras [4, 5].

Reinforcement learning (RL) has emerged as a powerful method for this problem, offering advantages such as reduced reliance on large, labeled datasets and the ability to improve through experience, like how humans learn tasks like driving [6]. RL has already demonstrated success in fields like gaming and autonomous vehicles, and its adaptive, sequential decision-making capabilities make it particularly well-suited for traffic signal control. This study focuses on the following objectives:

1. Design and Implementation of a Realistic Traffic Simulation Environment: Creating a simulation that closely replicates real-world traffic conditions, enabling robust and meaningful experimentation.
2. Development of RL-Based Traffic Signal Control Systems: Utilizing DQN and DDQN algorithms to learn optimal traffic signal timing policies through continuous interaction with the simulated environment.
3. Benchmarking Against Traditional Systems: Comparing the RL-based approaches against conventional fixed-cycle traffic control methods to provide clear, quantitative performance insights.
4. Training and Optimization of RL Agents: Enhancing the learning efficiency and generalization of RL agents by focusing on reducing overfitting and improving convergence within the simulation environment.
5. Evaluation Through Performance Metrics: Assessing system performance using metrics such as mean vehicle velocity, where results demonstrate that the DDQN-based system significantly outperforms traditional methods in minimizing congestion and improving overall traffic flow

2 Literature Review

The limitations of traditional fixed-cycle traffic signal control (TSC) systems have spurred the development of adaptive methods capable of responding dynamically to real-time traffic conditions. Among these, reinforcement learning (RL) has emerged as a promising approach for optimizing traffic signal control strategies. In the RL framework, five fundamental components are defined: agent, environment, state, action and reward function [6, 7]. The agent observes the current state of the environment, selects an action according to a policy, and receives a reward based on the resulting state transition. Over time, the agent refines its policy to maximize expected rewards through trial-and-error interaction.

The application of RL to traffic control can be traced back to early studies around 2001, where tabular methods such as Q-learning were primarily used [7]. A breakthrough occurred in 2015 with the introduction of the Deep Q-Network (DQN), which combined Q-learning with deep neural networks, enabling RL to handle high-dimensional, continuous state spaces [8]. Unlike rule-based systems that rely on predefined heuristics, RL models learn directly through interactions with their environment, discovering adaptive policies that maximize cumulative rewards, such as minimizing vehicle delay or maximizing traffic throughput [9] in complex traffic environments.

In DQN frameworks, the agent is typically implemented as a deep neural network and acts as the "brain," interacting with the environment to optimize traffic signal policies. Single-agent systems are commonly employed for isolated intersection signal control (IISC) [10–12], while multiple-agent systems have been developed for multi-intersection control (MIC) [13–15]. The environment consists of traffic intersections and their associated settings, typically simulated using tools such as SUMO, Vissim, CityFlow, and Aimsun [16]. These platforms enable the creation of realistic training scenarios, with traffic demand levels either fixed or dynamically varying over time to simulate real-world conditions [11, 15]. The definition of action in RL-based traffic control is critical and varies across studies. Actions may involve selecting a vehicle phase to display green for a specific duration [17], choosing the green time for the current phase [18], deciding whether to end the current phase [19], or adjusting green times for all phases in the next cycle [20]. State representations are also designer-defined and significantly impact learning outcomes. Vehicle-related features often form the basis of the state, which can be represented using point-based modes (e.g., gap time, occupancy), block-based modes (e.g., vehicle counts, queue lengths), or grid-based modes where the road network is divided into grids containing vehicle speed and presence information [21]. Grid-based states are particularly well-suited for convolutional neural networks (CNNs) [21]. Similarly, the reward function is a critical component that guides learning. Common reward designs include the number of vehicles served, the waiting time of served vehicles, the number of vehicles queued, and the waiting time of vehicles waiting to be served [14].

Beyond RL, other AI-based methods have also been employed in traffic signal control. These include supervised learning approaches, such as decision trees and support vector machines [22, 23], which predict traffic patterns and adjust signals accordingly. Fuzzy logic and genetic algorithms have been used to evolve rule-based systems for adaptive control [24]. Additionally, deep learning models like convolutional neural networks (CNNs) and recurrent neural networks (RNNs) have been applied to extract spatial and temporal traffic features from sensor data [25, 26]. While these methods provide improvements over traditional systems, RL remains uniquely suited for sequential decision-making under uncertainty, particularly in complex and dynamic traffic environments.

Despite its success, DQN is known to suffer from overestimation bias, where the agent tends to overestimate action values, leading to suboptimal policies. To address this issue, Van Hasselt et al. introduced the Double DQN (DDQN) algorithm [27]. The key difference between DQN and DDQN is that DDQN uses two neural networks: one for action selection and another for action evaluation. This decoupling significantly reduces overestimation and results in more stable and accurate learning. Applications of DDQN in traffic signal control have demonstrated improved convergence rates and policy performance compared to standard DQN models [28]. Despite these advancements, several challenges remain. Overfitting to specific simulated environments can limit the generalizability of trained policies, and ensuring the safety and robustness of RL controllers under unpredictable real-world conditions continues to be an active area of research [29]. Nevertheless, the progress achieved with DQN, DDQN, and multi-agent reinforcement learning methods underscores the growing potential of RL to revolutionize traffic signal control systems.

In this paper, we propose comparing RL-based approaches, i.e. specifically, DQN and DDQN against conventional fixed-cycle traffic control methods in a simulated traffic environment.

3 Methodology

The methodology of this project consists of designing the traffic simulation environment, selecting appropriate reinforcement learning (RL) components, and setting up the training and evaluation process.

3.1 Environment Setup

In this project, the traffic environment is created using the Simulation of Urban Mobility (SUMO) tool to simulate an intersection scenario. The intersection layout consists of six lanes in the north-south direction (three lanes per direction) and six lanes in the east-west direction (three lanes per direction), as illustrated in Fig. 1.

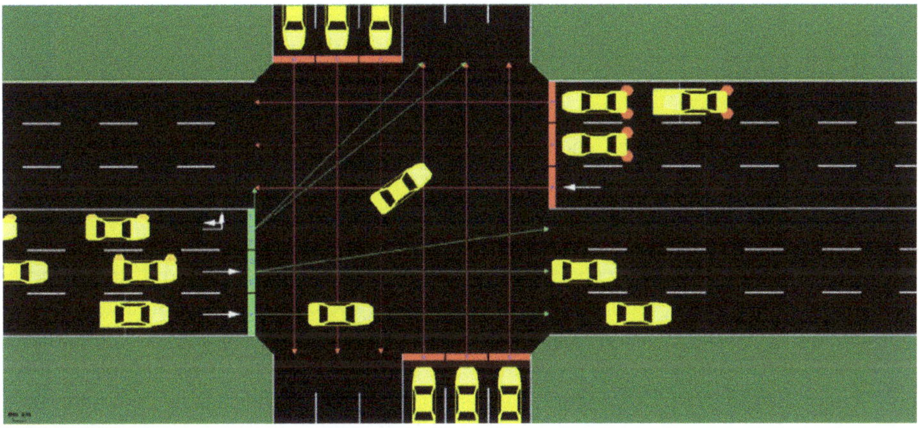

Fig. 1. Traffic environment for the DQN learning

The system models local roads with a speed limit of 35 miles per hour (mph). Vehicles enter and exit the system from all four directions, maintaining a minimum time headway of 3 s between any two vehicles. Each approach road leading to the intersection has a length of 400 yards. Vehicles follow standard traffic rules: proceeding during a green light and stopping at a red light. The system is designed so that vehicles continue moving until they reach the end of the road segment after passing through the intersection.

Vehicles are initially generated at the ends of the roads according to a predefined distribution pattern while maintaining a minimum inter-arrival time between cars to prevent congestion at entry points. The traffic lights at the intersection are organized into phases, corresponding to different light colors (green, yellow, red). For the baseline fixed-time traffic control method, a static light cycle with a duration of 30 s is employed.

3.2 Reinforcement Learning Setup

In this project, Deep Q-Network (DQN) and Double Deep Q-Network (DDQN) algorithms are applied to optimize the traffic signal control at the intersection. The RL framework is structured around five essential components: agent, environment, state, action, and reward.

a. Agent

   The agent is represented by a deep neural network that interacts with the SUMO environment. In the DQN and DDQN frameworks, the agent observes the current traffic state, selects an action (i.e., traffic signal operation), and receives a reward based on the resulting traffic conditions. DQN is recognized as a robust RL algorithm; however, it tends to overestimate Q-values during the learning process because the same network is used both to select and evaluate the next action. While Double DQN introduces a second identical neural network. In Double DQN, the target Q-value is computed using a separate network from the one selecting the action, effectively decoupling action selection and evaluation. This adjustment significantly mitigates the overestimation problem and leads to more stable and accurate learning. In this project, both DQN and Double DQN are explored and implemented as RL agents for optimizing traffic signal control.

b. Environment

   The environment consists of the simulated intersection described in Sect. 3.1, including all vehicles, lanes, and traffic signal infrastructure. The environment evolves based on the agent's actions and the internal SUMO traffic simulation dynamics.

c. State

   Selecting the appropriate length of the state observation zone involves a trade-off between training efficiency and the adequacy of information captured. If the observation zone is too large, the resulting state may include excessive, redundant information, leading to slower training and increased computational complexity. Conversely, if the observation zone is too small, critical traffic information necessary for effective decision-making may be omitted. In this project, the state is defined as the queuing length on each incoming lane at the intersection, providing a concise yet informative representation of traffic conditions for the RL agent.

d. Action

   The action selected by the agent is defined as the control of the traffic signal phase at the intersection. Specifically, the agent decides whether to maintain the current signal phase or switch to the next phase based on the observed traffic conditions. This action space allows the agent to dynamically adapt signal timing to minimize vehicle queuing and improve traffic flow.

e. Reward

To enhance the efficiency of traffic control, minimizing the vehicle queue length at the intersection is established as the primary objective. Therefore, the reward function is designed to directly reflect the queuing conditions: actions that lead to shorter queues yield higher rewards. In this project, the reward is designed to be proportional to the improvement in traffic conditions, thereby encouraging the agent to select actions that reduce congestion. A higher Q-value corresponds to actions that are better aligned with the current traffic state, guiding the agent to learn policies that optimize intersection throughput and minimize vehicle delays.

3.3 Neural Network and RL Training Settings Reinforcement

The DQN architecture used in this project consists of two convolutional layers designed to process the input traffic state data. The input data are randomly selected from a replay memory buffer, which stores experience samples collected during the traffic simulation process. The reinforcement learning training follows a sequential process: the agent interacts with the environment, consisting of a 6×6 lane traffic intersection, by selecting actions based on the current policy. The output of the neural network is the Q-value matrix corresponding to possible actions, while the input is the state matrix representing current traffic conditions. The network is trained by minimizing a loss function based on the difference between predicted Q-values and target Q-values. The Q-value update rule for DQN is expressed as:

$$Q(s_t, a_t) \leftarrow Q(s_t, a_t) + \alpha\left[r_t + \gamma \max_a Q(s_{t+1}, a) - Q(s_t, a_t)\right] \tag{1}$$

where s_t and a_t are the current state and action at time step t, r_t is the immediate reward, and γ is the discount factor. During training, the agent requires the reward at the current time step and the maximum Q-value of the next state from the memory to update the Q-value accurately.

DDQN employs a similar neural network structure as DQN but introduces an additional identical network called the target network. In Double DQN, the maximum Q-value at the next time step t + 1 is generated by the target network, rather than the evaluation network. This modification mitigates the overestimation bias observed in standard DQN. The Q-value update rule for Double DQN is:

$$Q(s_t, a_t) \leftarrow Q(s_t, a_t) + \alpha\left[r_t + \gamma Q_{\text{target}}(s_{t+1}, \arg\ \max_a Q_{\text{eval}}(s_{t+1}, a)) - Q(s_t, a_t)\right] \tag{2}$$

where Q_target denotes the value estimated by the target network, and Q_eval is estimated by the evaluation network.

The hyperparameters used in training DQN and DDQN are shown in Table 1. The parameters are set by trial-and-error as well as reference to the most used values in literature.

Table 1. Hyperparameters for DQN and Double DQN training

Hyperparameter	Value	Description
SGD Optimizer	Adam	Adam optimizer from TensorFlow for training the neural network
Gamma	0.9	Discount factor for future rewards
Mini-batch Size	32	Size of the batch randomly sampled from the replay memory
Replay Buffer Size	10,000	Size of the replay buffer that stores experience tuples
Final Epsilon	0.1	Final value of epsilon in the ε-greedy exploration strategy

4 Results

In this study, we conducted experiments to evaluate the performance of the DQN- and DDQN-based traffic control systems compared to a conventional fixed-cycle control system. The experiments were carried out in a simulated 6×6 lane intersection environment under varying traffic conditions.

4.1 Mean Delay Analysis

The mean delay of vehicles for both DQN-based and fixed-cycle systems over 20,000 training episodes is shown on Fig. 2. Initially, the mean delays of both systems are comparable. However, after approximately 1,800 episodes, the DQN-based system demonstrates a significant reduction in mean delay, while the fixed-cycle system exhibits a slight increase in delay over time. This indicates that the DQN agent successfully learns to adapt traffic signals to minimize congestion, outperforming the baseline system.

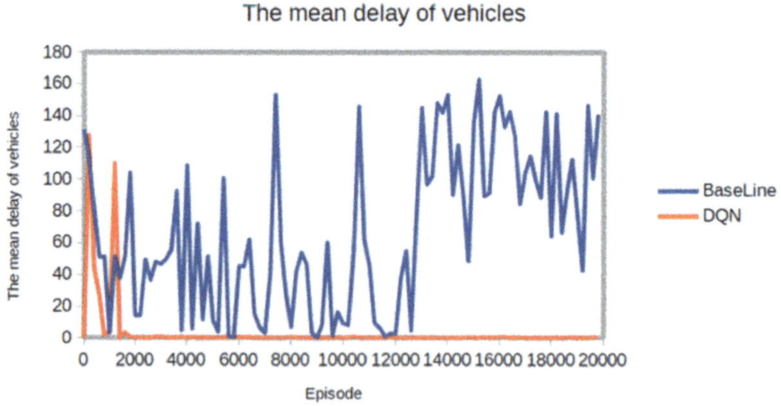

Fig. 2. The mean delays of the traffic control based on DQN (red), and fixed cycle(blue)

4.2 Mean Velocity Comparison

The comparison of mean vehicle velocity for the DQN, DDQN, and fixed-cycle systems is presented in Fig. 3. Throughout 10,000 episodes, the DDQN-based system consistently achieves the highest mean vehicle velocity, followed by the DQN system. The fixed-cycle system exhibits the lowest velocity, suggesting that RL-based approaches significantly improve overall traffic flow, with DDQN offering superior performance.

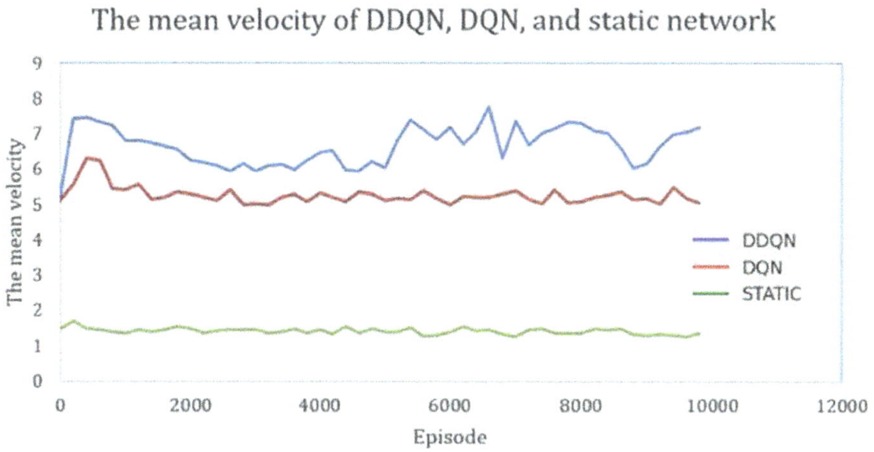

Fig. 3. The mean velocity of DQN based network and fixed cycle network

4.3 Reward Evolution During Training

Figures 4 and 5 illustrate the reward progression during the training processes of the DQN and DDQN networks, respectively. As shown in Fig. 4, the reward for the DQN agent increases rapidly from episode 0 to approximately episode 1,000. After this point, the reward continues to grow more gradually until around episode 8,000, indicating steady learning and policy improvement.

In comparison, Fig. 5 presents the reward trend for the DDQN agent. While the DDQN reward also increases sharply from episode 0 to approximately episode 500, it exhibits greater fluctuation throughout the subsequent training period up to episode 8,000. These fluctuations can be attributed to the structure of Double DQN, which utilizes two separate neural networks for action selection and evaluation. Although this design helps reduce overestimation bias and improves generalization, it can also introduce variability in the Q-value updates, resulting in less smooth reward growth compared to standard DQN. Nevertheless, both methods demonstrate positive learning trends over time.

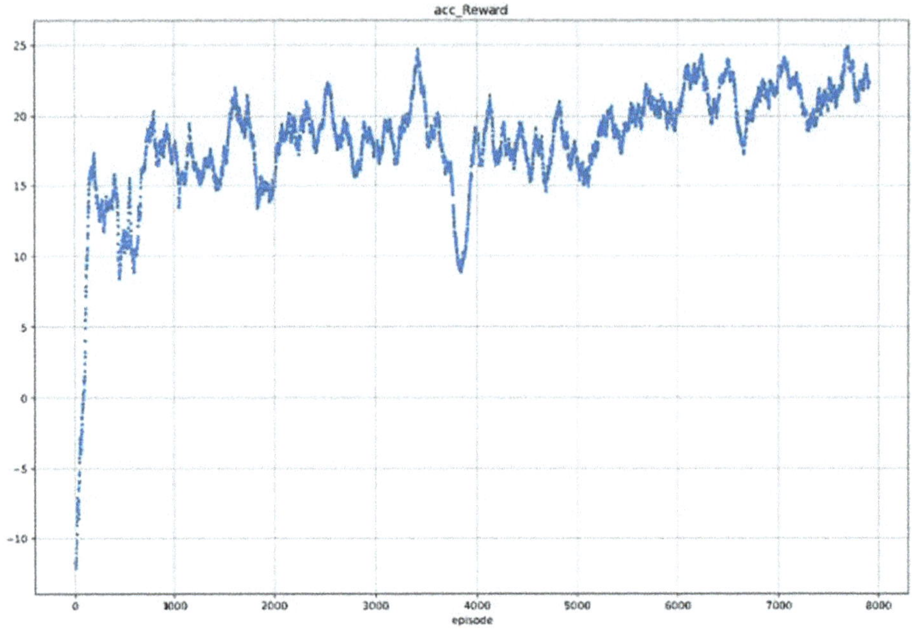

Fig. 4. The reward of the DQN based network

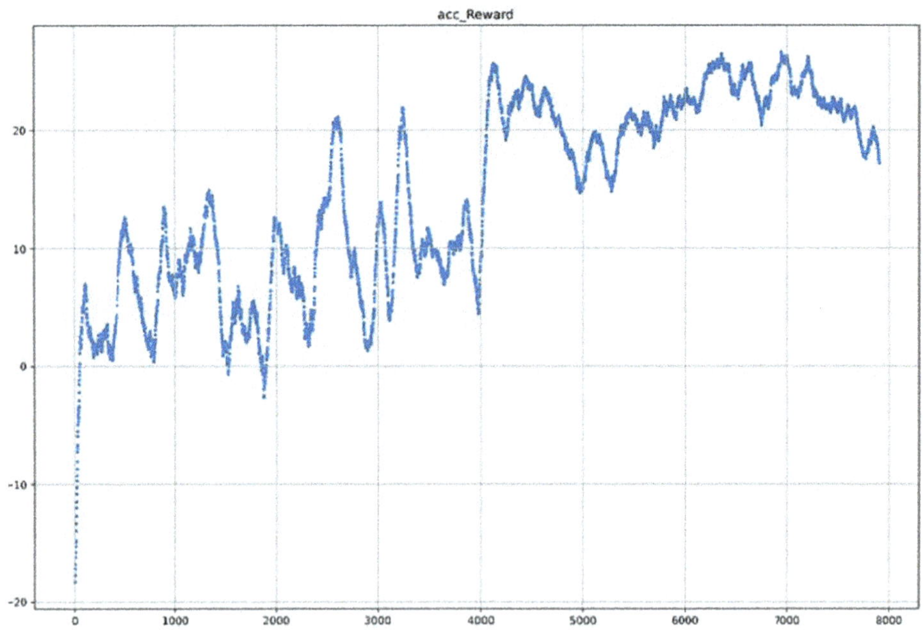

Fig. 5. The reward of the DDQN based network

4.4 Training Loss Analysis

The training costs (loss values) of the DQN and DDQN models are depicted in Fig. 6(a, b), respectively. Both models show a decreasing trend in loss overtime, confirming that the networks are learning effectively. The consistent reduction in training loss indicates convergence toward optimal policies for both DQN and DDQN models.

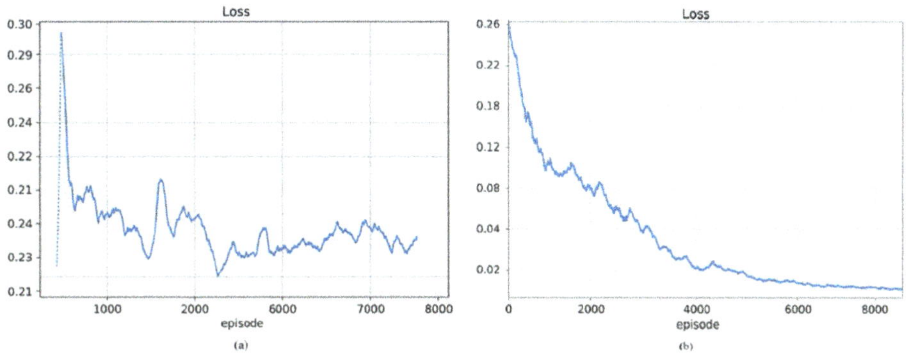

Fig. 6. Training costs (a) - DQN, (b) - DDQN

5 Conclusions and Recommendations for Future Work

This study employed machine learning techniques to develop an intelligent traffic signal control system capable of mitigating traffic congestion through adaptive signal timing. Specifically, reinforcement learning (RL) algorithms, including Deep Q-Networks (DQN) and Double Deep Q-Networks (DDQN), were utilized to design and implement the control system.

The performance of the RL-based adaptive traffic control system was benchmarked against a conventional fixed-cycle signal control strategy. A series of simulations were conducted using a 6×6 lane intersection environment generated through traffic simulation software. The results consistently indicated that the RL-based system significantly outperforms the fixed-cycle system, achieving notable improvements in traffic flow, average vehicle velocity, and overall congestion reduction.

While the simulation results demonstrate the benefits of RL-based methods, several limitations should be noted. First, the current study is based on simulation in a single intersection scenario and may not fully reflect the variability and noise in real-world traffic data. Second, the model's performance may degrade in highly irregular traffic patterns or under sensor failures. Third, the hyperparameters were tuned manually and might not generalize across scenarios. Addressing these limitations in future work through real-world pilot studies, robust policy training, and automated hyperparameter optimization will be essential for practical deployment.

Future research should include additional tests under varying traffic patterns, vehicle arrival rates, and network topologies. Such robust testing will help validate the model's generalizability and effectiveness in broader, more realistic urban settings. Furthermore, we will focus on extending the proposed approach to large-scale, networked traffic environments involving multiple interconnected intersections. Scaling the system to more complex and realistic urban traffic scenarios would enhance its practical applicability and provide further insights into the robustness and scalability of reinforcement learning-based traffic management solutions. Multi-agent reinforcement learning (MARL) could be employed to coordinate signal control among intersections, while techniques such as graph neural networks and hierarchical RL can help manage complexity and improve scalability.

References

1. INRIX: 2024 Global Traffic Scorecard: Employees & Consumers Returned to Downtowns, Traffic Delays & Costs Grew (2024). https://inrix.com/scorecard/
2. Ajayi, S.A., Adams, C.A., Dumedah, G., Adebanji, A.O.: The impact of vehicle engine characteristics on vehicle exhaust emissions for transport modes in Lagos City. Urban Plan. Transp. Res. **12**(1) (2024)
3. Nadrian, H., Taghdisi, M., Pouyesh, K., Khazaee-Pool, M., Babazadeh, T.: "I am sick and tired of this congestion": perceptions of Sanandaj inhabitants on the family mental health impacts of urban traffic jam. J. Transp. Health. **14**, 100587 (2019)
4. Miletić, M., Ivanjko, E., Gregurić, M., Kušić, K.: A review of reinforcement learning applications in adaptive traffic signal control. IET Intell. Transp. Syst. **16**, 1269–1285 (2022)
5. Siavash, A., Ramin, M., Ratnesh, S., Mehran, M.: Adaptive traffic control with deep reinforcement learning: towards state-of-the-art and beyond. arXiv:2007.10960 (2020)
6. Mnih, V., Kavukcuoglu, K., Graves, A.: Human-level control through deep reinforcement learning. Nature. **518**, 529–533 (2015)
7. Bingham, E.: Reinforcement Learning: An Introduction, 2nd edn. MIT Press, Cambridge (2018)
8. Sutton, R.S., Barto, A.G.: Reinforcement Learning: An Introduction, 2nd edn. MIT Press, Cambridge (2018)
9. Lai, X., Yang, Z., Xie, J., Liu, Y.: Reinforcement learning in transportation research: frontiers and future directions. Multimodal Transp. **3**(4), 100164 (2024)
10. Network Traffic Signal Control. IET Intell. Transp. Syst. **4**(2), 128–135 (2010)
11. Ma, D., Zhou, B., Dai, H.: A deep reinforcement learning approach to intelligent traffic signal control using the joint methods of multiagent reinforcement learning and artificial intelligence of things. IEEE Trans. Intell. Transp. Syst. **23**, 9335–9345 (2022)
12. Wei, H., et al.: CoLight: learning network-level cooperation for traffic signal control. In: Proceedings of the CIKM (2019)
13. Chu, C., Wang, B., Wu, X., Chen, X.: Multi-agent deep reinforcement learning for large-scale traffic signal control. IEEE Trans. Intell. Transp. Syst. **23**(5), 4513–4525 (2022)
14. Qi, Z., Kong, L., Wang, X.: Traffic signal control using a cooperative EWMA-based multi-agent reinforcement learning. Appl. Intell. (2022). https://doi.org/10.1007/s10489-022-03643-9
15. Mo, Z., Li, W., Vlight, C.: Decentralized learning for adaptive traffic signal control with connected vehicles. Transp. Res. C Emerg. Technol. **141**, 103728 (2022)
16. Krajzewicz, D., Hertkorn, G., Rössel, C., Wagner, P.: SUMO (simulation of urban mobility) – an open-source traffic simulation. In: Proceedings of the MESM (2002)
17. Jin, J., Ma, X.: Hierarchical multi-agent control of traffic lights based on collective learning. Eng. Appl. Artif. Intell. **68**, 236–248 (2018)

18. Rizzo, G.: Time critic policy gradient methods for traffic signal control in complex and congested scenarios. In: Proceedings of the KDD, pp. 1654–1664 (2019)
19. Gu, J.: Traffic signal optimization for multiple intersections based on reinforcement learning. Appl. Sci. **11**, 10688 (2021)
20. Liu, J.: Intelligent traffic light control by exploring strategies in an optimized space of deep Q-learning. IEEE Trans. Veh. Technol. **71**, 5960–5970 (2022)
21. Jenelius, E., Mattsson, L.-G.: Road network vulnerability analysis of area-covering disruptions: a grid-based approach with case study. Transp. Res. A Policy Pract. **46**(5), 746–760 (2012)
22. Zaklouta, F., Stanciulescu, B.: Real-time traffic-sign recognition using tree classifiers. IEEE Trans. Intell. Transp. Syst. **13**(4), 1507–1514 (2012)
23. Nagalla, R., Pothuganti, P., Pawar, D.S.: Analyzing gap acceptance behavior at unsignalized intersections using support vector machines, decision tree and random forests. Procedia Comput. Sci. **109**, 474–481 (2017)
24. Zhou, C., Mo, H., Chen, X., Wen, H.: Fuzzy control under time-varying universe and phase optimization in traffic lights (ICSSE 2020). Int. J. Fuzzy Syst. **23**(3), 692–703 (2020)
25. Feng, W., et al.: Traffic signal current prediction algorithm based on CNN and LSTM. Meas. Sci. Technol. **36**(1), 015032 (2024)
26. Baek, T., Lee, Y.G.: Traffic control hand signal recognition using convolution and recurrent neural networks. J. Comput. Des. Eng. **9**(2), 296–309 (2022)
27. Van Hasselt, H., Guez, A., Silver, D.: Deep reinforcement learning with double Q-learning. In: Proceedings of the AAAI (2016)
28. Shu, F., Feng, C., Li, H.: Dueling double deep Q-network for adaptive traffic signal control with low exhaust emissions in a single intersection. In: 2019 6th International Conferences Advanced Composite Materials and Manufacturing Engineering (ACMME), Yunnan, China (2019)
29. Sun, Y., Jiang, S., Zhang, Y.: Safe reinforcement learning for traffic signal control with model uncertainty. In: Proceedings of the ITSC (2021)

DP-UOTM: A Differentially Private Unbalanced Optimal Transport Based Approach for High Quality Medical Image Synthesis

Jinnan He, Ai Ran, Fengchi Yuan, and Wai Kin Chan(✉)

Institute of Data and Information, Tsinghua Shenzhen International Graduate School, Tsinghua University, Shenzhen, China
{hjn22,ra22,yfc22}@mails.tsinghua.edu.cn, chanw@sz.tsinghua.edu.cn

Abstract. Deep learning has transformed AI-driven healthcare services through advanced medical imaging analysis, yet privacy concerns over protected health information persistently hinder data sharing in clinical service ecosystems. Conventional anonymization methods, while compliant with healthcare data governance standards, irreversibly degrade diagnostic utility through quality loss or incomplete de-identification. We propose DP-UOTM, a differentially private Unbalanced Optimal Transport-based framework for medical service systems, generating regulatory-compliant synthetic images that preserve diagnostic fidelity via optimal transport, enforce provable (ϵ,δ)-DP guarantees against re-identification, and resolve class imbalances through label distribution calibration. Evaluated on benchmarks MNIST and Fashion-MNIST, DP-UOTM outperforms DP-GANs with ≥22% lower FID scores and ≥2% higher classification accuracy across privacy budgets. Clinical validation on chest X-ray services demonstrates synthetic data maintains ≥4% diagnostic accuracy improvement over raw data through class balancing. By enabling privacy-preserving substitutes for raw medical images, this framework directly supports federated healthcare services and multi-institutional collaborations while addressing critical needs in clinical AI service development—providing scalable, audit-ready synthetic datasets that comply with medical data regulations without compromising diagnostic utility, thereby bridging the gap between AI innovation and healthcare privacy requirements.

Keywords: private data synthesis · unbalanced optimal transport · differential privacy · medical decision making

1 Introduction

Recent advancements in artificial intelligence have revolutionized AI-powered medical service systems, enabling sophisticated diagnostic assistance through

X. Xie et al. (Eds.): ICSS 2025, LNOR, pp. 90–103, 2026.
https://doi.org/10.1007/978-3-032-13116-4_8

categrical data [1] and medical image analysis. However, critical challenges persist in healthcare services due to limitations in medical imaging datasets, primarily stemming from two issues: (1) inherent dataset constraints including severe class imbalance (particularly for rare diseases) and insufficient sample sizes, which degrade model performance in clinical service applications, and (2) stringent privacy protections governing sensitive patient information under regulations like HIPAA that restrict data sharing across healthcare systems. Conventional privacy approaches like data anonymization often compromise diagnostic utility; for instance, they may produce incorrect results that flag healthy patients for unnecessary treatment [2]. Differentially private data synthesis has emerged as a promising solution. This approach trains privacy-preserving generative models using DP-SGD[3], offering quantifiable privacy guarantees while maintaining three key advantages for medical services: enabling secure data sharing through synthetic models rather than raw patient data, generating balanced datasets to address class imbalance in clinical applications, and preserving diagnostic fidelity with theoretical privacy assurances - all crucial for developing robust AI systems in healthcare services.

Privacy-preserving synthetic data generation requires careful consideration of two key aspects: (1) privacy protection mechanisms and (2) generative models. Currently, GANs with differential privacy (DP) guarantees have emerged as the most prevalent approach due to their superior generation quality and flexible privacy-utility trade-offs. Among privacy protection methods, several distinct strategies exist. The Private Aggregation of Teacher Ensembles (PATE) framework [4] protects privacy through ensemble-based knowledge transfer, but suffers from complex implementation and substantial computational overhead. Alternative approaches like DP-MERF [5] and PEARL [6] employ feature anonymization and extraction techniques; however, their utility is fundamentally constrained by the limited expressive power of the anonymized features. In contrast, DP-SGD [3] stands out as particularly advantageous - it provides quantifiable privacy budgeting through (ε, δ)-DP guarantees and achieves an effective balance between privacy protection and data utility. More importantly, DP-SGD's model-agnostic nature makes it readily adaptable to various generative architectures. This flexibility has led to DP-SGD's widespread adoption across different generative models. When applied to variational autoencoders (VAEs), the inherent noise addition often results in overly smoothed and blurry samples [7]. For diffusion models (DMs), while DP-SGD enables training of privacy-preserving variants like DPDM [8], the approach demands careful hyperparameter tuning and suffers from computationally intensive noise sampling processes. DP-GANs [9,10] uniquely combine the strengths of both components: they maintain GANs' renowned generation quality while benefiting from DP-SGD's reliable privacy guarantees. Compared to alternatives, DP-GANs demonstrate: (1) higher sample fidelity than DP-VAEs, (2) more efficient training than DP-DMs, and (3) better utility preservation than feature-based methods like DP-MERF (Fig. 1).

However, the fundamental challenge in differentially private image synthesis is maintaining meaningful feature learning under constrained privacy bud-

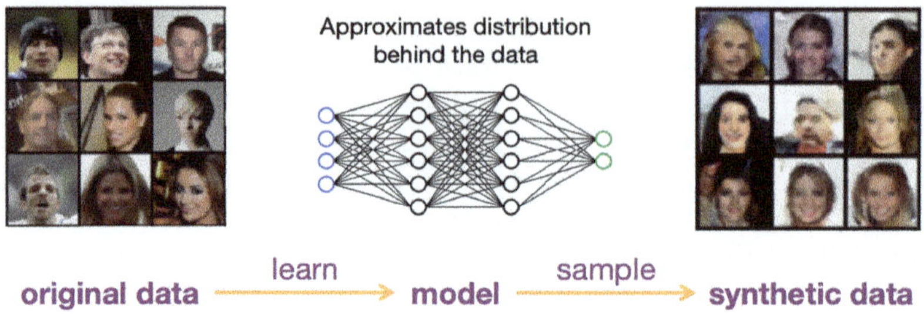

Fig. 1. The idea of privacy data synthesis.

gets (ϵ, δ)-DP. Unlike conventional approaches that apply DP through one-time noise injection[11,12], DP-SGD requires noise addition at every training iteration. This noise injection exacerbates inherent GAN instability, amplifying mode collapse and convergence issues. Unbalanced Optimal Transport based Generativce Model (UOTM) [13] addresses this through two key properties. Firstly, the UOTM objective's inherent stability (theoretically analyzed in [13]) mitigates gradient noise effects - a principle validated in DP-Sinkhorn [14] and DPDM [8]. Rapid convergence reduces required optimization steps, lowering cumulative noise. Secondly, as the first OT-based model matching SOTA generation quality, UOTM preserves GAN-level fidelity while overcoming mode collapse via its adversarial training stabilization. Empirical results across dataset scales confirm UOTM's superiority for privacy-constrained synthesis.

In this work, we make the following key contributions to privacy-preserving medical imaging: (1) First UOT-based Framework for Private Medical Synthesis. We propose DP-UOTM and its label-conditioned variant DP-CUOTM, establishing the first differentially private medical image synthesis framework grounded in Unbalanced Optimal Transport theory. This approach fundamentally addresses the privacy-utility trade-off in healthcare data generation. (2) Quantitative Performance Advancements. Extensive experiments on MNIST and Fashion-MNIST demonstrate consistent improvements over existing DP-GANs, achieving $\geq$22% reduction in FID and enhanced visual fidelity under equivalent privacy budget. (3) Clinical Validation with Diagnostic Utility. Our chest X-ray synthesis effectively balances imbalanced clinical labels, improving downstream diagnostic model accuracy by $\geq$4%. It Preserves pathological features while eliminating protected health information.

2 Differentially Private Unbalanced Optimal Transport-Based Generative Model

We propose a Differentially Private Unbalanced Optimal Transport-based generative model (DP-UOTM), UOTM [13] trained with rigorous DP guarantees based on DP-SGD. Assume that our source distribution consists of random Gaussian noise $\mathscr{X}$, which has the same image size as the target distribution $\mathscr{Y}$, which represents private data.

2.1 Data Synthesizer Based on Unbalanced Optimal Transport

Unbalanced Optimal Transport (UOT) [15] relaxes the hard constraints of marginal distribution in Kantorovich Optimal Transport Problem [16] with soft entropic penalties. Assume Ψ_1 and Ψ_2 are convex, lower semi-continuous, non-negative entropy functions, and $\Psi_1(0) = \Psi_2(0) = 1$, $\Psi_1(x) = \Psi_2(x) = \infty$ for $x < 0$. The problem of UOT is stated as follows:

$$Cost_{ub}(\mu, \nu) := \inf_{\pi \in \mathscr{M}_+(\mathscr{X} \times \mathscr{Y})} \left[\int_{\mathscr{X} \times \mathscr{Y}} c(x, y)\, d\pi(x, y) + \int_{\mathscr{X}} \Psi_1(\frac{\pi_0(x)}{d\mu(x)}) d\mu(x) + \int_{\mathscr{Y}} \Psi_2(\frac{\pi_1(y)}{d\nu(y)}) d\nu(y) \right]. \tag{1}$$

Here $\pi_0(x)$ and $\pi_1(y)$ represent the marginal distributions associated with $\mathscr{X}$ and $\mathscr{Y}$, respectively. By relaxing marginal constraint[15], UOT can effectively alleviate the influence of outliers, which is a notable limitation of traditional OT. We use its semi-dual formulation:

$$Cost_{ub}(\mu, \nu) = \sup_{v \in \mathscr{C}} \left[\int_{\mathscr{X}} -\Psi_1^* \left(-v^c(x)\right) d\mu(x) + \int_{\mathscr{Y}} -\Psi_2^*(-v(y)) d\nu(y) \right]. \tag{2}$$

The convex conjunction of function $f(y)$ is defined as $f^*(y) = \sup_{x \in \mathbb{R}} \{\langle x, y \rangle - f(x)\}$ for $f : \mathbb{R} \to [-\infty, \infty]$. Here $v \in \mathscr{C}(\mathscr{Y})$ is a continuous function on $\mathscr{Y}$, named as potential function on target space. And the c-transform of v is defined as $v^c(x) = \inf_{y \in \mathscr{Y}} (c(x, y) - v(y))$. Use a generative model $T_v \in \underset{y \in \mathscr{Y}}{\operatorname{arginf}} [c(x, y) - v(y)]$ to help approximate v^c, we have $v^c(x) = c\left(x, T_v(x)\right) - v\left(T_v(x)\right)$. Therefore the objective $J(v)$ becomes:

$$J(v, T) := \int_{\mathscr{X}} \Psi_1^* \left(- \left[\left(c\left(x, T_v(x)\right) - v\left(T_v(x)\right)\right)\right]\right) d\mu(x) + \int_{\mathscr{Y}} \Psi_2^*(-v(y)) d\nu(y).$$

In this objective function, we parameterize v and T_v by neural networks with parameters ϕ and θ. Note that the loss functions of v_ϕ and T_θ exhibit an adversarial relationship, requiring an alternating training procedure similar to that employed in GANs[17]. Building upon prior works [13], we introduce stochasticity into our generator T_θ by incorporating an auxiliary variable z as an additional input. Finally the objectives for v_ϕ and T_θ are:

$$\mathscr{L}_{v_\phi} = \frac{1}{X} \sum_{x \in X} \Psi_1^* \left(-c\left(x, T_\theta(x, z)\right) + v_\phi\left(T_\theta(x, z)\right)\right) + \frac{1}{Y} \sum_{y \in Y} \Psi_2^* \left(-v_\phi(y)\right) \tag{3}$$

$$\mathscr{L}_{T_\theta} = \frac{1}{|X|} \sum_{x \in X} \left(c\left(x, T_\theta(x, z)\right) - v_\phi\left(T_\theta(x, z)\right)\right) \tag{4}$$

Next, we introduce a Label-Conditional Unbalanced Optimal Transport-based Generative Model (CUOTM). To the best of my knowledge, our work is the first to propose the training of CUOTM, which constitutes one of our contributions. In many practical scenarios, generating labeled data requires the ability to condition image synthesis on specific labels. So a label $l \in L$ is incorporated as an additional conditioning variable, resulting in the formulations $T(x|l)$ and $v(y|l)$. Conditional Optimal Transport [18] has been explored and here it is extended to UOT problem. Let $\mathscr{X}$, $\mathscr{Y}$, $\mathscr{L}$ be polish spaces. And $\mu(\cdot|l)$, $\nu(\cdot|l)$ and u are two conditional distributions refer to source and target. Consider a cost function $c : (\mathscr{L} \times \mathscr{X}) \times (\mathscr{L} \times \mathscr{Y}) \to \mathbb{R}$ which is condituous and $\inf c > -\infty$, then the Conditional Unbalanced Optimal Transport problem is:

$$Cost_{cub}(\mu, \nu, l) = \inf_{\pi \in \Pi(\mu(x|l), \nu(y|l))} \int_{\mathscr{X} \times \mathscr{Y}} c^l(x, y) d\pi(x, y) + D_{\Psi_1}(\pi_0(\cdot|l)|\mu(\cdot|l)) + D_{\Psi_2}(\pi_1(\cdot|l)|\nu(\cdot|l)).$$

Here $c^l(x, y) = c(l, x; l, y)$, $\pi_0(\cdot|l)$ and $\pi_1(\cdot|l)$ are two conditional marginals. Then, as before, we present its semi-dual formulation:

$$Cost_{cub}(\mu, \nu, l) = \sup_{v \in \mathscr{C}} \left[\int_{\mathscr{X}} -\Psi_1^*\left(-v^c(x|l)\right) d\mu(x|l) + \int_{\mathscr{Y}} -\Psi_2^*(-v(y|l)) d\nu(y|l) \right], \tag{5}$$

where $v \in \mathscr{C}(\mathscr{Y} \times \mathscr{L})$ is a continuous function, named as conditional potential function. And the c-transform of v is defined as $v^c(x|l) = \inf_{y \in \mathscr{Y}}(c^l(x, y) - v(y|l))$. Use a generative model $T_v(\cdot|l) \in \operatorname*{arginf}_{y \in \mathscr{Y}}[c^l(x, y) - v(y|l)]$ to help approximate v^c, we have the objective:

$$J(v, T) = \int_{\mathscr{X}} \Psi_1^*\left(-\left[\left(c\left(x, T_v(x|l)\right) - v\left(T_v(x|l)\right)\right)\right]\right) d\mu(x|l) + \int_{\mathscr{Y}} \Psi_2^*(-v(y|l)) d\nu(y|l).$$

Algorithm 1. Training algorithm of DP-UOTM

Require: Source distribution μ and N private samples from target distribution ν. Functions Ψ_1^*, Ψ_2^*. Two networks: generator T_θ and potential v_ϕ. Privacy budget ϵ. Expected batch size B, gradient norm bound C. Privacy accountant $\mathscr{A}(\epsilon, C, B, N)$. Learning rate $\eta_{\phi t}$, $\eta_{\theta t}$

```
Use privacy accountant to get training steps T and noise level σ: (T, σ) = 𝒜(ε, C, B, N)
for t = 0, 1, 2, ..., T do
    Update φ:
    Sample a batch X ~ μ, and z ~ 𝒩(0, I). And sample a batch Y ~ ν with
  sampling rate B/N by Poission Sampling.
    ℒ_v = (1/B) Σ_{x∈X} Ψ₁*(−c(x, T_θ(x, z)) + v_φ(T_θ(x, z))) + (1/L) Σ_{y∈Y} Ψ₂*(−v_φ(y))
    Compute gradient: for each y ∈ Y, g_t(y) = ∇_{φ_t} ℒ_v
    Clip gradient: for each y ∈ Y, ḡ_t(y) = g_t(y) / max(1, ‖g_t(y)‖₂ / C)
    Add noise: g̃_t ← (1/B)(Σ ḡ_t(y) + 𝒩(0, σ²C²I))
    Descent: φ_{t+1} ← φ_t − η_{φt} g̃_t
    Update θ:
    Sample a batch X ~ μ, z ~ 𝒩(0, I).
    ℒ_T = (1/|X|) Σ_{x∈X} (c(x, T_θ(x, z)) − v_φ(T_θ(x, z))).
    Compute gradient for the batch: g̃_t ← ∇_{θ_t} ℒ_T
    Descent: θ_{t+1} ← θ_t − η_{θt} g̃_t
end for
```

2.2 DP-SGD for Privacy Protection

Differential Privacy (DP) seeks to describe information leakage resulting from multiple queries of a database. And Rényi Differential Privacy (RDP) offers a robust formulation that enables the integration of multiple queries while also being readily adaptable to standard DP:

Definition 1. (Rényi Differential Privacy) [19] A randomized Algorithm $\mathscr{A} : \mathscr{D} \to \mathscr{R}$ satisfies (α, ϵ)-RDP if for any adjacent inputs $d, d' \in \mathscr{D}$ it holds that:

$$D_\alpha \left(\mathscr{A}(d) || \mathscr{A}\left(d'\right)\right) \leq \epsilon \tag{6}$$

where D_α is the Rényi divergence of order α. Any $\mathscr{A}$ that satisfies (α, ϵ)-RDP also satisfies $\left(\epsilon + \frac{\log 1/\delta}{\alpha - 1}, \delta\right)$-DP.

DP-SGD[3], a modification of SGD, is a technique specifically designed to ensure that the trained model satisfies differential privacy. During the training process of a neural network, sensitive data information is inadvertently disclosed to the model parameters through the gradients computed during each backward propagation step. Therefore, DP-SGD updates parameters as $\theta \leftarrow \theta - \frac{\eta}{B}(\sum_{x_i \in B} \text{clip}_C(\nabla_\theta L(x_i; \theta) + Cz)$, where η is learning rate, B is a batch of data from a total of N training samples, C is a clipping bound, L is loss function and $z \sim \mathscr{N}(0, \sigma_{DP}^2 \mathbf{I})$. The clipping function $\text{clip}_C(g) = min\{1, \frac{C}{||g||_2}\}g$. Furthermore, privacy accounting methods can be utilized to calculate the cumulated privacy cost of multiple DP-SGD training steps. Based on training steps and subsampling rate B/N, it can determine the variance σ_{DP}^2 needed to satisfy (ϵ, δ)-DP (Fig. 2).

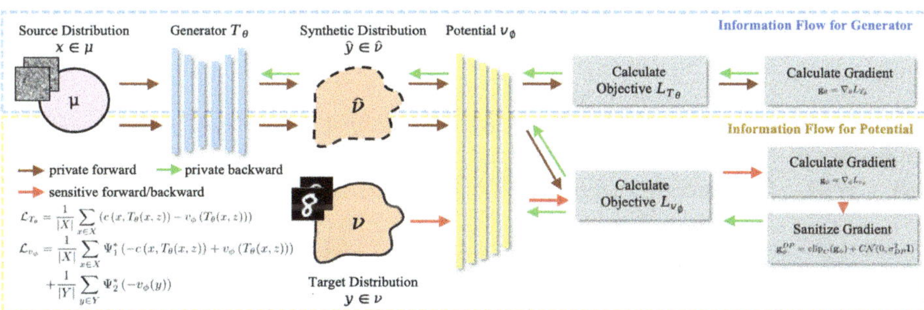

Fig. 2. Information flow during training in DP-UOTM. For the information flow of the generator (in the upper blue dotted-line box), both the forward and backward processes are private. For the information flow of the potential network (in the lower yellow dotted-line box), the forward process is sensitive. However, after the gradient is sanitized, the backward process becomes private.

2.3 Training Details

We present our efforts for privacy protection. Firstly, our methodology selectively applied DP-SGD to potential network training. Specifically, it is important to note that the loss computation for the potential function v_ϕ directly involves private data, whereas the generator loss does not depend on the private data explicitly. Although the generator indirectly interacts with private real data through its dependency on the potential network, it does not introduce additional privacy risks due to the post-processing property of differential privacy [20]. This ensures that no further privacy leakage occurs beyond what has already been accounted for in the potential function's training process. A detailed formal proof of this privacy-preserving mechanism will be provided in the following part.

Furthermore, for a fixed privacy budget, to incorporate less noise while ensuring comparable privacy protection, we employed the Sampled Gaussian Mechanism (SGM)[21]. In our framework, for each training iteration, we sample a mini-batch Y by Poisson Sampling[21] from target distribution ν. The sampling rate is $q = \frac{B}{N}$, where B is the expected batch-size and N is the total training samples. Other settings, such as large batch sizes (e.g., 2048 and 4096), have been recognized as effective in DP training. Regarding the clipping bound, previous work suggests that smaller bounds tend to yield more favorable outcomes. In line with this, we set the clipping bound to $C = 1$. Finally a RDP accountant[19] is adopted to track total privacy budget.

2.4 Theoretical Privacy Guareetee

We first provide a short proof that gradients of potential network released by Algorithm 1 are DP. Specifically, we commence by elucidating how the introduction of Gaussian noise can confer RDP in Theorem 1. And given that the parameters of the potential network are differentially private, it follows that the parameters of the generator themselves are also differentially private based on Lemma 2.

Lemma 1. *(RDP Gaussian mechanism [19]) For a real-valued function f with Sensitivity $S = max_{d,d'} ||f(d) - f(d')||_2$, the Gaussian mechanism that releases $f(d) + \mathcal{N}(0, \sigma_{DP}^2)$ satisfies $(\alpha, \alpha S^2/(2\sigma^2)) - RDP$.*

Lemma 2. *(Post-processing property [19]) If a real-valued function f is (α, ϵ)-RDP, then for any function or mechanism g, the composition $g \circ f$ is also (α, ϵ)-RDP.*

Theorem 1. *For noise scale σ, a batch Y sampled with an expected batch size B from N samples, releasing $\tilde{\mathbf{g}} = \frac{1}{B}(\sum clip(\mathbf{g}(y)) + \mathcal{N}(0, \sigma^2 C^2 \mathbf{I}))$ in Algorithm 1 satisfies $(\alpha, \alpha S^2/(2\sigma^2)) - RDP$.*

Proof. Note that $\text{clip}(g) = \mathbf{g}(y) / \max(1, \frac{\|\mathbf{g}(y)\|_2}{C})$. For adjacent mini-batches $\{y_i\}_{i\in\mathbb{B}}$ and $y' \cup \{y_i\}_{i\in\mathbb{B}}$, the difference of their gradients $\frac{1}{B}\sum \text{clip}(\mathbf{g}(y))$ can be bounded in L_2-norm as $||\frac{1}{B}\sum_{y_i\in Y} \text{clip}(\mathbf{g}(y_i)) - \frac{1}{B}[\text{clip}(\mathbf{g}(y')) +$

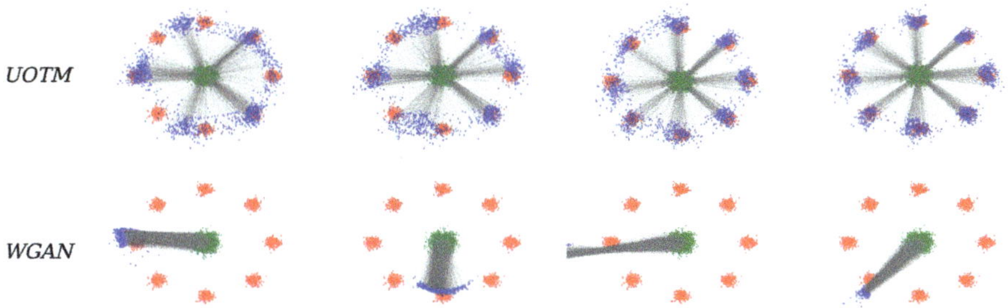

Fig. 3. Toy experiment under $\epsilon = 10$. UOTM and WGAN are trained under DP-SGD to learn 8 Gaussians (red dots). The training dynamics are plotted every 50 epochs, and the blue dots are sythetic data.

$\sum_{y_i \in Y} \text{clip}(\mathbf{g}(y_i))]||_2 = ||\frac{1}{B}\text{clip}(\mathbf{g}(y'))||_2 \leq \frac{C}{B}$. Thus we have sensitivity $S(\frac{1}{B}\sum_{y_i \in Y} \text{clip}(\mathbf{g}(y_i))) = \frac{C}{B}$. Furthermore, $\frac{1}{B}\mathcal{N}(0, \sigma^2 C^2 \mathbf{I})$ is equivalent to $\mathcal{N}(0, \sigma^2 \frac{C^2}{B^2}\mathbf{I})$, based on Lemma 1, releasing $\tilde{\mathbf{g}} = \frac{1}{B}(\sum \text{clip}(\mathbf{g}(y)) + \mathcal{N}(0, \sigma^2 C^2 \mathbf{I}))$ satisfies $(\alpha, \alpha S^2/(2\sigma^2)) - \text{RDP}$.

3 Experiment

Datasets. Our experiment employs two benchmark datasets in privacy data synthesis tasks, MNIST[22] and Fashion-MNIST[23], to accomplish the image synthesis task. Both datasets contain 70,000 grayscale images, each with dimensions of 28×28 pixels, encompassing 10 categories.

Architecture and Training Settings. We followed the architecture of the large model in UOTM[13], and an image size of 28×28. The source distribution μ is a standard Gaussian distribution $\mathcal{N}(0, \mathbf{I})$ with a dimension of 28×28. The cost function $c(x, y) = \frac{\tau}{2}||x - y||_2^2$ and $\Psi_1^* = \Psi_2^* = SP(x) = 2log(1 + e^x) - 2log2$. Code are written in PyTorch and we use Opacus[24], a DP-SGD library in PyTorch, for training and privacy accounting. We set $\delta = 10^{-5}$, which is a standard value[14] and a clipping bound $C = 1$.

Evaluation Metric. To evaluate the quality of the generated samples, we use the Fréchet Inception Distance (FID)[25]. Lower FID scores indicate higher-quality and more realistic generated images. Secondly, we assess the impact of the synthesized dataset on classification tasks by training classifiers - Logistic Regression (LR), Multilayer Perceptron (MLP), and Convolutional Neural Networks (CNN), and evaluating their accuracy on a standard test set.

3.1 Toy Experiment for Mode Collapse Problem

Mixture of Gaussians [26] serves as a benchmark dataset for visualizing the mode collapse problem in generative models. We conduct experiments in the

DP-SGD framework, comparing UOTM with WGAN—a well-known GAN variant designed to mitigate mode collapse. As illustrated in Fig. 3, under identical privacy budgets, the results demonstrate that UOTM achieves significantly faster convergence and effectively alleviates mode collapse. In contrast, WGAN exhibits noticeable deviation from the target distribution and suffers from pronounced mode collapse under the DP-SGD training regime.

3.2 Performance Under Moderate Privacy Budget

By set the privacy budget to a moderate level of $\epsilon = 10$, our model was trained for 300 epochs using a large batch size of 2048. Under these specified conditions, the noise added was $\sigma = 1.73$. Learning rates are $\eta_T = \eta_v = 5 \times 10^{-5}$ for MNIST, and $\eta_T = 1.4 \times 10^{-5}, \eta_v = 1 \times 10^{-5}$ for Fashion-MNIST.

Table 1. Class-conditional DP image generation performance (MNIST & Fashion-MNIST).

Method	DP-ϵ	FID↓		Acc(%)↑					
		MNIST	Fashion-MNIST	MNIST			Fashion-MNIST		
				LR	MLP	CNN	LR	MLP	CNN
DP-UOTM(*ours*)	10	**9.15**	**25.4**	**88.6**	**92.4**	**97.7**	**80.2**	**80.9**	**84.9**
DPGANr [27]	10	13.0	56.8	–	–	95.0	–	–	74.8
DP-CGAN [9]	10	179.2	243.8	60	60	63	51	50	46
GS-WGAN [10]	10	61.3	131.3	79	79	80	68	65	65

Best Generation Performance in GAN-like Models. We assess the quality of the generated data by examining its similarity to real private data. Under the same privacy budget of $\epsilon = 10$, as shown in table 1, our model achieved the lowest FID value in all GAN-like differentially private generative models. Notably, the advantage of DP-UOTM is particularly evident on the Fashion-MNIST dataset.

Sufficient Utility for Training Downstream Classifiers. Classifiers are trained by synthetic data from DP-CUOTM, and we compare their test accuracy using private data as a benchmark. Table 1 presents the results, which reveal that the classifiers trained on images generated by DP-CUOTM achieve remarkably high classification accuracy, outperforming other GAN-like differentially private generative models.

Table 2. Performance under low privacy budget on MNIST

Method	DP-ϵ	FID↓	DP-ϵ	FID↓
DP-UOTM(*ours*)	1	**18.06**	0.2	**38.25**
DPDM [8]	1	23.4	0.2	61.9
PEARL [6]	1	121	0.2	133
DPGANr [27]	1	56.2	0.2	-

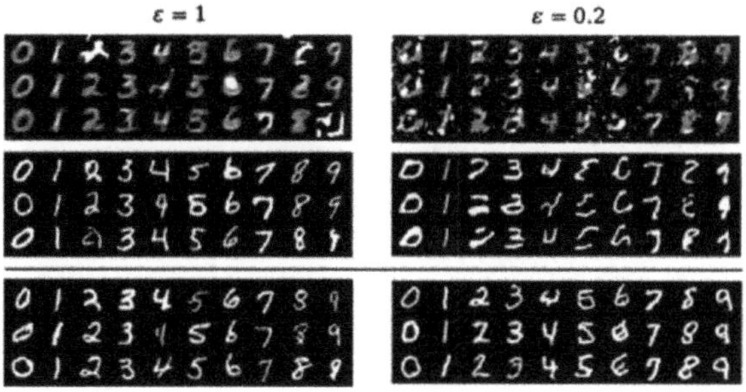

Fig. 4. MNIST images generated by, from top to bottom, PEARL, DPDM and our DP-UOTM under $\epsilon = 1$ and 0.2.

3.3 Performance Under Extremely Low Privacy Budgets

Under lower privacy budget $\epsilon = 1$ and $\epsilon = 0.2$, noise added are $\sigma = 2.9$ and $\sigma = 55.0$ respectively. When studying DP-GANs under extremely low privacy budgets, there are limited comparable baselines available in existing research. Therefore, we selected two alternative models for comparison: DPDM and PEARL. DPDM [8] is a diffusion model trained using DP-SGD, while PEARL [6] employs private features for model training rather than relying on DP-SGD. Table 2 demonstrates that on MNIST, our model achieved the lowest FID among these scenarios, surpassing even DPDM. As the privacy budget decreased, leading to an increase in noise level, our model performed even better. From the comparison of generated images (Fig. 4), we can also observe that DP-UOTM is capable of effectively learning the general morphology of images, while DPDM excels in the smooth processing of details.

3.4 Clinical Validation on Chest X-Rays Dataset

To rigorously evaluate the clinical applicability of DP-UOTM, we conducted experiments using a chest X-ray dataset [28] comprising 5,232 images (3,883 pneumonia cases vs. 1,349 normal cases), reflecting the significant global health burden of pneumonia which accounts for over 2.5 million annual deaths worldwide. Given chest radiography's critical role as a primary diagnostic tool, we specifically examined our model's performance on this imbalanced dataset (1:3 normal-to-pneumonia ratio) which presents both privacy preservation and diagnostic utility challenges.

We train DP-CUOTM with label condition (0: normal, 1: pneumonia). All images were standardized to 64×64 resolution and processed under strict differential privacy guarantees (ϵ=10) through 200 training epochs with a noise multiplier of 1.7. A comparative visualization between the synthetic images generated by our method and DP-WGAN is presented in Fig. 5. Our analysis reveals two key limitations in DP-WGAN's outputs: evident mode collapse in normal cases, where generated images show nearly identical skeletal structures, and poor

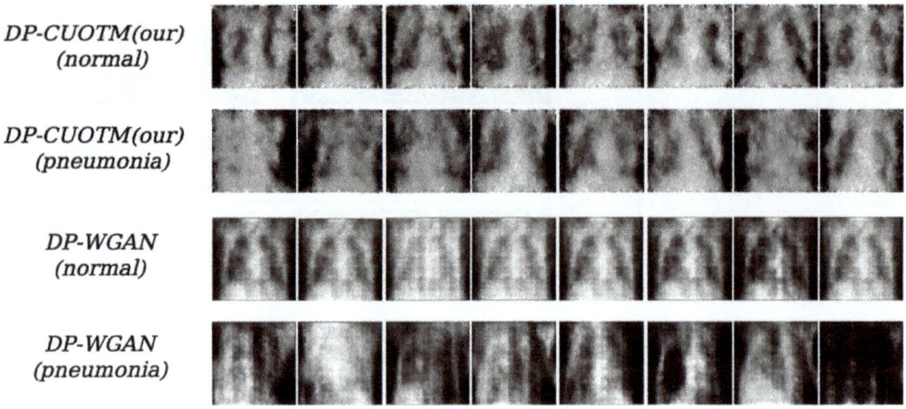

Fig. 5. Synthetic chest X-ray images

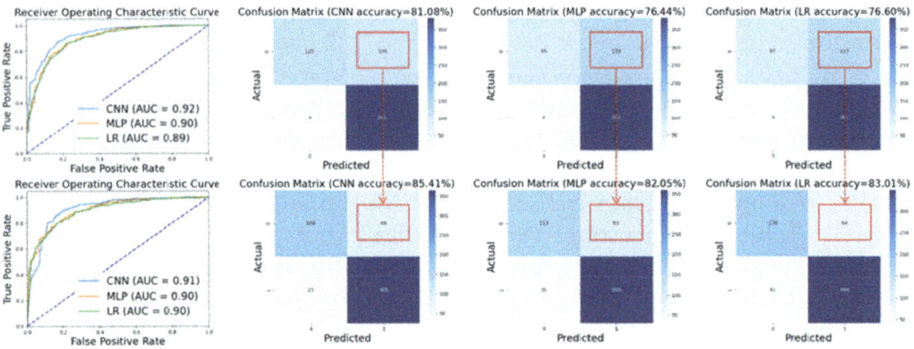

Fig. 6. Classification ROC curve and confusion matrix using original dataset (top) and synthetic dataset (bottom)

quality in pneumonia cases with blurred pathological features. These issues occur because pneumonia's characteristic lung shadows present complex patterns that are particularly challenging to learn under privacy constraints. In contrast, our DP-CUOTM generates more diverse and clinically realistic images - normal cases show natural anatomical variations, while pneumonia cases clearly display diagnostic features like asymmetric lung involvement and distinct pulmonary shadows.

To rigorously evaluate the clinical utility, we generated a balanced synthetic dataset comprising 10,000 medical images (5,000 normal and 5,000 pneumonia cases) and trained three diagnostic models - CNN, MLP, and LR - using both original and augmented datasets for comparative analysis. As evidenced by the ROC curves in Fig. 6, classifiers trained on synthetic-augmented data achieved comparable discriminative performance (AUC ≈ 0.9) to original dataset models, while demonstrating statistically significant accuracy improvements: +4.33% for CNN, +5.61% for MLP, and +6.41% for LR. Confusion matrix analysis revealed these gains stemmed primarily from enhanced specificity (i.e., the proportion of normal cases correctly identified as normal), as baseline models trained on the original imbalanced dataset exhibited pronounced false-positive rates due to underrepresentation of normal cases. Our synthetic data generation effectively mitigated this class imbalance by producing balanced label distributions, improving normal case discrimination without compromising overall discriminative power, thereby yielding more clinically reliable diagnostic models. The results substantiate that synthetic data augmentation can address critical class imbalance challenges in medical image analysis while maintaining diagnostic performance.

4 Conclusion

This work proposes DP-UOTM, the first Unbalanced Optimal Transport-based framework for differentially private medical image synthesis in clinical AI services, achieving superior performance under strict privacy constraints ($\geq$ 22% FID improvement over DP-GANs) while enhancing downstream diagnostic accuracy by $\geq$4% in healthcare applications. The UOT formulation's inherent noise stability enables high-fidelity synthetic data generation crucial for medical service systems, effectively addressing both class imbalance and privacy preservation challenges. Future work will explore DP-UOTM's integration with federated healthcare systems, extension to 3D imaging and complex diagnostic pipelines, and automated hyperparameter optimization for privacy-utility tradeoffs. A current limitation is that DP-UOTM's synthetic image quality, while superior to privacy-preserving GANs, lags behind diffusion-based methods due to intentional speed-quality tradeoffs. Despite this tradeoff, DP-UOTM's ability to simultaneously ensure diagnostic-grade data utility and rigorous privacy compliance represents a critical step toward deployable privacy-preserving AI in clinical infrastructures.

Acknowledgements. This research was funded by the Guangdong Pearl River Plan (2019QN01X890).

References

1. Jia, F., Zhu, H., Jia, F., Chan, W.K.V.: TRamCol-Diff: deepening diffusion model comprehension for extensive categorical medical service data. In: 2024 IEEE International Conference on Service Operations and Logistics, and Informatics (SOLI), pp. 1–7 (2024)
2. Cai, J., Zychlinski, N.: When AI is not enough: reducing diagnostic errors with radiologist oversight. Serv. Sci. (2025)
3. Abadi, M., et al.: Deep learning with differential privacy. In: Proceedings of the 2016 ACM SIGSAC Conference on Computer and Communications Security (2016)
4. Papernot, N., Abadi, M., Erlingsson, Ú., Goodfellow, I., Talwar, K.: Semi-supervised knowledge transfer for deep learning from private training data. In: International Conference on Learning Representations (2017)

5. Harder, F., Adamczewski, K., Park, M.: DP-MERF: differentially private mean embeddings with randomfeatures for practical privacy-preserving data generation. In: International Conference on Artificial Intelligence and Statistics, pp. 1819–1827. PMLR (2021)
6. Liew, S.P., Takahashi, T., Ueno, M.: PEARL: data synthesis via private embeddings and adversarial reconstruction learning. In: International Conference on Learning Representations (2022)
7. Acs, G., Melis, L., Castelluccia, C., De Cristofaro, E.: Differentially private mixture of generative neural networks. In: 2017 IEEE International Conference on Data Mining (ICDM) (2017)
8. Dockhorn, T., Cao, T., Vahdat, A., Kreis, K.: Differentially private diffusion models. Trans. Mach. Learn. Res. (2023)
9. Torkzadehmahani, R., Kairouz, P., Paten, B.: DP-CGAN: differentially private synthetic data and label generation. In: 2019 IEEE/CVF Conference on Computer Vision and Pattern Recognition Workshops (CVPRW) (2019)
10. Chen, D., Orekondy, T., Fritz, M.: GS-WGAN: a gradient-sanitized approach for learning differentially private generators. Adv. Neural. Inf. Process. Syst. **33**, 12673–12684 (2020)
11. Ran, A., Li, D., Ren, X., He, J., Yuan, F., Chan, W.K.V.: A novel differential privacy protection model for publishing time series of classification task. In: 2024 9th International Conference on Big Data Analytics (ICBDA), pp. 191–196 (2024)
12. Wang, J., Chan, W.V.: A design for private data protection combining with data perturbation and data reconstruction. In: 2021 International Symposium on Electrical, Electronics and Information Engineering, ISEEIE 2021, pp. 545–550. Association for Computing Machinery, New York (2021)
13. Choi, J., Choi, J., Kang, M.: Generative modeling through the semi-dual formulation of unbalanced optimal transport. Adv. Neural Inf. Process. Syst. **36** (2024)
14. Cao, T., Bie, A., Vahdat, A., Fidler, S., Kreis, K.: Don't generate me: training differentially private generative models with sinkhorn divergence. Adv. Neural. Inf. Process. Syst. **34**, 12480–12492 (2021)
15. Chizat, L., Peyré, G., Schmitzer, B., Vialard, F.-X.: Unbalanced optimal transport: dynamic and Kantorovich formulations. J. Funct. Anal. **274**(11), 3090–3123 (2018)
16. Kantorovich, L.K.: On a problem of Monge. J. Math. Sci. **133**(4) (2006)
17. Goodfellow, I., Pouget-Abadie, J., Mirza, M., Bing, X., Warde-Farley, D., Ozair, S., Courville, A., Bengio, Y.: Generative adversarial networks. Commun. ACM **63**(11), 139–144 (2020)
18. Hosseini, B., Hsu, A.W., Taghvaei, A.: Conditional optimal transport on function spaces. SIAM/ASA J. Uncertain. Quantif. **13**(1), 304–338 (2025)
19. Mironov, I.: Rényi differential privacy. In: 2017 IEEE 30th Computer Security Foundations Symposium (CSF), pp. 263–275 (2017)
20. Dwork, C., Roth, A.: The algorithmic foundations of differential privacy. Found. Trends® Theor. Comput. Sci. 211–407 (2013)
21. Mironov, I., Talwar, K., Zhang, L.: Rényi differential privacy of the sampled gaussian mechanism. CoRR, abs/1908.10530 (2019)
22. LeCun, Y., Bottou, L., Bengio, Y., Haffner, P.: Gradient-based learning applied to document recognition. Proc. IEEE **86**(11), 2278–2324 (1998)
23. Xiao, H., Rasul, K., Vollgraf, R.: Fashion-MNIST: a novel image dataset for benchmarking machine learning algorithms (2017)
24. Yousefpour, A., et al.: Opacus: user-friendly differential privacy library in pytorch. In: NeurIPS 2021 Workshop Privacy in Machine Learning (2021)

25. Heusel, M., Ramsauer, H., Unterthiner, T., Nessler, B., Hochreiter, S.: GANs trained by a two time-scale update rule converge to a local nash equilibrium. Neural Inf. Process. Syst. (2017)
26. Metz, L., Poole, B., Pfau, D., Sohl-Dickstein, J.: Unrolled generative adversarial networks. In: International Conference on Learning Representations (2017)
27. Bie, A., Kamath, G., Zhang, G.: Private GANs, revisited. Trans. Mach. Learn. Res. **2023** (2023)
28. Kermany, D.S., et al.: Identifying medical diagnoses and treatable diseases by image-based deep learning. Cell **172**(5), 1122–1131 (2018)

Analysis of Changes in Grey-White Matter Contrast in Healthy Older Adults

Jonas Swope[1], Mohammad Fili[1](✉), Guiping Hu[1], and Auriel A. Willette[2]

[1] School of Industrial Engineering and Management, Oklahoma State University, Stillwater, OK 74077, USA
{jonas.a.swope,mohammad.fili,guiping.hu}@okstate.edu

[2] Department of Neurology, Rutgers University, New Brunswick, NJ 08901, USA
aaw139@rwjms.rutgers.edu

Abstract. Grey-white matter contrast (GWC) is a valuable proxy for brain health. In this study, we analyzed regional changes in GWC using structural MRI data from the UK Biobank, collected between 2006 and 2019. Leveraging the Desikan-Killiany (DK) Freesurfer brain atlas, we identified two groups: "Contrast Gainers" and "Contrast Decliners". For each group, we computed Pearson correlations between GWC changes across brain regions and used Jennrich's test to assess statistical differences in correlation structures. Results showed that the correlation matrices corresponding to Contrast Gainers and Contrast Decliners were substantially different. To enhance interpretability, we employed hierarchically reordered heatmaps and network diagrams. The heatmaps revealed clusters of strongly correlated brain regions in both groups, while the network diagrams highlighted similarities and differences in regional patterns. Our analysis uncovers statistically significant differences in GWC correlation patterns between Contrast Gainers and Decliners, suggesting potential applications in understanding neurodegenerative processes and brain aging.

Keywords: Cognitive aging · Correlation patterns · Grey-white matter contrast · Structural Magnetic Resonance Imaging (sMRI)

1 Introduction

Grey-white matter contrast (GWC) reflects the integrity of tissue boundaries between grey and white matter, serving as a biomarker for microstructural brain changes [1, 2]. Its value is influenced by multiple biological factors: cholesterol in myelin [3], regional brain iron levels [4], and water content [5]. Unlike traditional measures of atrophy, GWC captures subtle tissue composition changes, such as those linked to aging. Brain atrophy progresses with age [6], and age-related shifts in regional signal intensity directly impact GWC [7]. This makes GWC a sensitive tool for studying both healthy aging and neurodegenerative disorders such as Alzheimer's Disease (AD). For instance, adjusting cortical thickness measurements for local GWC variability improves sensitivity in distinguishing AD from healthy aging [8], while worse baseline GWC predicts progression to Mild Cognitive Impairment, a precursor to all-cause dementia [9].

X. Xie et al. (Eds.): ICSS 2025, LNOR, pp. 104–113, 2026.
https://doi.org/10.1007/978-3-032-13116-4_9

In neurodegenerative and neuropsychiatric disorders, GWC reveals disease-specific patterns. In Parkinson's disease, GWC itself does not correlate with disease development [1], though white matter hyperintensities—another marker of brain tissue health—are tied to cognitive decline in Parkinson's disease patients [10]. By contrast, AD research highlights GWC's utility in enhancing cortical thickness sensitivity for early detection [8]. Schizophrenia, a condition with both neurodevelopmental and neurodegenerative components [11], shows elevated GWC in broad regions among first-episode patients compared to controls [12], suggesting early microstructural disruptions.

GWC's relationship with cognition further underscores its clinical relevance. A lack of contrast between grey and white matter on opposite sides of regional boundaries correlates with poorer performance in verbal comprehension and working memory [13], while deviations from age-predicted GWC are associated with IQ [14]. However, confounding factors like cognitive reserve, a measure of the brain's adaptability, are rarely accounted for in these studies [15], complicating interpretations. Together, these findings position GWC as a multifaceted biomarker, though its clinical application requires deeper mechanistic insights into how cholesterol, iron, and water interact, as well as longitudinal studies to clarify causality in aging and disease [3–5, 9].

The present study utilizes data from the UK Biobank to explore changes in GWC between two time points. Participants were divided into two groups based on their overall GWC change: Contrast Gainers, who showed an increase in GWC over time, and Contrast Decliners, who experienced a decrease over time. We calculate correlation matrices between brain regions for each group, and Jennrich's test was applied to assess the statistical significance of the differences between these groups. To facilitate the interpretation of the results, we also use hierarchical heatmaps and network diagrams, which provide visual insights into the regional interactions of GWC.

The remainder of this paper is organized as follows: We begin by describing the data processing and methodology used for statistical analysis and visualization in Sects. 2 and 3, respectively. Next, the results are discussed in Sect. 4. Finally, we conclude in Sect. 5 by highlighting the implications of our findings for understanding regional GWC in aging populations.

2 Data

The dataset for this study was obtained from the UK Biobank [16]. Baseline information for each participant was collected between 2006 and 2010, with imaging conducted between 2014 and 2019. Although UK Biobank collects data across various biological domains, only participants with complete structural MRI (sMRI) and demographic records were included. We used the Desikan-Killiany (DK) Freesurfer brain atlas for the sMRI data. Participants were excluded if they missed one or more time points (t_3 or t_4), lacked brain imaging data for a Region of Interest (ROI), or were identified as significant outliers (greater than six standard deviations from the mean).

The final processed dataset included 65 brain regions across 399 participants aged 55 years and above. Participants were grouped based on the mean change in grey-white matter contrast, calculated using Eq. (1).

$$O_i = \frac{1}{M}\sum\nolimits_{j=1}^{M}\left(C_{ij}^{t_4} - C_{ij}^{t_3}\right) \qquad \forall i = 1, 2, \ldots, N \qquad (1)$$

where, i denotes the participant index ($i = 1, 2, \ldots, N$), j denotes the brain region, O_i is the overall mean change in GWC from t_3 to t_4 across all regions; C_{ij}^{t3} and C_{ij}^{t4} are the grey-white matter contrast values for brain region j at visits t_3 and t_4, respectively; N is the total number of participants (i.e., $N = 399$); and M is the total number of neural regions used from the brain atlas ($M = 65$).

Participants were placed into one of three groups: "Contrast Decliners", "Neutral", and "Contrast Gainers". The grouping was accomplished using Eq. (2).

$$y_i = \begin{cases} \text{Contrast Gainer}, & O_i \geq D_8(O) \\ \text{Neutral}, & D_2(O) < O_i < D_8(O) \\ \text{Contrast Decliner}, & O_i \leq D_2(O) \end{cases} \quad \forall i = 1, 2, \ldots, N \tag{2}$$

where, $D_2(.)$ and $D_8(.)$ correspond to decile functions for 20th and 80th percentile for mean change in GWC, respectively. Here, O is the list of all values. Figure 1 illustrates the distribution of mean GWC change, and the resulting groupings.

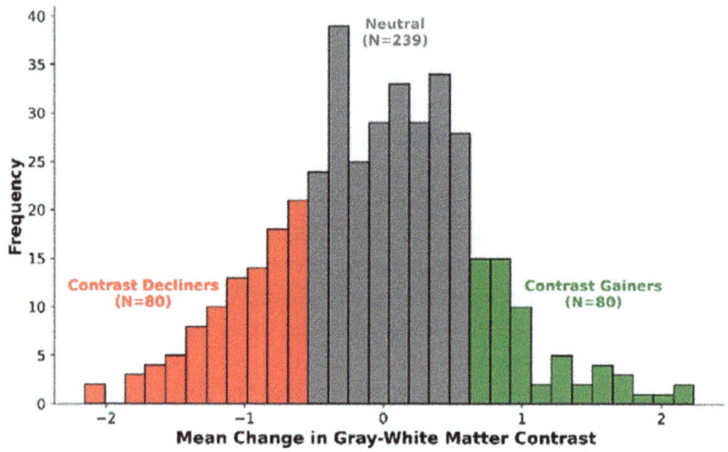

Fig. 1. Histogram of mean change in grey-white matter contrast.

3 Method

We computed the Pearson correlation coefficient between the mean changes of GWC for different areas of the brain, for "Contrast Gainer" and "Contrast Decliner" groups, separately. This process resulted in correlation matrices R_G and R_D, respectively. We then applied Jennrich's test to test for the similarity in the correlation matrices:

$$\begin{aligned} \boldsymbol{H_0} &: \boldsymbol{R_D} = \boldsymbol{R_G} \\ \boldsymbol{H_a} &: \boldsymbol{R_C} \neq \boldsymbol{R_G} \end{aligned} \tag{3}$$

A significance level of 0.05 was used for statistical analysis. Given the number of brain regions present in the dataset and the difficulty in finding sub-structures, we used Hierarchical Clustering to re-order the structure of the correlation matrices and group similar regions close to each other. Heatmaps were then employed to visualize the differences between the two matrices. First, for each entry in the correlation matrix, the corresponding distance was calculated using Eq. (4).

$$D = 1 - | r | \tag{4}$$

where, r is the Pearson correlation coefficient estimated for pairs of neural regions. We used the Ward linkage method as the merging strategy. A network graph was then constructed to visualize the similarities and differences in correlation patterns between Contrast Gainers and Contrast Decliners. In this network, nodes represent brain regions, and edges represent strong correlations between them. A threshold of 0.70 was applied to retain only edges corresponding to correlation coefficients with a magnitude greater than this value.

4 Results

To compare the patterns of correlation between "Contrast Decliners" and "Contrast Gainers", we first applied Jennrich's test to assess the similarity of the corresponding correlation matrices. Jennrich's test yielded a p-value less than 0.0001, providing strong evidence against the null hypothesis that the correlation matrices of GWC change across brain regions are identical between "Contrast Decliners" and "Contrast Gainers". While the test confirms a statistically significant difference, whether these differences are global or localized can be further examined through heatmaps and network diagrams.

Figures 2 and 3 display the heatmaps of correlation patterns in terms of the mean GWC change for "Contrast Gainers" and "Contrast Decliners", respectively. Regional grey-white matter contrast change appears to be mostly positively correlated.

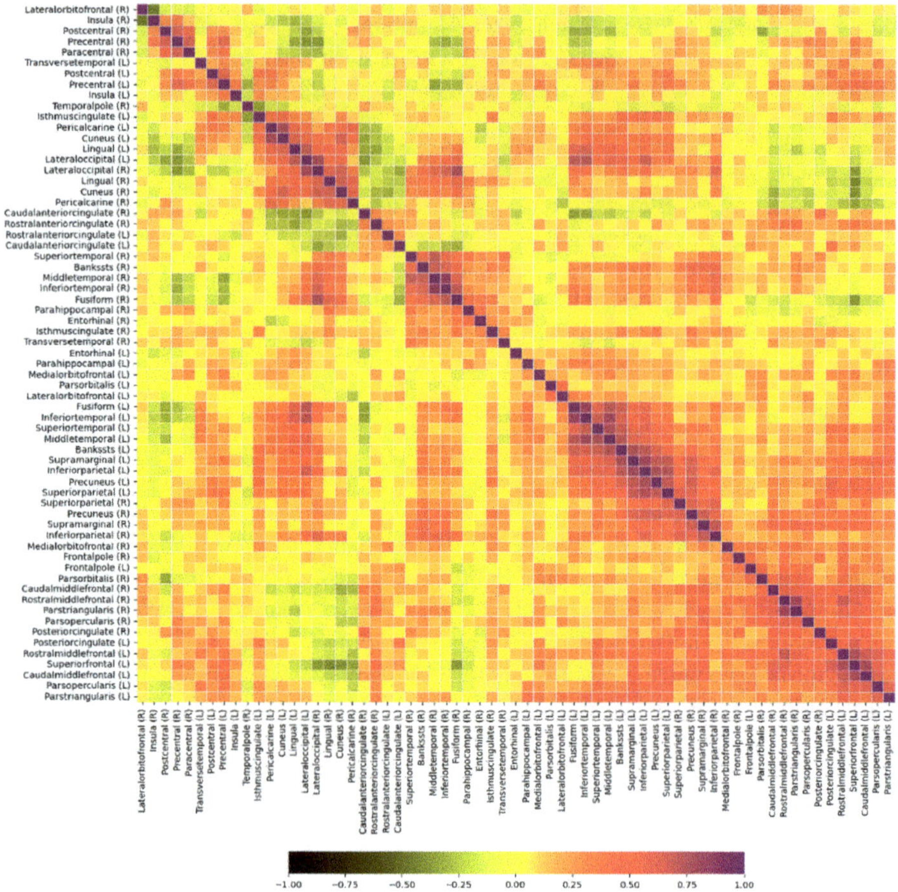

Fig. 2. Heatmap of correlation between regional GWC average change for Contrast Gainers.

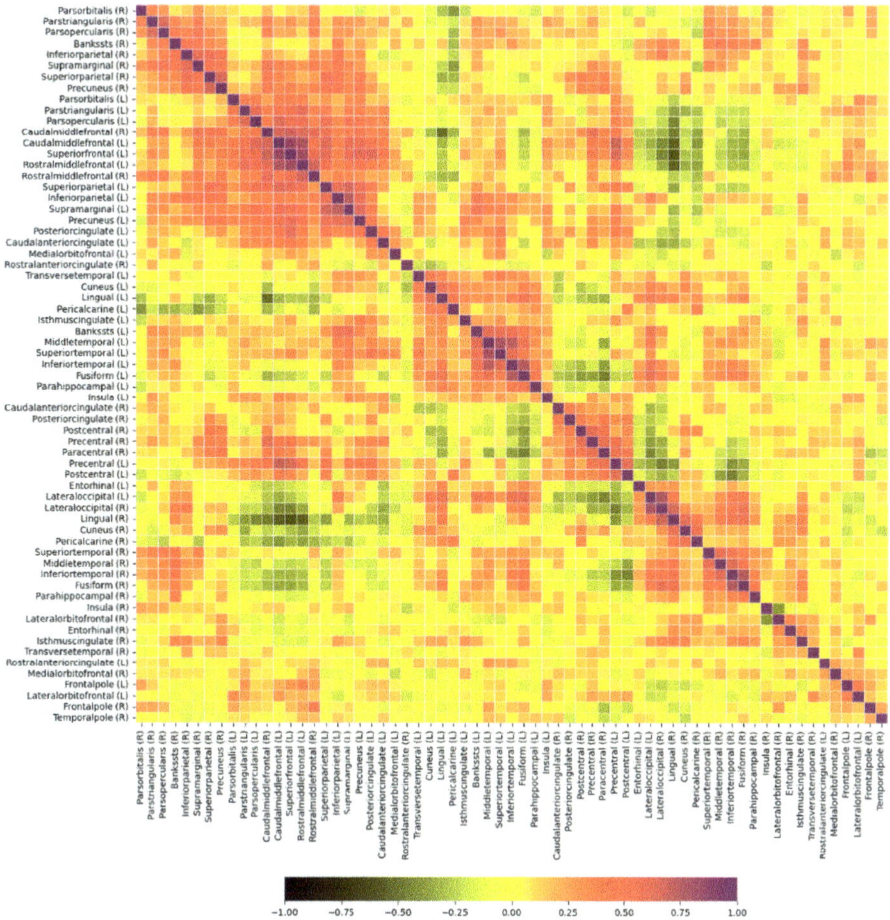

Fig. 3. Heatmap of correlation between regional GWC mean change for Contrast Decliners.

First, Contrast Gainers (Fig. 2) show stronger and more cohesive clusters of highly correlated regions, indicated by dark red color. These clusters suggest coordinated changes in GWC among specific brain areas. In contrast, Contrast Decliners (Fig. 3) display weaker and more diffuse correlations overall. The clustering is less pronounced, with smaller and more scattered high-correlation blocks, indicating less synchrony in regional GWC changes. Second, the overall distribution of higher correlations (orange and red) is denser in Gainers, whereas Decliners exhibit more yellow and light orange, reflecting lower correlation magnitudes across the matrix. Finally, we observe that in Contrast Decliners (Fig. 3), there is a more frequent and widespread presence of negative correlations, visible as green cells scattered throughout the matrix, while Contrast Gainers (Fig. 2) show fewer negative correlations overall. The increased presence and magnitude of negative correlations in Contrast Decliners may reflect a loss of coordinated regulation across brain networks during aging in the Contrast Decliners group. To better understand the clustering structure in these two groups, we plotted the dendrograms of each in Figs. 4 and 5.

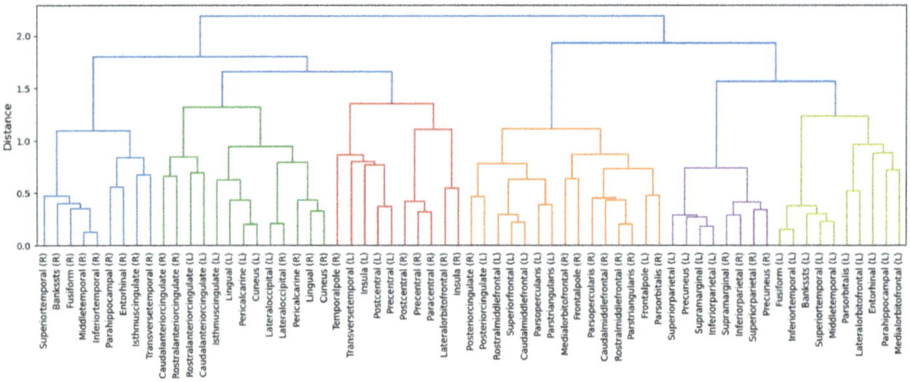

Fig. 4. Dendrogram of Grey-White Matter Contrast Changes in Brain Regions for Contrast Gainers.

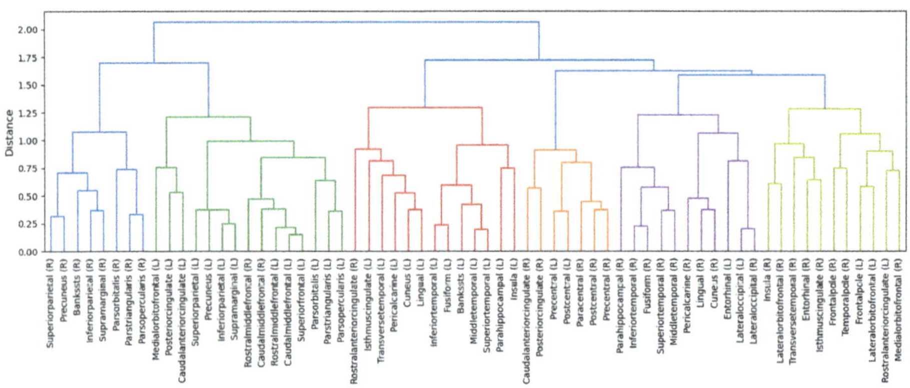

Fig. 5. Dendrogram of Grey-White Matter Contrast Changes in Brain Regions for Contrast Decliners.

Figure 6 presents a network diagram illustrating the similarities and differences in regional correlations of grey-white matter contrast (GWC) changes between "Contrast Gainers" and "Contrast Decliners". Only connections (edges) with an absolute Pearson correlation of 0.70 or higher are shown. (L) and (R) indicate the hemisphere (left or right) to which each brain region belongs.

Shared correlations between the two groups are shown in purple, highlighting overlapping patterns particularly in the left frontal region (e.g., Superior frontal – Caudal middle frontal and Superior frontal – Rostral middle frontal) and the ventral temporal-occipital pathway (Fusiform – Inferior temporal, Middle temporal – Superior temporal, and Lateral occipital L – R). Contrast Gainers (green edges) show distinct networks involving bilateral parietal and occipital regions, including strong links among Precuneus, Supramarginal, Inferior parietal, Superior parietal, and Bankssts. Other unique correlations are observed between Cuneus and Pericalcarine (both left hemisphere), and frontal-temporal connections such as Rostral middle frontal (R) and Pars triangularis (R). Additionally, exclusive right-hemisphere correlations appear in the ventral stream between Middle temporal, Inferior temporal, Fusiform, and Lateral occipital. Contrast Decliners (red edges), in contrast, show more restricted and localized exclusive correlations, including Caudal middle frontal (L) with both Precentral (L) and Rostral middle frontal (L), and Superior frontal (L) - Caudal middle frontal (R).

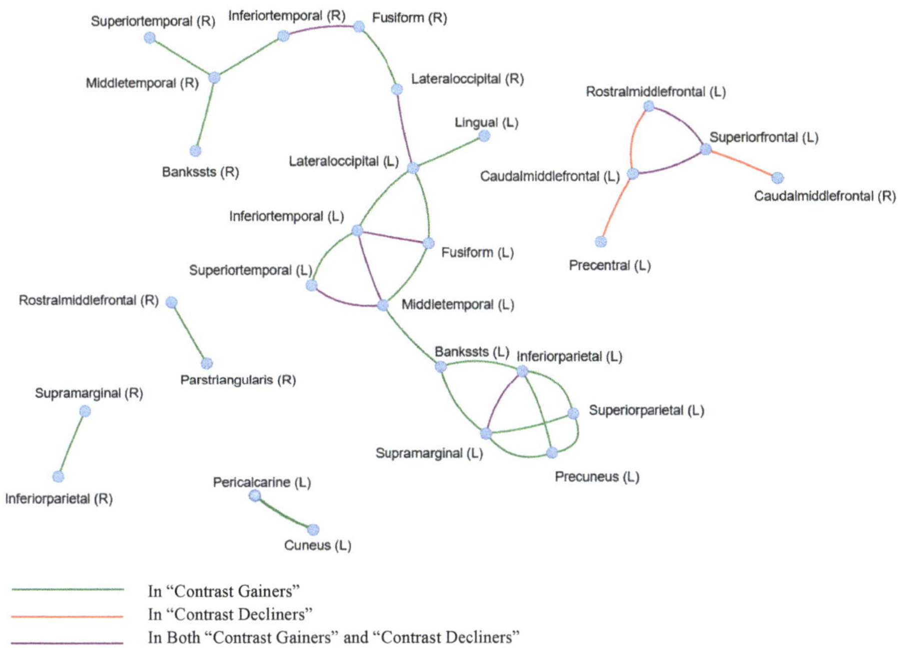

Fig. 6. Network diagram of the similarity and differences between correlations of regional GWC change.

5 Conclusion

This study demonstrates that grey-white matter contrast (GWC) changes in the brain exhibit distinct regional correlation patterns between individuals who experience gains versus declines in GWC over time. Using structural MRI data from the UK Biobank and leveraging the Desikan-Killiany brain atlas, we identified two groups, Contrast Gainers and Contrast Decliners, and systematically examined the correlation structures of GWC changes across brain regions. Jennrich's test revealed statistically significant differences between the correlation matrices of these two groups, indicating that the underlying patterns of regional interactions diverge meaningfully. Hierarchically clustered heatmaps and network diagrams further uncovered both shared and group-specific patterns. Contrast Gainers exhibited broader and more diverse connections, particularly involving parietal and occipital regions. In contrast, Contrast Decliners showed more localized interactions. These findings may suggest that the dynamics of GWC change, and their inter-regional coordination, may serve as indicators of brain aging trajectories and potential early markers of neurodegenerative processes.

Despite its contributions, this study has several limitations. First, the analysis is based on Pearson correlation, which captures linear associations and may overlook non-linear relationships between brain regions. Second, while the results demonstrate statistical differences between groups, the causal mechanisms driving these patterns remain unclear and warrant further investigation. Future work should consider integrating complementary biomarkers (e.g., cognitive scores and genetic risk factors) which may improve the interpretability and specificity of findings. While the current analyses are correlational, future research could employ longitudinal designs or interventional studies to better infer causal relationships between regional GWC changes and complementary biomarkers across contrast groupings. Given GWC's utility as a proxy for brain health, targeted experiments investigating the underlying causes of participant-level trends, whether a participant is a Contrast Gainer, Decliner, or neither, may yield valuable insights. Furthermore, expanding the analysis to include non-linear association measures and multivariate modeling could also uncover more complex inter-regional dependencies.

References

1. Uribe, C., et al.: Gray/white matter contrast in Parkinson's disease. Front. Aging Neurosci. **10** (2018). https://doi.org/10.3389/fnagi.2018.00089
2. Davatzikos, C.: Degenerative age changes in white matter connectivity visualized in vivo using magnetic resonance imaging. Cereb. Cortex. **12**(7), 767–771 (2002). https://doi.org/10.1093/cercor/12.7.767
3. Koenig, S.H.: Cholesterol of myelin is the determinant of gray-white contrast in MRI of brain. Magn. Reson. Med. **20**(2), 285–291 (1991). https://doi.org/10.1002/mrm.1910200210
4. Ogg, R.J., Steen, R.G.: Age-related changes in brain T1 are correlated with iron concentration. Magn. Reson. Med. **40**(5), 749–753 (1998). https://doi.org/10.1002/mrm.1910400516
5. Bansal, R., Hao, X., Liu, F., Xu, D., Liu, J., Peterson, B.S.: The effects of changing water content, relaxation times, and tissue contrast on tissue segmentation and measures of cortical anatomy in MR images. Magn. Reson. Imaging. **31**(10), 1709–1730 (2013). https://doi.org/10.1016/j.mri.2013.07.017
6. Giorgio, A., et al.: Age-related changes in grey and white matter structure throughout adulthood. NeuroImage. **51**(3), 943–951 (2010). https://doi.org/10.1016/j.neuroimage.2010.03.004
7. Salat, D.H., Lee, S.Y., van der Kouwe, A.J., Greve, D.N., Fischl, B., Rosas, H.D.: Age-associated alterations in cortical gray and white matter signal intensity and gray to white matter contrast. Neuroimage. **48**(1), 21–28 (2009). https://doi.org/10.1016/j.neuroimage.2009.06.074
8. Westlye, L.T., et al.: Increased sensitivity to effects of normal aging and Alzheimer's disease on cortical thickness by adjustment for local variability in gray/white contrast: a multi-sample MRI study. NeuroImage. **47**(4), 1545–1557 (2009). https://doi.org/10.1016/j.neuroimage.2009.05.084
9. Jefferson, A.L., et al.: Gray & white matter tissue contrast differentiates mild cognitive impairment converters from non-converters. Brain Imaging Behav. **9**(2), 141–148 (2014). https://doi.org/10.1007/s11682-014-9291-2

10. Sunwoo, M.K., et al.: The burden of white matter hyperintensities is a predictor of progressive mild cognitive impairment in patients with Parkinson's disease. Eur. J. Neurol. **21**(6), 922–922 (2014). https://doi.org/10.1111/ene.12412
11. Kulhara, P., Gupta, S.: What is schizophrenia: a neurodevelopmental or neurodegenerative disorder or a combination of both? A critical analysis. Indian J. Psychiatry. **52**(1), 21 (2010). https://doi.org/10.4103/0019-5545.58891
12. Kobayashi, H., et al.: The relationship between gray/white matter contrast and cognitive performance in first-episode schizophrenia. Cereb. Cortex (New York, N.Y. : 1991). **34**(2), 1–9 (2024). https://doi.org/10.1093/cercor/bhae009
13. Blackmon, K., et al.: Individual differences in verbal abilities associated with regional blurring of the left gray and white matter boundary. J. Neurosci. **31**(43), 15257–15263 (2011). https://doi.org/10.1523/jneurosci.3039-11.2011
14. Lewis, J.D., Evans, A.C., Tohka, J.: T1 white/gray contrast as a predictor of chronological age, and an index of cognitive performance. Neuroimage. **173**, 341–350 (2018). https://doi.org/10.1016/j.neuroimage.2018.02.050
15. Pettigrew, C., Soldan, A.: Defining cognitive reserve and implications for cognitive aging. Curr. Neurol. Neurosci. Rep. **19**(1) (2019). https://doi.org/10.1007/s11910-019-0917-z
16. UK Biobank. About our data. www.ukbiobank.ac.uk, 2024. https://www.ukbiobank.ac.uk/enable-your-research/about-our-data

Theorizing Generative AI's Role in Strategic Decision-Making: From Automation to Augmentation

Md Abu Toha[1,2], Parvez Alam Khan[3(✉)], and Md Salah Uddin[4]

[1] Faculty of education, University of Cambridge, Cambridge CB2 1TN, UK
mat80@cam.ac.uk

[2] Business Studies Group, National University, Gazipur 1704, Bangladesh

[3] Department of Management, University Technology PETRONAS, Persiaran UTP, 32610, Seri Iskandar, Perak, Malaysia
parvez.khan@utp.edu.my

[4] Department of Accounting and Information Systems, Jagannath University, Dhaka 1100,, Bangladesh
salahuddin@ais.jnu.ac.bd

Abstract. Generative artificial intelligence (GenAI) is profoundly reconfiguring strategic decision-making models within large-scale organizations. However, its theoretical implications within service science and organizational science remain emerging. The prime purpose of this study is to synthesize service science doctrines with the management theory domain to develop a conceptual framework that positions GenAI as an agent of value co-creation in strategic processes, excelling its conventional position as a decision-support instrument. Grounded in service-dominant logic and bounded rationality theory, this study theorizes GenAI's dual capacity role to augment human cognitive faculties through real-time scenario simulation, heterogeneous data synthesis, and predictive analytics while concurrently familiarizing systemic risks, comprising the perpetuation of latent biases inherent in training data and organizational over-reliance dynamics. By combining empirical case analyses with theoretical modelling, this study reveals how GenAI-mediated decision structural design reconfigures service systems, prompting decision velocity at the expense of ethical and epistemic accountability. This proposed conceptual framework suggests that GenAI's transformative potential lies in its capability to operationalize augmented intelligence, wherein human-AI symbiosis fosters adaptive, data-driven strategic agility while imposing robust governance instruments to alleviate algorithmic opacity and stakeholder dissonance. The study practically contributes to the interdisciplinary discourse on AI-enabled service innovation by recommending testable hypotheses regarding GenAI's impact on strategic foresight, resource orchestration, and organizational resilience, thereby projecting a research agenda for the responsible integration of generative technologies in enterprise strategic decision ecosystems.

Keywords: Generative artificial intelligence (GenAI) · strategic decision-making · service-dominant logic · augmented intelligence · algorithmic governance · digital transformation · AI ethics

X. Xie et al. (Eds.): ICSS 2025, LNOR, pp. 114–125, 2026.
https://doi.org/10.1007/978-3-032-13116-4_10

1 Introduction

Strategic decision-making is considered a critical component of corporate success as organizations consistently rely on human expertise, historical data, and analytical frameworks to traverse complex challenges such as market entry, mergers and acquisitions (M&A), and sustainability transitions (Toha et al., 2020; Toha, 2021; Hussain et al., 2023). Traditional decision-making approaches acknowledge that individuals often face cognitive and informational limitations. These constraints lead decision-makers to rely on incremental adjustments and heuristics, allowing them to strike a balance between the need for speed and the desire for accuracy in their choices. (Simon, 1957; Eisenhardt and Zbaracki, 1992). The launch of Coca-Cola's "New Coke" in 1985 exemplifies the pitfalls of human-driven decision-making, particularly highlighting the dangers of overconfidence and limited consumer testing. The CEO's strong belief in the product, combined with a narrow focus on consumer feedback, ultimately led to a miscalculation that resulted in a significant backlash (Greising, 1998). Although there are several tools, such as SWOT analysis and Porter's Five Forces, to reduce the biases, they are still constrained by reactive approaches and incomplete data. Additionally, a 2022 survey conducted by Deloitte indicated that 73% of the executives doubt their organization's data infrastructure for strategic decisions, identifying siloed systems and delayed insights as significant obstacles (Deloitte, 2022).

However, the advent of generative artificial intelligence (GenAI) is challenging conventional paradigms. GenAI is revolutionizing this field by offering capabilities like natural language processing, predictive modelling, and real-time data analysis that challenge traditional methods. Systems like ChatGPT, Claude, and Gemini are now practical tools, not just experiments, as evidenced in McKinsey reports that 60% of Fortune 500 firms are testing GenAI for strategic planning, with early adopters like Siemens and Unilever achieving 30–50% reductions in decision cycle times (Chui, Hazan, et al., 2023a; Majumder et al., 2024).

For instance, Unilever is utilizing Generative AI to analyse consumer trends across 50 markets, which has effectively shortened their product launch timelines from 12 months to just 3 months. Meanwhile, JPMorgan Chase has adopted AI for mergers and acquisitions due diligence, resulting in a 70% reduction in document review times and revealing risks that may have been missed by human analysts (Chase, 2024; Hu and Zeng, 2024). Notwithstanding these tactical advancements, the comprehensive strategic implications of GenAI, particularly how it redefines the decision-making authority, transforms the organizational learning, and affects the competitions remain undertheorized in the field of service science, which studies value creation in socio-technical systems (Vargo and Lusch, 2008).

This paper addresses two critical gaps. First, though service science underscores the collaborative interactions among human participants, the incorporation of nonhuman agents such as GenAI into strategic processes is not yet adequately examined. Specifically, this study raises the questions of whether algorithms can act as "co-creators" of value or if they are merely functional tools. Further, the boundary between human and machine agency is blurred when Microsoft's Azure OpenAI formulates market entry strategies by synthesizing data from LinkedIn, GitHub, and customer feedback loops, raising concerns about resource integrations and accountability. Second, practitioners confront a conflict between efficiency gains and emergent risks since GenAI speeds up the decision velocity while also bringing ethical concerns (e.g., biases in succession planning algorithms), reduces human skills (as seen in aviation automation), and creates uncertainties about accountability (Carr, 2014; Bender et al., 2021). A MIT study in 2024 revealed that 41% of executives using GenAI for strategy cannot explain how the tool prioritizes risks, pointing to a lack of transparency that weakens trust (Reed, 2024).

To address these gaps, we propose a GenAI-Augmented Decision Loop, where GenAI acts as a strategic partner that boosts human judgment, accelerates value co-creation, and reshapes organizational service systems. For instance, Siemens's "Industrial Copilot" has been instrumental for executives in designing sustainable manufacturing processes by synthesizing energy consumption patterns, supplier ESG scores, and regulatory data, while human leaders apply this information to real-world contexts (e.g., labor practices in emerging markets). This symbiosis resulted in reducing planning errors by 22% in 2023, showing the potential of human-AI teamwork (Gašević et al., 2023)

This study makes several significant contributions. First, we integrate service-dominant logic with AI theory, advancing the concept of algorithmic co-creation where human decision-makers and AI systems collaboratively generate strategic value through iterative feedback mechanisms. Second, this study provides practical insights by proposing a governance framework to mitigate risks, advocating for bias audits, human-in-the-loop protocols, and transparency standards aligned with the EU's AI Act (Evas, 2024). Finally, it offers a methodological contribution by introducing metrics to assess the impact of GenAI on decision quality, such as agility (time-to-insight), stakeholder satisfaction, and long-term adaptability. These contributions collectively address the growing need for interdisciplinary research at the intersection of AI, management, and service science, specifically in response to the societal implications of AI-driven digital transformation.

The structure of this paper is organized as follows: Sect. 2 provides a comprehensive review of the literature; Sect. 3 outlines the proposed theoretical framework; Sect. 4 critically examines the implications for service systems; Sect. 5 addresses the key limitations of the study and provides directions for future research; and finally, Sect. 6 concludes with actionable insights for scholars and practitioners.

2 Literature Review

2.1 Strategic Decision-Making: From Bounded Rationality to Dynamic Capabilities

Strategic decision-making has often relied on bounded rationality, a concept introduced by Herbert Simon (1957) explaining how human decision-makers navigate complexity under cognitive limitations and incomplete information. Traditional models underscore intuition (Kahneman and Klein, 2009) and incrementalism (Lindblom, 1989), where executives rely on heuristics and iterative adjustments to strike a balance between speed and accuracy. For example, the "garbage can model" of organizational choice (Cohen et al., 1972) emphasizes that decisions often result from chaotic interactions of problems, solutions, and participants rather than following a clear, logical process. However, these frameworks fail to address the velocity and scale of modern markets, where global disruptions such as geopolitical conflicts or climate-related supply chain shocks necessitate real-time adaptability. Enter Generative AI (GenAI) disrupts these models by quickly turning unstructured data such as social media sentiment, regulatory filings, and satellite imagery into actionable insights. For example, IBM's Watsonx helps executives in simulating the outcomes of mergers and acquisitions (M&A) by analysing thousands of global regulatory documents and predicting antitrust risks, shortening decision cycles from months to weeks (Majumder et al., 2024). Further, this capability is consistent with the dynamic capabilities theory (Teece, 2018), which recommends that firms continuously sense, seize, and transform their resources to thrive in rapidly changing environments. Although GenAI improves the ability to sense changes through predictive analytics, its role in fostering adaptive learning, such as refining decision-making heuristics based on real-time feedback, still remains underexplored, particularly in contexts requiring ethical trade-offs or long-term sustainability.

Recent research extends dynamic capabilities to AI-augmented agility, where algorithms enhance human abilities in recognizing patterns and planning for different scenarios. For example, a 2023 study by Accenture on Fortune 500 companies found that organizations using GenAI for strategic foresight achieved 40% faster response than those relying on traditional business intelligence tools. However, his increased agility comes with a risk of overconfidence, as executives may conflate algorithmic speed with strategic wisdom—an issue known as "automation bias" (Carr, 2014). Moreover, GenAI's reliance on historical data may unintentionally reinforce path dependencies, inhibiting innovation in industries undergoing radical transformation (e.g., energy transition). Thus, while GenAI overcomes bounded rationality by expanding cognitive capacity, it creates new conflicts between efficiency and adaptability that call into question established decision-making theories.

2.2 Generative AI: Capabilities, Limitations, and Organizational Adoption

GenAI marks a paradigm shift in how business organizations make decisions, offering tools that transcend traditional analytics. Additionally, its ability to generate scenarios using advanced models like GPT-4 helps predict possible outcomes of market disruptions, such as supply chain breakdowns or price wars with competitors. This allows leaders to test their strategies in uncertain environments. For instance, Unilever reduces supply chain planning errors by 35% by employing GenAI to model the impact of climate change on raw materials costs across over 100 global suppliers (Nkpurukwe et al., 2025). Simultaneously, natural language interfaces democratize strategic insights to make them more accessible. For example, Microsoft's Copilot converts complex datasets into easily comprehensible summaries, allowing nontechnical leaders to interact with predictive analytics. This democratization aligns with resource-based view theory (Barney, 1991) as GenAI turns data into a valuable strategic asset by combining siloed information (e.g., CRM systems, IoT sensors, R&D pipelines) into cohesive decision-making frameworks.

However, the adoption of GenAI comes with significant limitations. One major limitation is the bias amplification, which signifies that models trained on past data may codify outdated strategies, as reflected in a 2023 Harvard Business School study, which found that GenAI recommended gender-biased leadership appointments in 68% of simulated scenarios (Bender et al., 2021). Further, the explainability gap is another critical concern, highlighting the inability to trace how algorithms prioritize certain risks over others, eroding trust. For instance, JPMorgan Chase stopped using an AI tool for loan approval because executives failed to explain why it was rejecting creditworthy applicants from marginalized communities (Rai, 2020). Additionally, computational costs and data quality pose a significant threat to scalability. For instance, training enterprise-grade GenAI models can cost mid-sized companies over $10 million each year (Lamarre et al., 2023). To address these challenges, companies like Unilever adopt a hybrid approach, which requires the practice of manually validating AI-generated supply chain scenarios to avoid overfitting to past crises.

2.3 Service Science and Value Co-creation

Service science, grounded in service-dominant logic (Vargo and Lusch, 2008), views organizations as dynamic ecosystems where value is cocreated through the interactions between actors. However, with the emergence of GenAI, the paradigm is reshaping as it integrates non-human participants into the value creation process. Additionally, GenAI serves as a resource integrator, bridging stakeholders (e.g., executives, customers, suppliers) and data streams to co-create strategic value. A clear example of this is Siemens's "Industrial Copilot": the tool combines GenAI with human engineers to co-design sustainable manufacturing processes, balancing factors like technical feasibility, regulatory constraints, and ESG goals (Gašević et al., 2023). In this setup, data shifts from a passive asset to an active operand resource, manipulated by GenAI to generate insights (e.g., real-time carbon footprint simulations) that guide stakeholder negotiations.

This paradigm shift questions traditional hierarchies, portraying GenAI as a value cocreator rather than just a passive tool. For instance, Starbucks's Deep Brew platform employs GenAI to evaluate customer sentiment across 30000+ stores, permitting regional managers to collaborate with AI to develop highly localized marketing strategies based on AI-generated insights (Forbes, 2023). Further, this shift brings up critical issues regarding algorithmic agency: who will be responsible for the outcome if GenAI affects the strategic decisions? Service science focuses on value-in-context, where the beneficiaries in each unique situation define the value. Additionally, this concept becomes more complex when algorithms play a role in mediating human interactions. A 2024 MIT study found that 52% of executives view AI-generated strategies as collaborative outputs rather than purely human decisions, which makes it harder to determine who is accountable for the results (Reed, 2024). Moreover, the role of GenAI in service ecosystems creates power imbalances, as companies with advanced AI capabilities (such as Amazon's demand forecasting) can dominate markets, marginalizing smaller players. These challenges highlight the need for governance frameworks that strike a balance between innovation and equity, ensuring GenAI augments, rather than displaces, human judgment in value co-creation.

3 Theoretical Framework: the GenAI-Augmented Decision Loop

This study suggests the *GenAI-Augmented Decision Loop* (Fig. 1), a cyclical model that reframes strategic decision-making as a symbiotic process between human cognition and generative AI. Based on service science and dynamic capabilities theory, this framework asserts that GenAI does not replace human judgment but rather enhances it through an iterative feedback mechanism, helping organizations traverse complexity while maintaining ethical and contextual awareness. The model consists of four interconnected phases:

3.1 Phase 1: Input

Strategic goals and unstructured data, such as market reports, stakeholder feedback, real-time IoT sensor streams, and geopolitical risk alerts, act as key inputs. Unlike traditional decision systems that focus on structured data (like financial metrics), this phase embraces diversity, acknowledging that 80% of corporate data is unstructured and often overlooked (Gartner and Côté, 2023). For example, a multinational retailer could incorporate social media sentiment analysis, supplier sustainability scores, and climate projections to develop a net-zero supply chain strategy.

3.2 Phase 2: GenAI Synthesis

GenAI processes inputs through three core functions. Firstly, Scenario Generation is the process of Simulating probabilistic outcomes of strategic choices such as entering a regulated market under evolving trade policies. Secondly, identifying hidden risks, such as geopolitical instability or supplier dependencies, using transformer-based models trained on global news and regulatory databases. Thirdly, ranking strategic pathways based on multi-criteria optimization, such as balancing profitability, ESG compliance, and stakeholder alignment. For example, Siemens uses this approach to analyse patent databases and R&D pipelines, helping prioritize innovation investments and reducing time-to-insight by 65% (Gašević et al., 2023)

3.3 Phase 3: Human Judgment

Executives refine AI-generated options by applying their tacit knowledge, ethical judgment, and contextual expertise. This phase helps address the limitations of GenAI, such as algorithmic bias or the risk of overfitting to past patterns. For example, Coca-Cola uses GenAI for regional market entry, where the system predicts taste preferences based on demographic data. However, human managers adjust these recommendations when they conflict with cultural nuances, such as avoiding pork-derived ingredients in Muslim-majority markets (Forbes, 2023).

3.4 Phase 4: Feedback

In this phase, post-decision outcomes such as customer feedback, operational performance, and stakeholder sentiment are fed back into the system, allowing for continuous learning. For example, a financial firm using GenAI for M&A due diligence retrains its models every quarter with data from closed deals, improving prediction accuracy by 12% each year (Deloitte, 2022). This closed-loop process fosters adaptive evolution, consistent with Teece's (2018) dynamic capabilities framework.

The model is anchored in three constructs. Firstly, Augmented Intelligence, GenAI complements human cognition by addressing blind spots such as long-term climate risks in supply chains while preserving human agency. A healthcare firm reduced medication errors by 40% by using GenAI to cross-reference patient histories with clinical trials, while doctors retained final authority over prescriptions (Cooper and Rodman, 2023). Secondly, Decision Velocity, GenAI accelerates strategic cycles through real-time data processing. A Fortune 500 manufacturer cut product launch timelines from 18 to 6 months by automating competitor analysis and pricing simulations (Chui, Yee, et al., 2023b). Thirdly, Cognitive Load Reduction, by automating repetitive tasks (e.g., regulatory compliance checks), GenAI frees executives to focus on creative problem-solving. A European bank reported a 30% increase in executive productivity after deploying GenAI for risk reporting (Gavade, 2023). These constructs inform three testable propositions:

P1: GenAI reduces strategic myopia by simulating long-term scenarios (e.g., AI-driven carbon footprint projections for 2050), fostering proactive adaptation.
P2: Over-reliance on GenAI erodes tacit knowledge, risking organizational fragility in crises where historical data is irrelevant (e.g., novel pandemics).
P3: Human-AI collaboration enhances value co-creation in service systems, as seen in Starbucks's AI-human co-designed loyalty programs, which boosted customer retention by 25% (Meilani and Ellitan, 2023)

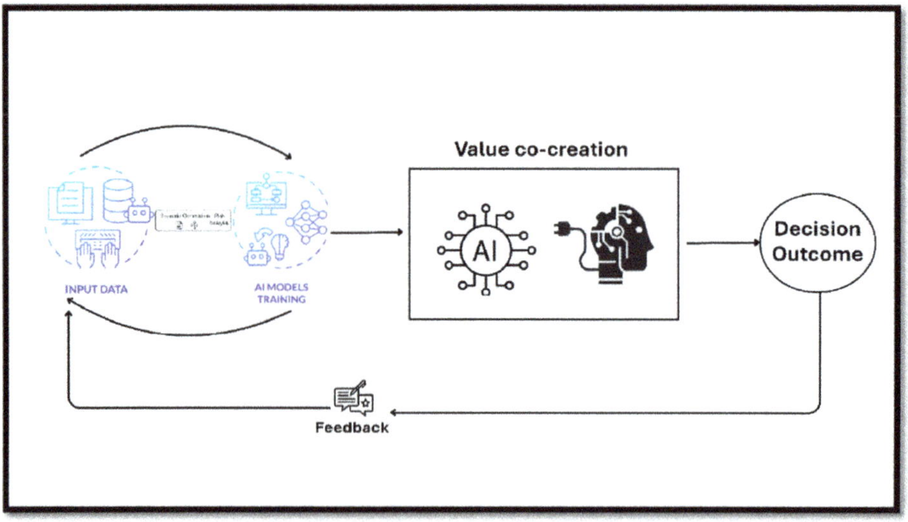

Fig. 1. GenAI-Augmented Decision Loop (GADL)

This study adopts a design science research (DSR) paradigm (Hevner et al., 2004) to develop and evaluate the *GenAI-Augmented Decision Loop*, a conceptual framework that redefines strategic decision-making through the lens of service science. DSR emphasizes the creation of practical, theoretically grounded artifacts in this case, the framework to address complex organizational challenges posed by generative AI (GenAI).

4 Implications for Service Science

GenAI is transforming traditional service systems into AI-augmented ecosystems where human and algorithmic agents co-create value through real-time, data-driven interactions. Platforms like Siemens's Industrial Copilot exemplify this shift by integrating GenAI with human engineers to collaboratively design manufacturing systems that balance technical, regulatory, and sustainability goals (Gašević et al., 2023). This reconfiguration also underpins the emergence of Decision Support as a Service, enabling mid-sized firms to simulate complex scenarios using platforms like Azure OpenAI without heavy infrastructure costs though these risks deepening digital divides (Lamarre et al., 2023). Tools such as Accenture's SynOps automate low-level tasks, freeing up human capacity for strategic and ethical decision-making, but necessitate widespread reskilling (Singh, 2023).

The deployment of GenAI raises ethical concerns, particularly regarding bias amplification and accountability. For instance, a Stanford study revealed geographic bias in AI-generated expansion strategies, disproportionately favouring North America (Bender et al., 2021). Despite initiatives like Unilever's bias audits, most responses remain reactive. Legal ambiguity further complicates governance, as shown in a 2024 antitrust lawsuit involving AI-generated pricing strategies, which questioned whether liability rests with the AI provider or the user (Mollick et al., 2024). Meanwhile, mandates like the EU's AI Act push for transparency, but GenAI's black-box nature resists full explainability, requiring interdisciplinary frameworks to ensure ethical deployment (Reed, 2024).

GenAI is redefining stakeholder engagement by enabling hyper-personalized services and participatory governance. Tools like Bank of America's Erica automate most customer service interactions, yet a 2023 error affecting thousands underscored the importance of trust and transparency (Newman et al., 2024). Companies like Starbucks now disclose how AI customizes offers to address privacy concerns. Regulators are also adopting proactive roles e.g., the EU's AI Act and Singapore's Monetary Authority stress tests reflecting AI's systemic significance. In sectors like healthcare, GenAI collaborations such as Med-PaLM 2 enhance diagnostic capabilities but also challenge human agency, demanding a recalibration of service science to preserve stakeholder co-authorship in decision-making.

5 Discussion and Future Research

Service science is a multidisciplinary field that studies how complex configurations of people, technology, organisations, and shared information are designed, improved, and innovated to provide value (Maglio et al., 2009; Spohrer and Kwan, 2009; Spohrer and Maglio, 2010). The focus on value co-creation, in which service interactions between both providers and consumers produce reciprocal benefits rather than unilateral value transfer, is a fundamental tenet of service science. A customer-focused context-sensitive adaptation, technology-mediated collaboration, as well as the dynamic reconfiguration of capabilities and resources are additional distinguishing traits (Chesbrough and Spohrer, 2006; Spohrer et al., 2007; Lusch and Nambisan, 2015)

These fundamental ideas of service science are well-aligned with the proposed GenAI-Augmented Decision Loop (GADL). First, it stresses the co-creation of strategic decisions, in which generative AI collaborates with human agents to frame issues, produce alternatives, and assess possible outcomes rather than just serving as a tool. This is in line with the co-creation processes Lusch and Vargo (2014) outline, in which technological and human resources combine to create value (Lusch and Vargo, 2014). Second, the GADL's iterative feedback structure provides contextual responsiveness, which is essential for dynamic service systems because it enables AI to continuously modify recommendations in response to changing data inputs, stakeholder feedback, and organisational objectives (Maglio and Spohrer, 2008).

Furthermore, the incorporation of generative AI into decision loops is a prime example of the socio-technical interaction that service science emphasises, especially when it comes to digital transformation (Barile et al., 2016). By combining extensive, imaginative data synthesis with human reasoning, generative AI improves the ability of service systems to share knowledge and innovate. Therefore, in unpredictable and data-intensive environments, the GADL can be viewed as a strategic enabler that strengthens the dynamic, cooperative, and adaptable nature of service systems.

Although generative AI has a lot of promise as a co-creative partner in strategic decision-making, there are a few drawbacks that should be carefully considered. Existing models frequently have algorithmic and data biases that could unintentionally affect decision results (Bender et al., 2021). Furthermore, the "black box" problem the opaqueness of generative models' output production raises questions regarding trust and accountability in strategic contexts (Cowls et al., 2019). Additionally, deep contextual reasoning, ethical judgement, and domain-specific nuance all crucial for making critical managerial decisions are absent from generative AI. Furthermore, its implementation in corporate environments is made more difficult by unresolved concerns about data security, privacy, and intellectual property. Given these difficulties, it is recommended that GenAI be viewed as an auxiliary tool that complements human decision-makers rather than taking their place, even though it can improve strategic thinking.

GenAI provides immense value in the realm of strategic decision-making, especially within the turbulent and unpredictable landscapes of emerging markets and the fast-evolving fintech sector. Here, it shines by harnessing unconventional data sources, delivering insights that traditional tools often miss. However, its inherent limitations become starkly apparent in situations steeped in moral complexity or characterized by a scarcity of data such as layoffs or unprecedented crises where the nuances of human judgment are irreplaceable. These critical boundary conditions underscore the importance of establishing clear governance frameworks to guide when GenAI should complement human intuition rather than supplant it.

Future research must embark on a journey to redefine success metrics, extending beyond the traditional confines of return on investment (ROI). It should aim to capture the elusive yet critical intangible benefits, such as the cultivation of stakeholder trust and the organization's ability to adapt in an ever-evolving landscape. Delving into the nuances of cross-cultural adoption patterns is essential; after all, diverse regulatory frameworks and cultural contexts play a pivotal role in shaping the integration of Generative AI. We also need to conduct longitudinal studies that explore the long-term impacts of AI utilization, investigating whether it serves as a catalyst for enhanced learning or, conversely, leads to the gradual erosion of human expertise. Bridging these knowledge gaps necessitates a rich tapestry of interdisciplinary collaboration, weaving together insights from AI ethics, management principles, and behavioural sciences. Only through such collective efforts can we ensure the responsible and effective deployment of these transformative technologies.

6 Conclusion

GenAI marks a paradigm shift in strategic decision-making by serving as a co-creative partner that enhances decision speed, reduces cognitive strain, and supports scenario planning within dynamic service systems. This paper proposed a GenAI-Augmented Decision Loop, integrating service science and management theory to show how GenAI transforms value co-creation. However, these gains are tempered by ethical challenges, accountability gaps, and risks to human agency. Emphasizing augmented intelligence, the framework reaffirms the centrality of human judgment in guiding AI outputs, especially in sensitive or uncertain contexts. Looking forward, responsible GenAI integration requires interdisciplinary collaboration, long-term impact studies, and reskilling efforts to ensure ethical, context-aware decision-making. Ultimately, success lies not in AI's capabilities alone, but in building systems where technology amplifies rather than replaces human values.

Acknowledgments. We sincerely thank the Conference organizing committee, reviewers, and our session chair for their valuable contributions. Additionally, we would like to express our gratitude to the Faculty of Education, University of Cambridge; The Department of Management and Humanities, Universiti Teknologi PETRONAS, Malaysia; and The Department of Accounting and Information Systems, Jagannath University, Bangladesh, for their support and collaboration.

Disclosure of Interests The authors have no competing interests to declare that are relevant to the content of this article.

References

Barile, S., Lusch, R., Reynoso, J., Saviano, M., Spohrer, J.: Systems, networks, and ecosystems in service research. J. Serv. Manag. **27**(4), 652–674 (2016)

Barney, J.: Firm resources and sustained competitive advantage. J. Manag. **17**(1), 99–120 (1991)

Bender, E.M., Gebru, T., McMillan-Major, A., Shmitchell, S.: On the dangers of stochastic parrots: can language models be too big?? In: Proceedings of the 2021 ACM Conference on Fairness, Accountability, and Transparency (2021)

Carr, N.: The Glass Cage: Automation and Us. WW Norton & Co (2014)

Chase, J.M.: Powering Growth with Curiosity and Heart Annual Report 2023. JP Morgan Chase & Co. (2024)
Chesbrough, H., Spohrer, J.: A research manifesto for services science. Commun. ACM. **49**(7), 35–40 (2006)
Chui, M., Hazan, E., Roberts, R., Singla, A., Smaje, K.: The Economic Potential of Generative AI. McKinsey & Company (2023a)
Chui, M., Yee, L., Hall, B., Singla, A.: The State of AI in 2023: Generative AI's Breakout Year. McKinsey & Company (2023b)
Cohen, M.D., March, J.G., Olsen, J.P.: A garbage can model of organizational choice. Adm. Sci. Q. **17**, 1–25 (1972)
Cooper, A., Rodman, A.: AI and medical education-a 21st-century Pandora's box. N. Engl. J. Med. **389**(5), 385–387 (2023)
Cowls, J., King, T., Taddeo, M., Floridi, L.: Designing AI for Social Good: Seven Essential Factors. Available at SSRN 3388669 (2019)
Deloitte, S.: Deloitte. Deloitte (2022)
Eisenhardt, K.M., Zbaracki, M.J.: Strategic decision making. Strateg. Manag. J. **13**(S2), 17–37 (1992)
Evas, T.: The EU artificial intelligence act. J. AI Law Regul. **1**(1), 98–101 (2024)
Forbes, D.: AI, Andy, Alan, algorithms, and Boden: the A's and B's of making art with aesthetic meaning. In: Liberal Education: Analog Dreams in a Digital Age, vol. 85. Vernon Press, Willington, DE (2023)
Gartner, J.-B., Côté, A.: Optimization of care pathways through technological, clinical, organizational and social innovations: a qualitative study. Health Serv. Insights. **16**, 11786329231211096 (2023)
Gašević, D., Siemens, G., Sadiq, S.: Empowering Learners for the Age of Artificial Intelligence, vol. 4, p. 100130. Elsevier (2023)
Gavade, D.: Navigating economic fluctuations in smart cities: a fusion of BCG matrix and AI-driven business analysis. In: 7th IET Smart Cities Symposium (SCS 2023) (2023)
Greising, D.: I'd Like the World to Buy a Coke: the Life and Leadership of Roberto Goizueta. Wiley (1998)
Hevner, A.R., March, S.T., Park, J., Ram, S.: Design science in information systems research. MIS Q., 75–105 (2004)
Hu, Y., Zeng, Y.: Achieving sustainable operations: challenges, countermeasures, and the case of Unilever. SHS Web of Conferences. **181** (2024)
Hussain, A., Abid, M.F., Shamim, A., Ting, D.H., Toha, M.A.: Videogames-as-a-service: how does in-game value co-creation enhance premium gaming co-creation experience for players? J. Retail. Consum. Serv. **70**, 103128 (2023)
Kahneman, D., Klein, G.: Conditions for intuitive expertise: a failure to disagree. Am. Psychol. **64**(6), 515–526 (2009)
Lamarre, E., Smaje, K., Zemmel, R.: Rewired: the McKinsey Guide to Outcompeting in the Age of Digital and AI. Wiley (2023)
Lindblom, C.E.: The science of muddling through. Readings in managerial psychology. **4**, 117–131 (1989)
Lusch, R.F., Nambisan, S.: Service innovation. MIS Q. **39**(1), 155–176 (2015)
Lusch, R.F., Vargo, S.L.: Service-dominant logic: Premises, perspectives, possibilities. Cambridge University Press (2014)
Maglio, P.P., Spohrer, J.: Fundamentals of service science. J. Acad. Mark. Sci. **36**(1), 18–20 (2008)
Maglio, P.P., Vargo, S.L., Caswell, N., Spohrer, J.: The service system is the basic abstraction of service science. IseB. **7**(4), 395–406 (2009)
Majumder, S., Bhattacharjee, A., Kozhaya, J.N.: Enhancing AI Governance in Financial Industry through IBM Watsonx. Governance. Authorea Preprints (2024)

Meilani, T., Ellitan, L.: Supply chain management processes and competitive advantage: the case of Starbucks. Int. J. Multidiscip. Res. Publ. **6**(8), 76–88 (2023)
Mollick, E., De Cremer, D., Neeley, T., Sinha, P.: Generative AI: the insights you need from Harvard Business Review. Harvard Business Press (2024)
Newman, N., Fletcher, R., Robertson, C.T., Ross Arguedas, A., Nielsen, R.K.: Reuters Institute Digital News Report 2024. Reuters Institute for the Study of Journalism (2024)
Nkpurukwe, O.I., Nwokah, E., Arumdeben, G.: Diversity management strategies and employee performance in Unilever plc. RUN J. Manage. **1**(1) (2025)
Rai, A.: Explainable AI: from black box to glass box. J. Acad. Mark. Sci. **48**, 137–141 (2020)
Reed, W.: MIT SMR's 10 AI must-reads for 2024. MIT Sloan Manage. Rev. (Online), 1–3 (2024)
Simon, H.A.: Models of Man: Social and Rational; Mathematical Essays on Rational Human Behavior in Society Setting. Wiley (1957)
Singh, V.K.: Leveraging Artificial Intelligence in sales management and HRM: a case study of Accenture's strategic implementation and impact. Change Manage. Int. J. **24**(2) (2023)
Spohrer, J., Kwan, S.K.: Service science, management, engineering, and design (SSMED): an emerging discipline-outline & references. Int. J. Inf. Syst. Serv. Sect. **1**(3), 1–31 (2009)
Spohrer, J., Maglio, P.P., Bailey, J., Gruhl, D.: Steps toward a science of service systems. Computer. **40**(1), 71–77 (2007)
Spohrer, J.C., Maglio, P.P.: Toward a science of service systems: value and symbols. In: Handbook of Service Science, pp. 157–194. Springer (2010)
Teece, D.J.: Business models and dynamic capabilities. Long Range Plan. **51**(1), 40–49 (2018)
Toha, M.A.: Eco-innovation, Board of Directors and Firms Financial Performance: A Study on Malaysian Energy Listed Companies. Universiti Teknologi Petronas (2021)
Toha, M.A., Johl, S.K., Khan, P.A.: Firm's sustainability and societal development from the lens of fishbone eco-innovation: a moderating role of ISO 14001-2015 environmental management system. Processes. **8**(9), 1152 (2020)
Vargo, S.L., Lusch, R.F.: Service-dominant logic: continuing the evolution. J. Acad. Mark. Sci. **36**, 1–10 (2008)

Static vs. Dynamic Scheduling in Teleconsultation Systems: Managing Uncertainty and Walk-Ins in Teleconsultation

Mengbo Zhai[1], Yan Qiao[1](✉), Lun Ran[2], and Yunkai Zhai[1]

[1] School of Management, Zhengzhou University, Zhengzhou, China
zzuqiaoyan@zzu.edu.cn

[2] School of Management and Economics, Beijing Institute of Technology, Beijing, China

Abstract. Teleconsultation in China, primarily based on appointment systems, often face walk-ins who have not scheduled in advance, leading to delays for scheduled patients. To minimize waiting and overtime costs, we propose a two-stage static scheduling model for optimal patient allocation. Additionally, a dynamic update model is developed for scheduled patients, integrated with a rolling horizon optimization strategy, resulting in a Greedy-based Rolling Horizon Optimization (GRHO) approach. We introduce two insertion principles for walk-ins: GRHO and GRHO*. The SAA-VNS-Integer L-Shaped (SVILS) algorithm is employed to solve the static model, and results show that GRHO* outperforms both GRHO and SVILS. GRHO* reduces waiting time by up to 67.86%, and increases teleconsultation room resource utilization to 89.29%, offering valuable insights for teleconsultation management.

Keywords: appointment scheduling · dynamic scheduling · rolling horizon optimization · SVILS algorithm

1 Introduction

Telemedicine includes services such as teleconsultations, monitoring, treatment, and diagnosis. It was first proposed by American scholar Thomas in the 1970s and is often referred to as "long-distance medical care" [1]. In recent years, telemedicine has rapidly developed worldwide. In particular, it plays a key role in expanding healthcare access, improving outcomes, and reducing costs, especially in developing countries [2]. Research firm Stratistics MRC predicts the global telemedicine market will reach $394.4 billion by 2028.[1] Telemedicine has become an essential part of healthcare systems, especially in areas with uneven resource distribution or during public health crises. Post-pandemic, telemedicine services continue to grow. In 2024, the National Health Commission of China announced that telemedicine networks covered all cities

[1] https://www.strategymrc.com/report/telehealth-market [accessed in March 2025].

X. Xie et al. (Eds.): ICSS 2025, LNOR, pp. 126–142, 2026.
https://doi.org/10.1007/978-3-032-13116-4_11

and counties.[2] Moreover, healthcare professionals continue to strongly support recommending telemedicine to patients [3], which drives further expansion and adoption. As a result, the demand for telemedicine is expected to rise.

In China, the core of telemedicine services is teleconsultation. Teleconsultations require prior scheduling and involve two types of patients: those with appointments (Patient A) and walk-in patients (Patient W) who arrive on the day. As the use of teleconsultation grows, many hospitals from higher level face increasingly uncertain environments. Patient A is more likely to be affected by the unscheduled arrival of Patient W, often leading to disruptions. Our research was inspired by a study conducted at the National Telemedicine Center (NTCC) from September 2024 to February 2025. We observed various uncertainties, including the duration of the service, patient unpunctuality, expert delays, and the random arrival of W patients. We also observed that the scheduling of Patient A's appointments at the telemedicine center was still handled manually. For example, when a patient W arrived, staff members adjusted the schedule by manually assigning an appropriate slot and updating the system for the subsequent Patient A appointments. While this approach offers some flexibility, it is not optimal as the demand for teleconsultation continues to grow. Over time, this method leads to inefficiencies and delays. We studied similar issues in other industries, such as dynamic scheduling in relay satellites [4] and vehicle trajectory control [5], and found that dynamic models, such as rolling time domain optimization, are practical in resource allocation and scheduling. This further encouraged our exploration of the dynamic scheduling performance in teleconsultation. Unlike outpatient visits, teleconsultation involves uncertainties for both grassroots patients and senior experts. Since teleconsultations have no geographical limitations and can occur during working hours, walk-in situations are more prominent. Therefore, considering dynamic scheduling in teleconsultations, particularly in scenarios where walk-ins disrupt the schedule, is essential.

To implement our idea, we first constructed a two-stage static scheduling model as a baseline for comparison. This model incorporated walk-in scenarios in advance, which, from a management perspective, served as a feedforward control when assigning appointment times to Patient A. Based on the static scheduling model from the first stage, we then developed a dynamic scheduling model. This model built a global update system tailored to the walk-in scenario. On the day of the teleconsultation, Patient A participated according to the scheduled appointment. If a Patient W arrived, the telemedicine center assigned a new slot and adjusted the subsequent appointments of affected Patients A. To enhance the timeliness of dynamic adjustments, we further developed a rolling horizon optimization model, combined with a greedy strategy to allocate slots to patients W. In the rolling horizon optimization process, each adjustment only affected teleconsultations within the current optimization window, and changes were made progressively forward in time. Throughout this process, we continued to consider uncertainties such as actual service duration, unpunctuality of Patient A, and the randomness of Patient W's arrival. Based on our research findings, when a Patient W was inserted into a certain slot, the Patient A scheduled for that slot experienced a delayed teleconsultation.

[2] http://www.nhc.gov.cn/wjw/tia/202407/5175661d63df4449919a8308a9b5c9da.shtml [accessed in March 2025].

Most previous studies on the scheduling of medical service appointments have focused on outpatient studies [6–9]. Although some studies also have dealt with the problem of the combination of online medical care and outpatient services [10, 11], the number of studies examining teleconsultation scheduling is still relatively limited. Our research is divided into two parts: static scheduling and dynamic scheduling. For research in this area, refer to [8–13]. Static scheduling is widely used across various research fields, aiming to optimize resource allocation and enhance operational efficiency. Common approaches include Mixed Integer Programming (MIP) models [14–16], For example, Erdogan et al. [17] developed a two-stage stochastic linear programming model for rural tele-cystoscopy in the United States, considering uncertainties such as random examination duration, equipment cleaning time, and patient no-shows. In a similar vein, Qiao et al. [18] proposed a multi-room teleconsultation scheduling model that incorporates service time variability. By comparing its performance against deterministic models and staffing plan. Furthermore, Qiao et al. [15] built a two-stage stochastic programming model that describes workflows under various random scenarios and identifies scheduling plans that minimize patient waiting times. And Wan et al.[19] also focusing on reducing wait time. Young et al. [20] evaluated different scheduling strategies for teleconsultation and recommended longer time blocks to reduce patient delays. These studies generally adopt MIP-based modeling approaches, combined with solution techniques such as Sample Average Approximation (SAA), L-Shaped methods, and Variable Neighborhood Search (VNS). However, a common challenge with SAA lies in its long computation time [14, 15]. To address this, we take SAA as a baseline and integrate it with L-Shaped and VNS methods to develop an efficient solution approach for solving our two-phase SMIP model. These studies provide valuable references. However, teleconsultation presents different challenges. Uncertainties come from both patients and remote experts. Since geography is not a limiting factor, teleconsultations can occur at any time during working hours. As a result, walk-ins are even more common in teleconsultation scenarios. And, experts are significantly more mobile. After one expert finishes a teleconsultation, another may take over, each expert is flexible and their arrival time is inherently uncertain We believe that addressing dynamic allocation under walk-in disruptions is essential for improving service performance in teleconsultation systems. Dynamic scheduling has been widely adopted in industrial systems such as satellite operations [4] and vehicle routing [5]. In the healthcare domain, outpatient dynamic scheduling has attracted significant attention, with Markov Decision Processes (MDPs) and game-theoretic approaches being among the most commonly used methods. For example, Geng et al. [21] proposed a finite-horizon MDP model to optimize outpatient scheduling, aiming to maximize resource utilization and expected revenue. In the context of teleconsultation, Chen et al. [22] applied a combination of MDP and deep Q-learning. Through extensive simulations and comparative analysis, they demonstrated that their model could effectively reduce patient waiting times. Wen et al. [23] focused on real-time scheduling for semi-urgent patients in emergency departments. They proposed a real-time MDP model along with an approximate dynamic programming approach, which proved effective in reducing waiting costs. Beyond MDP-based methods, game-theoretic approaches are also common in real-time scheduling research. For instance, Wang et al. [24] considered walk-in arrivals and developed an optimization model to determine the best appointment strategy under potential registration scenarios. Similarly, Liu et al. [25] investigated walk-in behavior by designing both real-time and asynchronous online scheduling systems, and found no universally superior solution between the two. Few studies in the healthcare field have explored Rolling Horizon Optimization (RHO) models. And existing studies often focus on either real-time or dynamic scheduling in isolation, with limited efforts to systematically integrate both static and dynamic scheduling approaches. Therefore, our research incorporated walk-in behavior and applied RHO-based optimization to develop a dynamic scheduling framework for teleconsultation, aiming to support timely appointment updates for patients in primary care settings.

The contributions of this paper are as follows:

1. We systematically develop both static and dynamic scheduling models for teleconsultation resource allocation and scheduling, providing a comparative analysis of the two approaches.
2. Using SAA as a benchmark, we employ a combination of SAA, VNS, and the integer L-Shaped method to solve the static scheduling model, improving computational efficiency.
3. We are the first to apply a rolling time-domain optimization strategy in teleconsultation scheduling research, incorporating a greedy strategy for patient W allocation, thereby enhancing the adaptability of the dynamic model.

2 Modeling Framework

2.1 Assumptions and Notations

Based on our preliminary research at NTCC, we developed a two-phase stochastic Mixed Integer Programming (MIP) model. In the first phase, experts, consultation rooms, and start times are allocated to scheduled patients, and pre-allocations are made for patients W. In the second phase, we integrate the uncertainties of actual consultation, including the tardiness of both patients and experts. The overall optimization process yields a reasonable allocation of patients. Before establishing the model, we made the following assumptions:

Assumption 1: The service duration for scheduled patients is random. Based on the service duration range in the actual data, we assume it to be a random variable sequence within a certain interval.

Assumption 2: The arrival of scheduled patients relative to their scheduled start time is random. The delay in their arrival is assumed to be a random variable sequence within a certain interval.

Assumption 3: Experts will not arrive earlier than the scheduled start time. A buffer is used during scheduling to mitigate the impact of expert delays on the consultation.

Assumption 4: We assume that all patients will be consulted in the order of their scheduled appointments.

2.2 Model Description

2.2.1 TATIC Appointment Scheduling

Based on the aforementioned assumptions and notations, we construct a two-stage stochastic mixed-integer programming model based on the SAA. By introducing the variable $Z_{ijm1} = U_i X_{ijm1}$, and applying SAA to linearize the nonlinear constraints, the SMIP can be transformed into the following deterministic model (SAA-MIP). The objective function is as follows:

$C_m \bullet y_m$ and $C_A \bullet X_{ijmt}$ represent the opening cost of consultation rooms and the assignment cost for patients, respectively. $C_W \bullet W_{ijmt}$ and $C_O \bullet O_m$ denote the waiting cost for patients A and overtime cost of consultation rooms in the second stage.

$$\begin{aligned} \min \textstyle\sum_{m\in M} C_m \bullet y_m + \sum_{m\in M}\left(C_u \bullet D \bullet y_m - C_u \sum_{i\in I}\sum_{j\in J}\sum_{t\in T} d_i \bullet X_{ijmt}\right) + \sum_{i\in I} C_A \bullet X_{ijmt} \\ + \tfrac{1}{K}\textstyle\sum_{k\in K}\left[\sum_{m\in M}\left(\sum_{i\in I}\sum_{j\in J}\sum_{t\in T} C_W \bullet W^k_{ijmt} + C_O \bullet O^k_m\right)\right] \end{aligned} \tag{1}$$

Constraints for Static Phase I: During the allocation, it is essential to ensure that each patient is assigned to only one position in a single teleconsultation room and is assigned to only one expert. At any given time, each expert can only consult one patient, and each teleconsultation room can only accommodate one patient. All teleconsultations are scheduled sequentially based on the assigned positions.

$$\textstyle\sum_{j\in J}\sum_{m\in M}\sum_{t\in T} X_{ijmt} = 1, \forall i \in I \tag{2}$$

$$\textstyle\sum_{i\in I}\sum_{m\in M} X_{ijmt} \le 1, \forall j \in J, \forall t \in T \tag{3}$$

$$\textstyle\sum_{i\in I}\sum_{j\in J} X_{ijmt} \le 1, \forall m \in M, \forall t \in T \tag{4}$$

$$\textstyle\sum_{i\in I}\sum_{j\in J} X_{ijmt} \le \sum_{i\in I}\sum_{j\in J} X_{ijm(t-1)}, \forall m \in M, \forall t \in \{1\ldots T\} \tag{5}$$

To ensure the effective availability of teleconsultation rooms, we introduce a big-*M* method. Additionally, all teleconsultations in each room are scheduled to be completed before the end of the working day.

$$\textstyle\sum_{i\in I}\sum_{j\in J}\sum_{t\in T} X_{ijmt} \le M \bullet y_m, \forall m \in M \tag{6}$$

$$\textstyle\sum_{i\in I}\sum_{j\in J}\sum_{t\in T} X_{ijmt} \ge y_m, \forall m \in M \tag{7}$$

$$\textstyle\sum_{i\in I}\sum_{j\in J}\sum_{t\in T} d_i \bullet X_{ijmt} \le D, \forall m \in M \tag{8}$$

We describe the relationship between the teleconsultation start time, end time, and service duration. To mitigate the impact of expert tardiness, we introduce a buffer threshold $\Delta\rho_j$.

$$\mathcal{B}_{ijm1} = 1 \tag{9}$$

$$\mathcal{E}_{ijmt} = \mathcal{B}_{ijmt} + d_i + \Delta\rho_j, \forall i \in I, \forall j \in J, \forall m \in M, \forall t \in T \tag{10}$$

Constraints for Static Phase II: In the second phase, we describe the allocation of patients W and detail the construction of the waiting time for scheduled patients A and the overtime duration for the teleconsultation rooms. The waiting time for a patient scheduled for the first teleconsultation is denoted by W_{ijmt}, which is defined as:

$$W_{ijm1} = \left(\sum_{i=1}^{I} U_i X_{ijm1} \right)^+, \forall j \in J, \forall m \in M, \forall t \in T \tag{11}$$

$(A)^+$ denotes a positive value; if $A > 0$, the value is equal to A; if $A = 0$, the value is equal to 0; if $A < 0$, the value is taken as 0. In this study, we do not consider the case that the remote expert arrives early, so $A \geq 0$.

$$\sum_{m=1}^{M} \sum_{t=1}^{T} X_{i-Njmt} = 1, \forall i-N \in I-N, \forall j \in J \tag{12}$$

$$\sum_{m=1}^{M} X_{i-Njmt} \leq 1, \forall i-N \in I-N, \forall j \in J, \forall t \in T \tag{13}$$

$$\sum_{i-N=1}^{I-N} X_{i-Njmt} \leq \sum_{i=1}^{I} X_{ijmt}, \forall m \in M, \forall j \in J, \forall t \in T \tag{14}$$

for K samples, we transform the constraints in the second phase into the following form:

$$\begin{aligned} &\Theta_{ijm(t-1)} X_{ijm(t-1)}(\xi_{i,k} + \Delta\rho_j) - \left(X_{ijmt}\mathcal{B}_{ijmt} - X_{ijm(t-1)}\mathcal{E}_{ijm(t-1)} \right); \\ &- \sum_{i-N \in I-N} \sum_{m \in M} \sum_{t \in \{1...T\}} X_{i-Njm(t-1)} d_{i-N}, \\ &\forall i \in I, \forall i-N \in I-N, \forall j \in J, \forall m \in M, \forall t \in \{1 \ldots T\} \end{aligned} \tag{15}$$

$$W_{ijm1}^{k} \geq \sum_{i=1}^{I} Z_{ijm1}^{k}, \forall j \in J, \forall m \in M, \forall t \in T \tag{16}$$

$$\begin{gathered} W_{ijmt}^{k} \geq \sum_{m \in M} \sum_{t \in \{1...T\}} Z_{ijm(t-1)}^{k} + \theta_{ijm(t-1)} - Z_{ijmt}^{k}, \\ \forall i \in I, \forall j \in J \end{gathered} \tag{17}$$

$$\begin{gathered} O_{m}^{k} \geq \sum_{i \in I} \sum_{j \in J} \sum_{i \in T} \sum_{i-N \in I-N} \Big(W_{ijmt}^{k} + \xi_{i,k} X_{ijmt} \\ + \xi_{i-N,k} X_{i-Njmt} - D \Big), \forall m \in M \end{gathered} \tag{18}$$

$$W_{ijm1}^{k} \geq 0, \forall i \in I, \forall j \in J, \forall m \in M, \forall t = 1 \tag{19}$$

$$W_{ijmt}^{k} \geq 0, \forall i \in I, \forall j \in J, \forall m \in M, \forall t \in T \tag{20}$$

$$O_{m}^{k} \geq 0, \forall m \in M \tag{21}$$

$$\sum_{i \in I} \sum_{j \in J} \sum_{t \in T} d_i \bullet X_{ijmt} \leq D + O_{m}^{k}, \forall m \in M \tag{22}$$

2.2.2 Dynamic Scheduling

We develop a dynamic scheduling model based on the first phase of static scheduling. Initially, we consider only the patient W, and after a global update, we adjust the appointments for all affected scheduled patients. We then further optimize the model using the rolling horizon optimization strategy. Figures 1 and 2 demonstrates the global update model considering only the insertion of patient W. $\mathcal{A}_{i-N}$ represents the arrival time of the patient W. After the arrival of the patient W, the greedy strategy is used to select the nearest available position with the shortest waiting time in different teleconsultation rooms. The original scheduled patient at that position is delayed, and the appointment information for this patient and all subsequent patients is updated accordingly. The dynamic scheduling model is built on the first phase of the SMIP, with the initial scheduling goal as given in Eq. (23).

$$\min \sum_{m\in M} (C_m \bullet y_m + C_O \bullet O_m) + \sum_{m\in M} \left(C_u \bullet \mathcal{D} \bullet y_m - C_u \sum_{i\in I} \sum_{j\in J} \sum_{t\in T} d_i \bullet X_{ijmt}\right) + \sum_{i\in I} C_A \bullet X_{ijmt} \tag{23}$$

The allocation constraints for Patient A remain unchanged, still governed by constraints (2) to (7). Equation (24) indicates that the initial scheduling only requires the completion of Patient A's appointments before the end of the workday. Equations (25) to (27) represent the start and end times for Patient A's teleconsultations (Fig. 2).

$$\sum_{i\in I} \sum_{j\in J} \sum_{t\in T} d \bullet X_{ijmt} \le D + O_m, \forall m \in M \tag{24}$$

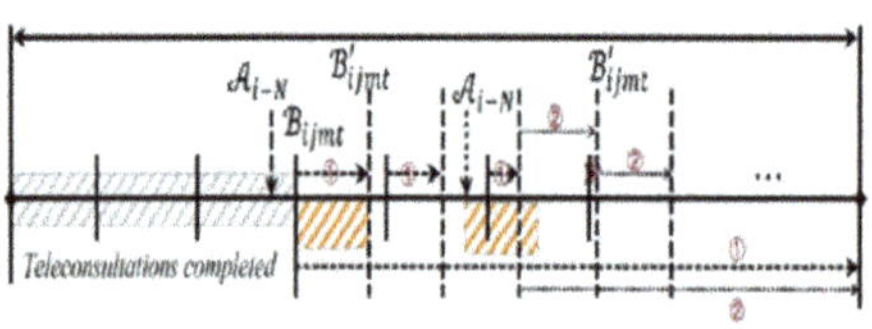

Fig. 1. Update on Start Times of Appointment

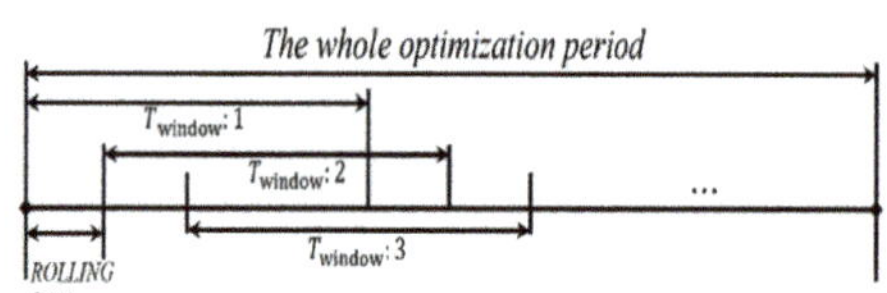

Fig. 2. Illustration of Rolling Horizon Optimization Strategy

Patients Considering Only Walk-ins

$$\mathcal{B}_{ijm1} = 1, \forall i \in I, \forall j \in J, \forall m \in M, \forall t \in T \tag{25}$$

$$\mathcal{B}_{ijmt} = X_{ijmt} \bullet (d_i \bullet t + 1), \forall i \in I, \forall j \in J, \forall m \in M, \forall t \in T \tag{26}$$

$$\mathcal{E}_{ijmt} = \mathcal{B}_{ijmt} + d_i + \Delta\rho_j, \forall i \in I, \forall j \in J, \forall m \in M, \forall t \in T \tag{27}$$

The service duration for a patient W is denoted as d_{i-N}. When a patient W is inserted, the appointment start time for Patient A is updated accordingly. If only one patient W is inserted before Patient A, the start time needs to be updated once. However, if multiple patients W are inserted, the update will occur multiple times. The updated start time for Patient A is represented by $\mathcal{B}\prime_{ijmt}$.

$$\mathcal{B}'_{ijmt} = \mathcal{B}_{ijmt} + X_{i-Njmt} d_{i-N}, \forall i \in I, \forall j \in J, \forall m \in M, \forall t \in T \tag{28}$$

$$\mathcal{B}'_{ijm(t+1)} = \mathcal{B}_{ijmt} + X_{i-Njmt}(d_{i-N} + d_{i-N}), \forall i \in I, \forall j \in J, \forall m \in M, \forall t \in T \quad (29)$$

$$\mathcal{B}'_{ijm(t+n)} = \mathcal{B}_{ijmt} + X_{i-Njmt}(d_{i-N} + nd_{i-N}), \forall i \in I, \forall j \in J, \forall m \in M, \forall t \in T \quad (30)$$

1) *GREEDY-BASED RECEDING HORIZON OPTIMIZATION*

Figure 3 illustrates two types of update-driving modes in the rolling horizon optimization model: (a) describes how the unpunctuality of both A and W patients and the random service duration of A patients may impact subsequent teleconsultations; (b) describes how the allocation of W patients, in random scenarios, can also affect A patients. Overall, our model employs a hybrid-driven rolling horizon optimization.

We first use static scheduling to allocate appointments for scheduled patients A, determining the initial appointment information. The specific objectives and constraints are given by (23), (2) ~ (7), (24) ~ (27).

We define$\mathcal{A}_i$ as the actual arrival time of scheduled patient *i*, $\mathcal{S}_i$as the actual service duration for scheduled patient *i*, and Δ_i as the arrival delay or advance of scheduled patient *i*. All optimizations are performed within each rolling window $[t, t + T_{window}]$.

$$\mathcal{A}_i = \mathcal{B}_{ijmt} + \epsilon_i, \epsilon_i \sim U(-\sigma, \sigma), \forall i \in I, \forall j \in J, \forall m \in M, \forall t \in T \quad (31)$$

$$\mathcal{S}_i \sim U(d_i - \eta, d_i + \eta), \forall i \in I \quad (32)$$

$$\Delta_i = \mathcal{A}_i - \mathcal{B}_{ijmt} + \Delta\rho_j, \forall i \in I, \forall j \in J, \forall m \in M, \forall t \in T \quad (33)$$

For all scheduled patients, calculate their actual arrival time $\mathcal{A}_i$and actual service duration$\mathcal{S}_i$, and adjust their teleconsultation start time $\mathcal{B}'_{ijmt}$.

$$\mathcal{B}'_{ijm(t+1)} = \mathcal{B}_{ijm(t+1)} + X_{ijmt}[\max(0, \Delta_i) + (\mathcal{S}_i - d_i)], \forall i \in I, \forall j \in J, \forall m \in M, \forall t \in T \quad (34)$$

Combining with the adjustment formula mentioned earlier for when no patient W arrive, we obtain the teleconsultation time adjustment formula under rolling horizon optimization with a hybrid driving mode.

$$\mathcal{B}'_{ijmt} = \mathcal{B}_{ijmt} + [\max(0, \Delta_i) + \mathcal{S}_i - d_i] + d_{i-N}, \forall i \in I, , \forall i - N \in I - N, \forall j \in J, \forall m \in M, \forall t \in T \quad (35)$$

$$\mathcal{B}^{frozen}_{ijmt} = \mathcal{B}'_{ijmt}, \forall t \in [t, t + T_{step}] \quad (36)$$

$$t \leftarrow t + T_{step} \quad (37)$$

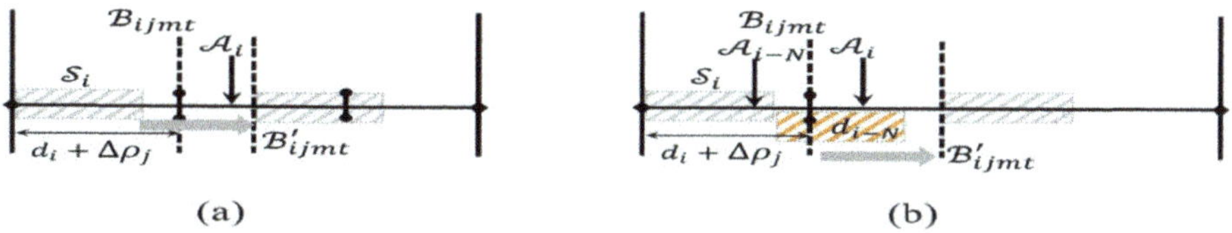

Fig. 3. Illustration of Two Types of Dynamic Update Factors

3 Methodology

3.1 SVILS Algorithm

Based on SAA, we use the L-Shaped method to decompose the SAA-MIP into a master problem and a subproblem. To avoid getting trapped in local optima and to improve global search capability, we introduce the VNS method when solving the master problem, perturbing the patient and teleconsultation room allocations. The specific solution procedure is as follows:

Step 1: First, we generate K random scenarios Ψ for the random variables to approximate the uncertainty distribution. We then set the maximum number of iterations MaxIterations and the stopping threshold for no improvement MaxNoImprove, and initialize the Cuts set.

Step 2: Next, we solve the master problem using the L-Shaped method to determine the initial patient allocations. The decision variables involved include X_{ijmt}, X_{i-Njmt}, y_m, θ_{sub}, where θ_{sub} represents the objective value of the subproblem, and θ_{sub}>cut. After solving the master problem, we solve the subproblem, where the optimal cuts generated vary across different scenarios.

Step 3: In combination with the VNS algorithm, we address the patient-room-location scheduling problem. We construct three perturbation strategies using three different neighborhood structures to generate perturbed solutions. In each search iteration, we select the appropriate neighborhood for further search. We define $\imath$ as the currently selected neighborhood structure, starting from $\imath = 1$, and attempt up to 3 neighborhood structures. The specific perturbation strategies are illustrated in Fig. 4.

Neighborhood Structure 1 (N1): The order of scheduled patients in different positions within the same teleconsultation room can be swapped.

Neighborhood Structure 2 (N2): Scheduled patients in the same position across different teleconsultation rooms can swap rooms.

Neighborhood Structure 3 (N3): If neither N1 nor N2 provides a reasonable allocation for some scheduled patients, these patients are randomly reallocated to new teleconsultation rooms and time slots.

To prevent omissions and conflicts (such as two patients being allocated to the same position) after perturbation, we incorporate conflict resolution and constraint handling in the algorithm. It ensures that each scheduled patient is assigned to a unique teleconsultation room and time slot. If a scheduled patient is not assigned, an optimal available position is selected. If a patient is assigned to multiple positions, the most optimal position is chosen as the final assignment.

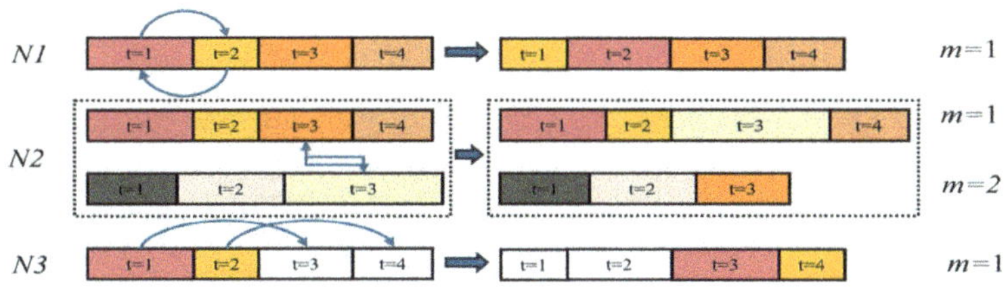

Fig. 4. VNS Perturbation Strategy

3.2 Greedy-Based Rolling Horizon Optimization

When dynamically adjusting the consultation arrangements for scheduled patients using the GRHO method, we first input the algorithm parameters based on the static scheduling solution, including the number of positions T, the rolling step *ROLLING STEP*, the window size T_{window}, the number of W patients *N_walk-in* randomly inserted at each position, and the number of available consultation rooms M. The static scheduling solution is marked as the current frozen schedule, *frozen_B_values*, and the arrival times of randomly generated patients W, *t_arr(p)*, are created.

Minimum Consultation Room Load Strategy (GRHO): For W patients, we need to find the least crowded teleconsultation room, m^*, and, if possible, assign them to positions close to the W patient's arrival location.

Minimum Disruption Strategy (GRHO):* When allocating W patients, we must simultaneously consider the adjustment magnitude for A patients, avoiding excessive changes to A schedules.

Algorithm 1. Rolling Horizon Scheduling with Greedy New Patient Insertion (GRHO*)

Algorithm 1 Rolling Horizon Scheduling with Greedy New Patient Insertion (GRHO*)

Input: T, T_{window}, *ROLLING STEP*, η, *B_values, N_walk-in*
$t \leftarrow 0$
inserted patients $\leftarrow \emptyset$
  while $t + T_{\text{window}} \leq T$ **do**
    for $t\_opt = t$ to $(t + T_{\text{window}} - 1)$ **do**
    if $B\_values[i, j, m, t\_opt] > 0.5$ **then**
      actual arrive $\leftarrow \max(0, B\_values[i, j, m, t\_opt]) + \epsilon_i$
      actual service $\leftarrow \mathcal{S}_i \sim U(d_i - \eta, d_i + \eta)$
      delay $\leftarrow \Delta_i \leftarrow (\mathcal{A}_i - \mathcal{B}_{ijmt} + \Delta\rho_j)$
      for all (*i-N, arrive_t*) in *I-N* **do**
        if i-N $\notin$ inserted patients & $t \leq arrive\_t < t + T_{\text{window}}$ **then**
          actual rooms $\leftarrow \{m \mid \exists(i', j', m, arrive\_t): B\_values[i', j', m, arrive\_t] > 0\}$
          if actual rooms $\neq \emptyset$ **then**
            $m^* \leftarrow \text{argmin}_{m' \in actual\ rooms}$ compute interference $(m', arrival\_t)$
            affected $i \leftarrow \{(i', j') \mid B\_values[i', j', m^*, arrive\_t] > 0\}$
            continue
          end if
      end for
      $adjustment \leftarrow \mathcal{B}_{ijmt} + [\max(0, \Delta_i) + \mathcal{S}_i - d_i] + d_{i-N}$
        if $|adjustment| > \beta$ **then**
          $\mathcal{B}'_{ijmt} \leftarrow B\_values[i, j, m, t\_opt] + adjustment$
  end while
Output: Optimized scheduling *B_values*

4 Numerical Analysis

In this section, we perform a numerical analysis of the static and dynamic scheduling models proposed earlier. Data were obtained from the NTCC, which contains 5,288 teleconsultation records from 2 January 2024 to 16 May 2024. Teleconsultation records contain basic information such as the time the lower-level hospital requested teleconsultation, length of the requested teleconsultation, actual start time, actual end time, patient's department, patient's condition, and experts.

4.1 Empirical Statistics and Parameter Setting

For example, we found that the earliest teleconsultation start time is 8:30 AM, while the typical end time is 12:00 PM. Based on this practical experience, we set the scheduling period $D = 210$. Additionally, we performed statistical analysis on the actual service duration data of teleconsultations, the number of the total statistics is 5,288. The mean service time is 16.88 min, with the maximum and minimum service times being 199.15 min and 0.44 min, respectively, which shows a significant difference. The coefficient of dispersion is 0.59, indicating a high degree of dispersion in the service time. Zhai et al. [26] predicted the actual service duration of teleconsultations using NTCC data, with the predicted service duration of 15 min being the most reasonable. Based on this statistical analysis and previous predictive studies using real data, we set the consultation allocation duration in the dynamic scheduling model as d_i=15 min, and the service duration for W patients as 10 min. For typical teleconsultations, we select $I = 25$ and $I - N = 6$. According to previous studies [18, 19], similar models have controlled the ratio of overtime costs to allocation costs to be greater than 3, with other cost coefficients aligned with standard hospital practices. T = 14 and $C_m = 20$, $C_u = 30$, $C_A = 20$, $C_W = 50$, $C_O = 110$. All computational experiments are implemented in PyCharm on a laptop with an Intel(R) Core(TM) i7-7500U CPU @ 2.70 GHz. The solver used is Gurobi 11.0.3.

4.2 Performance of Proposed Static Schedule Algorithm

We present some numerical studies based on the SAA and SVILS methods. Figure 5 shows the solution time and objective values obtained using the SVILS method. We observe that the solution time increases smoothly, with the solution for 100 samples requiring only 13,010.73 s. Due to the varying number of samples and the random distribution not being completely fixed, there is noticeable variation in the objective values. However, most values fall between 20,000 s and 23,000 s, which is not significantly different from the exact solution. Table 1 compares the performance of SAA and SVILS. We can clearly see that when the sample size is 30, SAA has not yet found the optimal solution at 30,000 s, with a gap value of 9.5%. Overall, regardless of the sample size, the solution speed of SVILS is superior to that of SAA. Therefore, the algorithm we proposed is effective. We then proceed with static scheduling for further numerical analysis using the SVILS method.

Figure 6 shows the boxplot of the waiting time distribution for Patient A across different slots. From the waiting time data, we can observe the impact of W patients on the waiting times for A patients in the static scheduling model. As shown in Fig. 6, most patients' waiting times fall within the range of 5 to 25 min, with the maximum waiting time approaching 30 min. Although static scheduling incorporates prior control, longer waiting times still occur. Two outliers are present in the figure, which can be attributed to multiple teleconsultation rooms conducting teleconsultations at the same time, each subject to uncertainties. These outliers are reasonable and can be considered as part of the inherent variability in the system.

4.3 Performance of Proposed Dynamic Schedule

Figures 7 and 8 show the initial allocation and subsequent updates of scheduled patients in two teleconsultation rooms. A total of 6 patients W are considered, requiring 6 updates. The blue lines represent the initial allocation, while the red arrows indicate the insertion of patients W. We observe that if only the random insertion of patient W is considered for updates, the teleconsultation start times for the scheduled patients may be pushed beyond the end of the working day. This is a cumulative process, meaning that with each update, the teleconsultation start time for a scheduled patient is delayed. However, in actual teleconsultations, service durations vary and patient arrivals are random. Therefore, the updates should reflect three possible states: migration, delay, or no change. As a result, the baseline full-process update plan needs further improvement.

Building upon the baseline dynamic model, we consider the random arrival of A patients, the random arrival of W patients, and the random service durations, while using a fixed buffer threshold to mitigate the impact of expert unpunctual. We first solve and analyze the initial arrangement of GRHO, followed by an analysis of GRHO and GRHO*. In Figs. 9 and 10, the final updated schedules for GRHO and GRHO* show that all appointments include various states, such as early, delayed, and on-time, with all consultations completed before the end of the workday. We also observed that, despite considering multiple uncertainties due to the impact of W patients, all teleconsultations after the adjustments were completed within the two consultation rooms. This indicates that we can maximize teleconsultation room utilization, which highlights the effectiveness and rationality of dynamic scheduling.

4.4 Comparative Analysis

After analyzing the proposed model, we first compare GRHO and GRHO* to explore the performance of the rolling optimization model under different greedy strategies. We continue to evaluate the timeliness of the updated results based on the waiting times of Patient A.

Table 1. Comparison of Solution Algorithm Performance

Number of Samples	Computational time (s)		
	SVILS	SAA	SAA:Gap (%)
1	18.65	178.05	0.00
10	219.90	1200.55	0.00
20	706.44	17550.43	0.00
30	1502.62	30000.00	0.095
...	...	–	–
100	13010.73	–	–

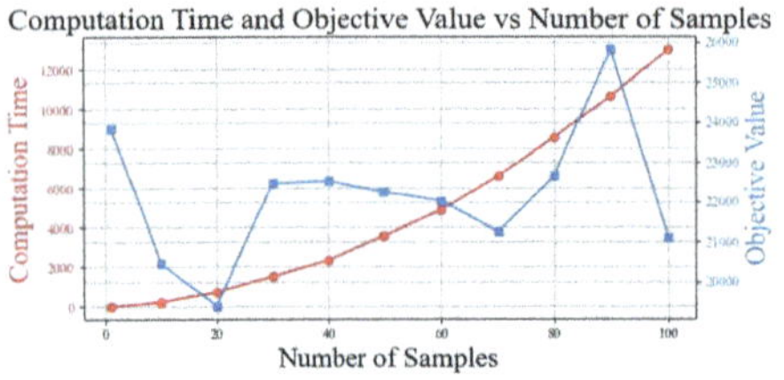

Fig. 5. Performance of SAA_VNS_L-Shaped

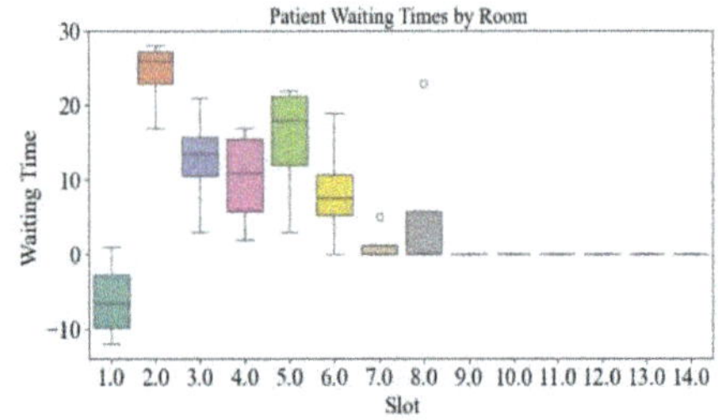

Fig. 6. Patient Waiting Times in Static Scheduling

The waiting times for Patient A at different positions are presented using boxplots. Figure 11 show the waiting times for Patient A after solving using GRHO and GRHO*, respectively. We observe that in GRHO*, the waiting times for Patient A are generally shorter, with a more uniform overall distribution and no significant outliers. In contrast, the waiting times for Patient A in GRHO show a more dispersed distribution, with a larger range of waiting times. This indicates that when we apply the greedy allocation strategy for W patients, focusing on minimizing disruption results in smaller adjustments. In other words, by prioritizing timeliness, we can make appropriate adjustments without excessive delays.

To further compare the performance of GRHO, GRHO*, and static scheduling (SMIP), we introduce two performance metrics: the adjustment rate and resource utilization rate. The adjustment rate primarily observes the changes in dynamic adjustments. We define the adjustment rate for Patient A as: $ADJ = \sum_i 1\left(B\prime_{ijmt}\right)/\sum_i X_{ijmt}$. The resource utilization rate (RUR) is defined as: $RUR = \left(\sum_i t \bullet X_{ijmt}\right)/(T \bullet y_m)$. Table 2 compares the performance of SVILS with GRHO and GRHO*. From the results, we can see that in the pre-allocation scheduling of SVILS, the start time for the first patient is relatively late, while the start time for the last patient is earlier. However, the maximum waiting time for Patient A can reach up to 28 min, with an average waiting time of 5.40 min, and the resource utilization rate is only 44.64%. This suggests that there is significant room for improvement in static scheduling. In contrast, whether considering the average deviation, maximum deviation, maximum waiting time, or average waiting time for Patient A, GRHO* consistently performs better than both GRHO and SVILS. Specifically, GRHO* achieves an adjustment rate (ADJ) of 80% and a resource utilization rate (RUR) of 89.29%.

5 Conclusion

This study, based on the research conducted at NTCC, explores the advantages and disadvantages of static scheduling that considers patients W W in advance, and dynamic scheduling that adjusts based on the arrival of W patients. The primary goal is to provide suitable methods for handling the walk-in scenarios and the uncertainties related to patients and experts in actual teleconsultations. By providing more timely appointment information, this study aims to enhance the operational efficiency of the China National Telemedicine Center (NTCC) and reduce the workload. We analyze the performance of the static model, including SVILS, and the dynamic models GRHO, and GRHO*. The results demonstrated the overall advantage of the dynamic scheduling model, particularly in the rolling time-domain optimization under the minimum disruption strategy.

This study also provides the following insights for teleconsultation professionals: (1) From the perspective of primary healthcare patients: If primary healthcare centers do not want later appointment times and prefer not to have frequent rescheduling, static scheduling methods can be used. By pre-allocating possible patients W, the appointment times for primary healthcare centers are typically earlier. (2) From the perspective of NTCC staff: To avoid patient complaints, staff should focus on resolving consultation queue blockages and ensuring timely updates to appointment information. The GRHO and GRHO* methods can facilitate this process without requiring manual adjustments, saving considerable administrative effort. (3) To prevent excessive adjustment of appointment information, which may lead to long waiting times for scheduled patients, staff can use the GRHO* method to minimize disruption to scheduled patients while maintaining timely updates.

While this study considers various scenarios in both static and dynamic scheduling, there is room for further development, especially in dynamic scheduling. Future research could focus solely on dynamic scheduling, combining Markov Decision Processes (MDP) and machine learning methods to further refine the decision-making process for patient W allocations.

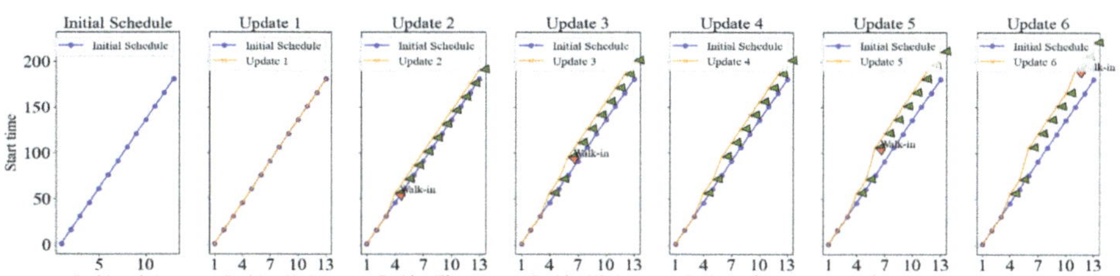

Fig. 7. Update on Start Times of Appointment Patients Considering Only Walk-ins (Room 1)

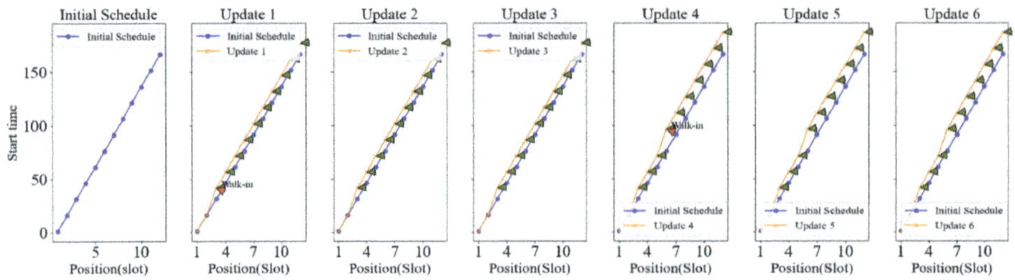

Fig. 8. Update on Start Times of Appointment Patients Considering Only Walk-ins (Room 2)

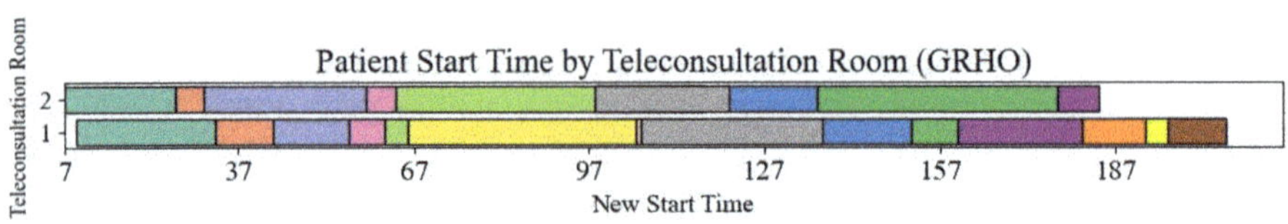

Fig. 9. Updated Start Time in GRHO

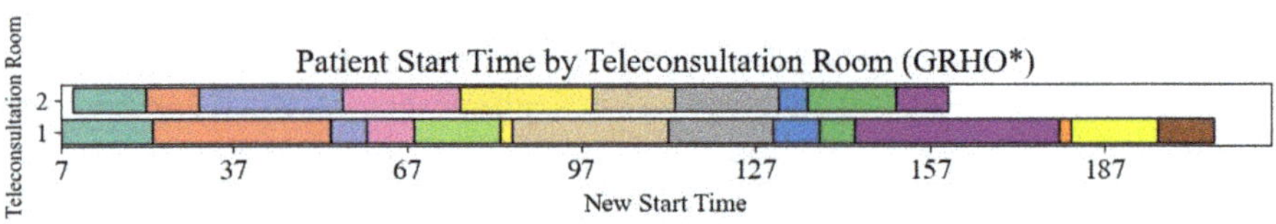

Fig. 10. Updated Start Time in GRHO*

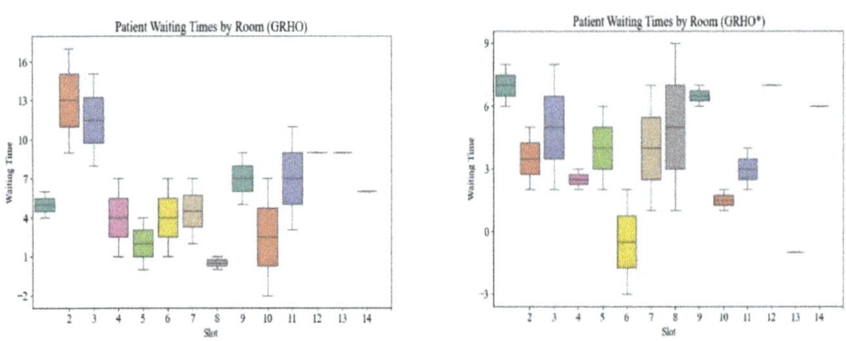

(a) (b)

Fig. 11. Patient Waiting Times in Dynamic Scheduling (GRHO and GRHO*)

Table 2. Comparison of Static and Dynamic Scheduling

	Static appointment scheduling	Dynamic scheduling	
	SVILS	GRHO	GRHO*
Min new start time	11	7	7
Max new start time	179	196	196
Mean deviation amplitude	–	8.80	6.12
Max deviation amplitude	–	26	23
Average waiting time	**5.40**	5.84	**3.80**
Max waiting time	**28**	17	**9**
Adjustment ratio	–	80%	80%
Resource utilization rate	**44.64%**	89.29%	89.29%

References

1. Trehle, E., Shabde, N.: One hundred years of telemedicine: does this new technology have a place in pediatrics? Arch. Dis. Child. **91**(12), 956–959 (2006)
2. Denton, B., James, V., Andrea, V.: Optimization of surgery sequencing and scheduling decisions under uncertainty. Health Care Manag. Sci. **10**(1), 13–24 (2007)
3. Wang, W., He, X., Zhang, X., et al.: Willingness of healthcare professionals in China to continue participating in and recommend telemedicine post COVID-19 pandemic. Sci. Rep. **15**, 9581 (2025)
4. Li, H.W., Luo, Q.Z., Gu, Y., et al.: Multi-objective dynamic scheduling optimization method for relay satellites based on rolling horizon strategy. Acta Aeronaut. Astronaut. Sin. **45**(16), 329706 (2024)
5. Xie, X.Y., Zhao, X., Jin, L.S., et al.: Trajectory tracking control of intelligent vehicles based on deep reinforcement learning and rolling horizon optimization. J. Traffic Transp. Eng. **24**(6), 259–272 (2024)
6. Zhou, S., Yue, Q.: Sequencing and scheduling appointments for multi-stage service systems with stochastic service durations and no-shows. Int. J. Prod. Res. **60**(5), 1500–1519 (2022)
7. Xue, L., Li, Y., Wang, Z., Chung, S.H., Wen, X.: Distributed appointment assignment and scheduling under uncertainty. Int. J. Prod. Res. **62**(1–2), 318–335 (2024)
8. Wang, K., Qin, H., Huang, Y., Luo, M., Zhou, L.: Surgery scheduling in outpatient procedure center with re-entrant patient flow and fuzzy service times. Omega. **102**, 102350 (2021)
9. Shehadeh, K.S., Cohn, A.E.M., Jiang, R.: A distributionally robust optimization approach for outpatient colonoscopy scheduling. Eur. J. Oper. Res. **283**(2), 549–561 (2020)
10. Balasubramanian, S., Shukla, V., Islam, N., Upadhyay, A., Duong, L.: Applying artificial intelligence in healthcare: lessons from the COVID-19 pandemic. Int. J. Prod. Res. **63**(2), 594–627 (2025)
11. Guo, H., Xie, Y., Jiang, B., Tang, J.: When outpatient appointment meets online consultation: a joint scheduling optimization framework. Omega. **127**, 103101 (2024)
12. Gupta, D., Denton, B.: Appointment scheduling in healthcare: challenges and opportunities. IIE Trans. **40**(9), 800–819 (2008)

13. Sun, D., Liao, Q.: Real-time coordination of multiple robotic arms with reactive trajectory modulation. IEEE Trans. Robot. **41**, 200–219 (2025)
14. Pan, X., Geng, N., Xie, X., Wen, J.: Managing appointments with waiting time targets and random walk-ins. Omega. **95**, 102062 (2020)
15. Qiao, Y., Zhai, Y., Ma, R., Ji, M., Lu, W.: Optimizing teleconsultation scheduling to make healthcare greener. J. Clean. Prod. **422**, 138569 (2023)
16. Anvaryazdi, S.F., Venkatachalam, S., Chinnam, R.B.: Appointment scheduling at outpatient clinics using two-stage stochastic programming approach. IEEE Access. **8**, 175297–175305 (2020)
17. Erdogan, S.A., Krupski, T.L., Lobo, J.M.: Optimization of telemedicine appointments in rural areas. Service Sci. **10**(3), 261–276 (2018)
18. Qiao, Y., Ran, L., Li, J., Zhai, Y.: Design and comparison of scheduling strategy for teleconsultation. Technol. Health Care. **29**(5), 939–953 (2021)
19. Wan, M., Shukla, N., Li, J., Pradhan, B.: Optimization of teleconsultation appointment scheduling in National Telemedicine Center of China. Comput. Ind. Eng. **183**, 109492 (2023)
20. Young, K., Gupta, A., Palacios, R.: Impact of telemedicine in pediatric postoperative care. Telemed. E-Health. **25**(11), 1083–1089 (2019)
21. Geng, N., Xie, X.: Optimal dynamic outpatient scheduling for a diagnostic facility with two waiting time targets. IEEE Trans. Autom. Control. **61**(12), 3725–3739 (2016)
22. Chen, W., Li, J.: Teleconsultation dynamic scheduling with a deep reinforcement learning approach. Artif. Intell. Med. **149**, 102806 (2024)
23. Wen, J., Geng, N., Xie, X.: Real-time scheduling of semi-urgent patients under waiting time targets. Int. J. Prod. Res. **58**(4), 1127–1143 (2019)
24. Wang, S., Liu, N., Wan, G.: Managing appointment-based services in the presence of walk-in customers. Manag. Sci. **66**(2), 667–686 (2019)
25. Liu, N., van Jaarsveld, W., Wang, S., Xiao, G.: Managing outpatient service with strategic walk-ins. Manag. Sci. **69**(10), 5904–5922 (2023)
26. Zhai, Y., Jia, Q., Qiao, Y., et al.: Prediction of remote consultation service duration based on the ATT-FC-LSTM model. J. Manag., 1–9 (2025)

The Impact of AI on Job Polarization in Egyptian Banking: Patterns, Implications and Associated Risks

Nada Ali(✉), Lamia Guseinova, and Chuanmin Mi

College of Economics and Management, Nanjing University of Aeronautics and Astronautics, Nanjing, Jiangsu 211106, China
{nadaali,lamiaguseinova,cmmi}@nuaa.edu.cn

Abstract. The expanding use of artificial intelligence (AI) in Egypt's banking and finance services industry has significantly impacted employment trends, leading to job polarization. This study investigates the relationship between AI adoption and industry job polarization using data from Egyptian financial institutions and national statistical agencies between 2015 and 2024. Employing principal component analysis and multiple regression analysis, the research reveals three key findings: (1) AI applications are driving job polarization, characterized by a decline in middle-skilled jobs (e.g., clerks, loan officers) alongside an increase in high-skilled (e.g., data scientists, AI engineers) and low-skilled employment (e.g., AI support staff); (2) Regional differences in AI adoption influence job polarization patterns, with urban areas like Cairo and Alexandria experiencing higher demand for highly skilled jobs compared to rural regions; (3) AI implementation creates new opportunities in high-skilled jobs focused on system development and low-skilled positions related to infrastructure maintenance, while simultaneously reducing demand for middle-skilled jobs that involve routine tasks. While growing social and economic gaps, regional disparities, and function-specific risk factors that call for strategic management represent some of the major risks related to implementation highlighted in the study.

Keywords: Artificial Intelligence · Job Polarization · Banking Sector · Multiple Regression Analysis · Employment Structure · Egypt · Risk Assessment

1 Introduction

The world is witnessing a profound shift in every sector due to the sprouting development of artificial intelligence (AI), especially in the financial and banking services. Customer service AI chatbots, automated loan processing systems, fraud detection systems, and robo-advisors are altering the functionality and service delivery in every business. These tools offer clear benefits: They reduce operational costs, increase the efficiency of work processes, and improve customer satisfaction. However, a critical question arises from the implementation of new technology; what happens to employment? With the automation of processes, there is a phenomenon termed as job polarization. This denotes

X. Xie et al. (Eds.): ICSS 2025, LNOR, pp. 143–157, 2026.
https://doi.org/10.1007/978-3-032-13116-4_12

the increase in demand for high-skill and low-skill jobs with a decrease in availability of mid-skill roles. All over the world, researchers are looking into this problem more closely, particularly in terms of AI's impact on employment structures in Egypt's banking and finance sector, there is still a lot to uncover. Considering the significance of this industry to the economy of the country, it becomes essential to study the impact of AI on employment patterns. This research seeks to bridge this gap by analyzing the effect of job polarization and AI technology uptake in the Egyptian banking and finance services industry. The research addresses critical data collection considerations through a comprehensive multi-source approach, combining employment records from the Central Agency for Public Mobilization and Statistics (CAPMAS) with detailed institutional data from Egyptian financial institutions. Data collection challenges were systematically addressed through validation protocols, cross-referencing multiple sources, and implementing quality control measures to ensure accuracy and completeness of the dataset spanning 2015–2024.

This study's importance results from its contribution to the understanding of how technological transformation impacts labor markets in developing economies. By identifying potential risks and opportunities associated with the implementation of AI, the study offers valuable insights for developing strategic responses. The findings will direct financial institutions in their workforce development initiatives, assist educational institutions in coordinating training programs with changing market demands, and inform policymakers in designing regulatory frameworks.

2 Literature Review

2.1 Artificial Intelligence in Banking and Finance

Finance and banking around the world are undergoing severe changes thanks to the implementation of Machine Learning Artificial Intelligence. Financial services processes have been transformed with the help of technologies like Natural Language Processing, Computer Vision and even Robotic Process Automation [1]. The AI comes with innovations such as algorithmic trading, automated customer support, more advanced credit scoring, fraud detection, and bespoke financial counseling. These technologies have been underutilized by banks and financial institutions until recently when they became compelled to change due to increased competition, heightened client demands, and the rapid speed of technological advancement. Ibrahim et al. [1] Innovations like algorithmic trading, automated customer support, smarter credit scoring, fraud detection, and personalized financial advice are now possible thanks to AI. In recent years, banks and financial institutions have been adopting these technologies more rapidly than ever pushed by growing competition, rising customer expectations, and the fast pace of technological progress. In the Egyptian context specifically, Ramadan Ali [4] found that AI has helped boost the performance of banks and improved the quality of financial services.

2.2 Job Polarization and Technological Change

The routine-biased technological change hypothesis is a common theoretical framework used to explain job polarization. It posits that routine, codifiable tasks that are primarily performed by middle-skilled workers are most easily automated by technologies such as artificial intelligence. [2]. While these technologies have little effect on non-routine manual tasks typically associated with low-skilled positions, they do complement the creative, abstract tasks carried out by high-skilled workers.

Research by Tariq [3] shows that the use of AI in financial services results to the automation of repetitive cognitive tasks that were previously handled by middle-skilled workers like loan officers, clerks, and customer support agents. The need for low-skilled workers who perform support tasks for AI-enabled services as well as high-skilled workers who can create and oversee AI systems (data scientists, AI engineers, financial analysts) is growing at the same time [2]. This restructuring of employment patterns has significant implications for career development, wage inequality, and social mobility. As Tariq [3] observes that middle-skilled workers have a difficult time making this shift and frequently need to reskill significantly either upskill into technical roles or adjust to middle-skilled jobs that still emphasis non-routine and interpersonal skills.

2.3 Analytical Methodologies for Studying AI Impact

2.3.1 Principal Component Analysis in Technology Adoption Studies

A useful methodological approach for examining intricate technological phenomena, like the adoption of AI, is Principal Component Analysis (PCA). However, this study acknowledges important limitations of PCA in the context of AI adoption measurement. While PCA effectively reduces dimensionality and identifies underlying patterns in the nine AI adoption indicators, it assumes linear relationships between variables and may not capture complex, non-linear interactions inherent in technological adoption processes. The current methodology addresses these limitations by conducting robustness checks and validating PCA results against alternative dimensional reduction techniques, though future research should explore non-linear approaches to better capture the complexity of AI integration patterns. Numerous interdependent variables represent adoption patterns are encountered by researchers looking at technological implementation across organizations. These multidimensional datasets can be reduced into manageable parts that capture key variation patterns using PCA. Marukatat [7] mentioned how complex technological indicators can be properly reduced by PCA to create composite indices that underlying adoption trends. PCA enables researchers to integrate factors like technological sophistication, workforce training intensity, implementation breadth, and investment levels into strong metrics that capture organizational commitment to AI integration. [6]. For regional studies, PCA proves especially valuable. Zounemat-Kermani [8] demonstrates how this method can analyze component loadings across geographic units to find spatial patterns in the diffusion of technology.

2.3.2 Regression Techniques in Employment Structure Analysis

The principal analytical tool for investigating the connections between employment constructions and technological change is Multiple Regression Analysis (MRA). Using this method, researchers can measure correlations between AI adoption and employment trends while accounting for potential confounding variables. Alita et al. [9] show how, by including suitable controls for organizational characteristics, economic conditions, and workforce attributes, regression models can isolate the effects of technological variables on employment outcomes. Institution size, capitalized levels, market competition intensity, and pre-existing workforce compositions are a few examples of these controls in the banking and finance industry. The study acknowledges methodological limitations regarding causation and potential impacts. While the regression analysis demonstrates strong correlations between AI adoption and employment changes, establishing definitive causal relationships remains challenging due to potential reverse causality and omitted variable bias. The research addresses these limitations through fixed-effects models and robustness checks, but future studies should consider instrumental variable approaches or natural experiments to strengthen causal inference. Additionally, the study recognizes that AI's potential impacts may extend beyond the measured employment categories, potentially affecting job quality, working conditions, and wage structures in ways not captured by the current skill-based classification system. MRA's strength is its capacity to pinpoint conditional relationships between the use of AI and particular job categories. Chang et al. [10] used this methodology to investigate the effects of blockchain adoption on employment in the financial services industry, identifying varying effects for different job categories according to task composition and skill requirements. Analytical frameworks of a similar nature can shed light on how AI technologies are changing employment trends in Egypt's banking industry across skill levels.

2.4 Research Gaps in the Literature

Current literature on AI and labor markets contains five significant gaps this study addresses:

1. **Sector-specific analysis**: While job polarization has been studied in manufacturing, media, and telecommunications, limited research examines the banking sector's unique characteristics (skill requirements, regulatory constraints, customer interaction models).
2. **Emerging economy context**: Existing research focuses primarily on advanced economies (Germany, Finland, UK), overlooking the distinct conditions in emerging economies like Egypt where digital infrastructure, workforce literacy, and institutional capacities differ substantially.
3. **Regional variation**: The literature inadequately addresses within-country regional differences in AI adoption, particularly relevant in Egypt where significant disparities exist between metropolitan financial centers and provincial regions.
4. **Longitudinal perspective**: Most studies use cross-sectional designs that capture only single time points, whereas this research's 2015–2024 timeframe reveals how adoption patterns and employment effects evolve over time.
5. **Mechanism specificity**: Previous research has focused either on technical implementation or broad employment outcomes without examining the specific mechanisms through which different AI applications affect various skill categories.

3 Research Hypotheses

Based on the literature review, this study proposes three hypotheses examining AI adoption and job polarization in Egypt's banking sector:

3.1 AI Adoption and Skill-Based Employment Shifts

H1: AI adoption in Egypt's banking sector reduces middle-skilled jobs while increasing both high-skilled and low-skilled roles.

This hypothesis draws on routine-biased technological change theory [2, 3], which suggests AI primarily automates routine cognitive tasks performed by middle-skilled workers (loan officers, customer service representatives, administrative clerks). As AI systems automate standardized processes, demand decreases for these middle-skilled positions while simultaneously increasing for:

- High-skilled workers (data scientists, AI engineers, financial technologists) who develop and manage AI systems [4]
- Low-skilled workers who support AI-enabled services, maintain infrastructure, and manage customer interactions where AI capabilities remain insufficient [3]

This pattern represents classic job polarization, with middle-skilled contraction alongside expansion in both high-skilled and low-skilled categories.

3.2 Regional Variation in AI Impact

H2: AI's effects on job polarization vary regionally, with urban areas experiencing more pronounced shifts than rural areas.

This hypothesis addresses geographic variation in AI's employment impacts based on regional differences across Egypt. Urban centers (Cairo, Alexandria) have advantages that accelerate AI adoption and employment restructuring:

- Superior technological infrastructure and internet connectivity
- Higher concentration of technology-oriented educational institutions
- Larger financial institutions with greater investment capacity [11]

Rural areas face implementation constraints (limited infrastructure, lower digital literacy, smaller institutions) that likely slow AI adoption [12]. The expected result is geographic divergence in employment polarization, with urban centers experiencing more pronounced effects than rural areas.

3.3 Differential Impact across Banking Functions

H3: Job polarization varies across banking functions, with greater effects in areas with high AI integration potential (fraud detection, loan processing) than in customer relationship management.

The hypothesis acknowledges that, depending on automation suitability, AI adoption varies across banking functions. Artificial intelligence (AI) can automate tasks that require structured data, predetermined rules, and little customer interaction, such as fraud detection, credit scoring, and loan processing. [11]. By contrast, functions requiring interpersonal skills, contextual understanding, and adaptive reasoning (customer relationship management, advisory services) present greater automation challenges [5].

4 Data and Methodology

4.1 Data Sources and Collection

This study shows panel data collected from 2015 to 2024, encompassing the period of accelerated AI adoption within Egypt's financial sector. The dataset draws from multiple sources to ensure comprehensive coverage of employment patterns, technological implementation, and contextual factors:

1. **Employment data**: Information on workforce composition and job categories across skill levels was obtained from the Central Agency for Public Mobilization and Statistics (CAPMAS) [13]. Employment records from Egyptian financial institutions were added to this, offering accurate job descriptions broken down by functional areas, skill levels, and educational requirements.
2. **AI adoption indicators**: The nine AI adoption indicators were tested by Principal Component Analysis (PCA) to produce a thorough measure of AI adoption that accounts for various aspects of technological implementation. [14] and industry associations.
3. **Regional indicators**: To better understand the factors influencing AI adoption and job structures, data on regional economic and technological development was gathered from CAPMAS regional statistics [15] and the Information and Decision Support Center (IDSC).

The comprehensive dataset addresses data collection challenges through systematic multi-source validation protocols. Employment data from CAPMAS was cross-validated with institutional records from individual banks to ensure accuracy in skill-level classifications. The nine AI adoption indicators (I1-I9) were verified through multiple industry sources including the Central Bank of Egypt and banking association reports. Quality control measures included systematic checks for missing values, outlier detection using statistical thresholds, and temporal consistency validation across the 2015–2024 period. Standardized data collection instruments were implemented to ensure uniform measurement across all 10 Egyptian cities, with regular quality assurance reviews conducted to maintain dataset reliability and completeness.

4.2 Variable Measurement

4.2.1 Dependent Variables: Employment Distribution by Skill Level

The dependent variables measure the distribution of employment across skill categories within Egypt's banking and finance sector:

1. **High-skilled employment (DH)**: Percentage of employees in positions requires advanced educational qualifications (bachelor's degree or higher) and specialized technical knowledge. This category includes data scientists, AI engineers, financial analysts, risk managers, and senior management positions.
2. **Middle-skilled employment (DM)**: Percentage of employees in positions requires secondary education or vocational training but not advanced degrees. This category includes clerks, customer service representatives, loan officers, and administrative staff.

3. **Low-skilled employment (DL)**: Percentage of employees in positions requiring minimal formal education or entry-level qualifications. This category includes support staff, facilities maintenance personnel, document processors, and basic customer service roles [2, 3].

4.2.2 Independent Variables: AI Adoption Measures

Multiple indicators to capture different dimensions of AI adoption within banking institutions:

1. **AI implementation rate (I1)**: Percentage of banking services utilizing AI technologies for core functions, measured on a service-by-service basis for each institution.
2. **AI investment intensity (I2)**: Percentage of technology budget allocated to AI development, implementation, and maintenance, normalized by institution size.
3. **AI-related staff training (I3)**: Percentage of employees who have completed formal training in AI-related technologies and applications.
4. **AI transaction processing (I4)**: Percentage of financial transactions (e.g., loan applications, account openings) processed using AI systems rather than manual procedures.
5. **Customer interaction automation (I5)**: Percentage of customer interactions handled through AI-powered systems such as chatbots and virtual assistants.
6. **AI risk management integration (I6)**: Extent of AI implementation in fraud detection, credit scoring, and risk assessment functions, measured on a standardized scale.
7. **Digital infrastructure quality (I7)**: Composite measure of technological infrastructure supporting AI implementation, including internet reliability, data center capabilities, and cloud computing access.
8. **AI governance framework (I8)**: Assessment of institutional policies, procedures, and standards governing AI deployment and utilization.
9. **AI-enabled product innovation (I9)**: Number of financial products and services developed or substantially modified using AI capabilities.

Table 1 presents descriptive statistics for these AI adoption indicators, showing substantial variation across institutions and regions.

Table 1. Descriptive Statistics of AI Adoption Indicators

Indicator	Count	Mean	Minimum	Maximum	Std. Error
I1	100	69.26	21.15	97.06	1.996
I2	100	66.41	19.97	86.07	2.143
I3	100	38.27	12.08	78.05	1.554
I4	100	54.52	1.11	92.43	3.027
I5	100	22.76	2.03	44.90	1.340
I6	100	11.84	4.00	91.44	2.328

(continued)

Table 1. (*continued*)

Indicator	Count	Mean	Minimum	Maximum	Std. Error
I7	100	28.10	1.00	33.23	0.718
I8	100	1.56	0.40	5.90	0.164
I9	100	4.51	0.41	27.88	0.557

4.2.3 Control Variables

To isolate the relationship between AI adoption and employment structure, the analysis incorporates several control variables addressing potential confounding factors:

1. **Institution size (C1)**: Total assets (in million EGP), logarithmically transformed to address skewness.
2. **Regional economic development (C2)**: GDP per capita by region (in thousand EGP), serving as a proxy for local economic conditions.
3. **Educational attainment (C3)**: Average years of formal education among employees, controlling for regional human capital development.
4. **Market competition (C4)**: Herfindahl-Hirschman Index measuring banking sector concentration in each region, controlling for competitive pressures that might independently influence employment structures.

Table 2 presents' descriptive statistics for the dependent and control variables, showing substantial variation across observations.

Table 2. Descriptive Statistics of Dependent and Control Variables

Variable	Count	Mean	Minimum	Maximum	Std. Error
DH	100	22.07	17.9	32.01	0.212
DM	100	56.72	40.54	71.21	0.793
DL	100	21.2	9.85	35.48	0.657
AI Index	100	−0.09	−4.85	4.21	0.26
C1	100	9,220	4,477	20,883	432.6
C2	100	6.19	2.4	20.93	0.503
C3	100	12.17	8.12	16.09	0.271
C4	100	0.267	0.129	0.446	0.041

4.3 Analytical Approach

4.3.1 Model Specification

To test the hypotheses regarding the relationship between AI adoption and employment distribution across skill levels, we employ a multivariate regression approach. The baseline model specification is as follows:

$$Employment_{ijt} = \beta_0 + \beta_j AI_{ijt} + \delta_j X_{ijt} + \lambda_{jt} + u_{ij} + \epsilon_{ijt}$$

Where $Employment_{ijt}$represents the employment proportion of banking labor at skill level j in region i during year t**.** The core explanatory variable AI_{ijt} represents the level of AI adoption in the banking and finance services sector in region i during year t. The variable $X_{ijt}$is a series of control variables such as AI investment**,** staff training in AI**,** bank size**,** and regional digital infrastructure. λ_{jt} is the time effect, u_{ij} is the individual effect, and ϵ_{ijt} is the random error. In addition, corner markers j = 1, 2, and 3 denote low-skilled, middle-skilled, and high-skilled labor, respectively, i denotes region, and t denotes year.

5 Results

This section presents empirical findings regarding the relationship between AI adoption and employment structure in Egypt's banking and finance services sector. The results are organized to address each research hypothesis systematically, beginning with overall employment effects and progressing to analyses of regional variation and functional domain differences.

Table 3. Regression Results of AI Impact on Employment Distribution by Skill Level

Variable	High-Skilled (DH)	Middle-Skilled (DM)	Low-Skilled (DL)
AI Index	5.85*** (1.82)	−2.94*** (−3.09)	0.32** (0.74)
Institution Size (C1)	0.0002*** (1.49)	−0.0004*** (−2.06)	0.0002*** (3.46)
Economic Development (C2)	0.077* (0.62)	−0.064** (−0.45)	0.014 (0.44)
Educational Attainment (C3)	−1.537* (−1.06)	−2.579 (−1.54)	1.038 (2.83)
Market Competition (C4)	−4.12*** (−0.42)	1.081*** (0.96)	−6.671** (−2.71)
Individual fixed effects	Yes	Yes	Yes
Time fixed effects	Yes	Yes	Yes
R^2	0.86	0.79	0.81
Observations	100	100	100

*Note: t-statistics in parentheses; *** p < 0.01, ** p < 0.05, * p < 0.1*

The results provide strong support for Hypothesis 1, which predicted that AI adoption would lead to a reduction in middle-skilled employment while increasing both high-skilled and low-skilled employment. Specifically:

1. **High-skilled employment**: The AI Adoption Index shows a positive and statistically significant relationship with high-skilled employment ($\beta = 5.85$, $p < 0.01$). This coefficient indicates that a one-unit increase in the AI Adoption Index is associated with a 5.85 percentage point increase in the proportion of high-skilled workers.

2. **Middle-skilled employment**: The results show a clear pattern: as AI adoption increases, the number of middle-skilled jobs goes down. Specifically, for every one-unit rise in the AI Adoption Index, there's a 2.94 percentage point drop in the share of middle-skilled workers ($\beta = -2.94$, $p < 0.01$).
3. **Low-skilled employment**: The findings also reveal that higher AI adoption is linked to a slight but statistically significant increase in low-skilled jobs ($\beta = 0.32$, $p < 0.05$). Taken together with the earlier results, this paints a clear picture of job polarization in Egypt's banking sector.

5.1 Regional Variation in AI Impact

Table 4 presents regression results examining how the relationship between AI adoption and employment structure varies across Egyptian regions. This analysis tests Hypothesis 2, which predicted more pronounced polarization effects in urban areas compared to rural regions.

Table 4. Regression Results of AI Impact on Employment Structure by Region

Region	High-Skilled (DH)	Middle-Skilled (DM)	Low-Skilled (DL)
Cairo	1.99*** (3.93)	−1.98*** (−4.19)	−0.007 (−0.05)
Alexandria	1.40*** (3.36)	−0.87*** (−2.52)	−0.04* (−0.36)
Giza	1.11*** (3.12)	−1.54*** (−4.47)	0.36** (1.04)
Shubra El-Kheima	1.04*** (3.34)	−1.21*** (−3.72)	0.56*** (2.60)
Port Said	1.53*** (3.95)	−2.11*** (−4.69)	−0.53 (−0.95)
Suez	0.36** (2.29)	−1.39*** (−3.29)	0.025 (0.07)
Luxor	0.99*** (3.12)	−1.09*** (−2.96)	1.09** (2.69)
Aswan	−0.35*** (−2.21)	1.34*** (3.50)	−0.13 (−1.74)
Mansoura	−0.52 (−1.28)	−0.56** (−1.97)	0.91*** (1.87)
Tanta	1.37*** (3.30)	−1.86*** (−4.20)	−0.56 (−0.92)
Individual fixed effects	Yes	Yes	Yes
Hausman test	Fixed effects	Fixed effects	Fixed effects
R^2	0.93	0.89	0.78
Observations	168	168	168

*Note: t-statistics in parentheses; *** $p < 0.01$, ** $p < 0.05$, * $p < 0.1$*

The results reveal substantial regional heterogeneity in how AI adoption affects employment structure, providing nuanced support for Hypothesis 2:

1. **Urban centers** (Cairo, Alexandria, Port Said): These metropolitan regions show the strongest polarization effects. Cairo demonstrates the largest positive effect of AI adoption on high-skilled employment ($\beta = 1.99$, $p < 0.01$) coupled with substantial negative effects on middle-skilled employment ($\beta = -1.98$, $p < 0.01$). Alexandria shows a similar but somewhat attenuated pattern.

2. **Secondary cities** (Giza, Tanta, Suez): These regions exhibit moderate polarization effects, with significant positive coefficients for high-skilled employment and negative coefficients for middle-skilled employment, but with generally smaller magnitudes than observed in primary urban centers.
3. **Rural regions** (Aswan, Luxor, Mansoura): These regions show distinct patterns that diverge from urban polarization trends. Aswan exhibits a negative relationship between AI adoption and high-skilled employment ($\beta = -0.35$, $p < 0.01$) coupled with a positive relationship with middle-skilled employment ($\beta = 1.34$, $p < 0.01$)—a pattern that contradicts typical polarization. Mansoura shows no significant relationship between AI adoption and high-skilled employment. The regional analysis also reveals interesting patterns regarding low-skilled employment. While some regions (Giza, Shubra El-Kheima, Luxor) show significant positive relationships between AI adoption and low-skilled employment, others (Alexandria, Port Said) show negative or non-significant relationships. This heterogeneity suggests that low-skilled employment effects depend on region-specific factors beyond simply urban-rural distinctions. The Hausman test results consistently support fixed effects specifications over ran dom effects models, indicating the presence of region-specific unobserved factors correlated with the explanatory variables. The high R^2 values (ranging from 0.78 to 0.93) indicate that the models explain substantial variation in employment patterns across regions.

5.2 Functional Domain Analysis

Table 5 presents regression results examining how AI adoption affects employment structure across different functional domains within Egypt's banking sector. This analysis tests Hypothesis 3, which predicted more pronounced polarization in functions with higher AI integration potential.

Table 5. Regression Results of AI Impact on Employment Structure by Banking Function

Function	High-Skilled (DH)	Middle-Skilled (DM)	Low-Skilled (DL)
Loan Processing	4.76*** (3.87)	−5.21*** (−4.56)	0.45 (0.79)
Fraud Detection	5.32*** (4.12)	−4.78*** (−3.89)	−0.54 (−0.92)
Credit Scoring	3.89*** (3.45)	−4.11*** (−3.76)	0.22 (0.43)
Customer Service	1.23** (2.21)	−0.98** (−2.33)	0.75** (2.14)
Financial Advisory	2.56*** (2.97)	−1.43** (−2.25)	−0.13 (−0.38)
Transaction Processing	3.12*** (3.32)	−3.54*** (−3.65)	0.42* (1.78)
Individual fixed effects	Yes	Yes	Yes
Time fixed effects	Yes	Yes	Yes
R^2	0.87	0.84	0.76
Observations	240	240	240

*Note: t-statistics in parentheses; *** $p < 0.01$, ** $p < 0.05$, * $p < 0.1$*

The results provide strong support for Hypothesis 3, revealing substantial variation in polarization effects across banking functions:

1. **Functions with high algorithmic potential** (Loan Processing, Fraud Detection, Credit Scoring): These domains exhibit the most pronounced polarization patterns. Loan Processing shows large positive effects on high-skilled employment ($\beta = 4.76$, $p < 0.01$) coupled with substantial negative effects on middle-skilled employment ($\beta = -5.21$, $p < 0.01$). Fraud Detection shows a similar pattern with even larger positive effects on high-skilled employment. These functions involve highly structured data and well-defined decision rules, making them particularly amenable to AI automation.
2. **Customer-facing functions** When it comes to customer service and financial advisory roles, the effects of AI-driven job polarization are more modest. In customer service, the impact on high-skilled jobs is positive but smaller ($\beta = 1.23$, $p < 0.05$), while middle-skilled jobs show a slight decline ($\beta = -0.98$, $p < 0.05$). Financial advisory roles see a bit more pronounced changes, though still not as significant as those in highly automated areas like algorithmic trading.
3. **Operational functions** (Transaction Processing): This domain shows intermediate polarization effects, with substantial impacts on both high-skilled and middle-skilled employment, reflecting moderate AI automation potential for transaction-related tasks.

The analysis also uncovers an interesting trend in low-skilled employment. In customer service, AI adoption is linked to the biggest rise in low-skilled jobs ($\beta = 0.75$, $p < 0.05$). This suggests that while AI handles many routine interactions, it also creates a need for support staff people who help manage the systems and step in when human judgment is needed for unusual or complex cases.

6 Discussion

6.1 AI-Driven Job Polarization

Our results confirm job polarization associated with AI adoption in Egypt's banking sector (Hypothesis 1), with high-skilled employment increasing by 5.85 percentage points and middle-skilled employment decreasing by 2.94 percentage points per unit of AI adoption. This pattern reflects routine-biased technological change theory [2, 3], but with distinctive Egyptian features. Unlike developed economies, Egypt shows only modest growth in low-skilled positions (0.32 percentage points), suggesting that AI primarily creates demand for technical expertise rather than support roles. This concentration of gains in high-skilled categories presents transition challenges for displaced middle-skilled workers, potentially exacerbating socioeconomic disparities without targeted interventions.

6.2 Regional Variation

Our analysis reveals significant regional differences in AI's employment effects (Hypothesis 2). Urban centers like Cairo demonstrate classic polarization (high-skilled: +1.99, middle-skilled: −1.98), driven by superior digital infrastructure, educational institutions, and larger financial firms with greater investment capacity [12]. Surprisingly, some rural areas like Aswan show inverse patterns (high-skilled: −0.35, middle-skilled: + 1.34), possibly reflecting "human-in-the-loop" implementations where AI complements rather than replaces middle-skilled workers. These regional disparities suggest differentiated policy approaches are needed: urban areas should focus on transition support for middle-skilled workers, while rural regions should address constraints on high-skilled employment growth.

6.3 Functional Differences

The analysis strongly supports Hypothesis 3, showing that polarization varies substantially across banking functions. The most pronounced effects occur in domains with structured data and defined rules loan processing (high-skilled: +4.76, middle-skilled: −5.21) and fraud detection (high-skilled: +5.32, middle-skilled: −4.78). This functional heterogeneity aligns with task-based frameworks [2], where automation potential depends on task characteristics rather than broad occupational categories. Functions requiring social intelligence and contextual understanding show more modest disruption. The positive effect on low-skilled employment in customer service (+0.75) suggests automation creates complementary roles in system support and exception handling, offering potential transition paths for some displaced middle-skilled workers with strong interpersonal capabilities.

7 Conclusion

This study examined AI adoption and job polarization in Egypt's banking sector during 2015–2024, revealing three key findings: (1) AI drives job polarization by expanding high-skilled employment (+5.85%), reducing middle-skilled positions (−2.94%), and modestly increasing low-skilled roles (+0.32%); (2) significant regional differences exist, with urban centers experiencing more pronounced polarization than rural areas; and (3) polarization varies by banking function, with greater impact in algorithmic-friendly domains than in areas requiring interpersonal skills.

Our comprehensive analysis of Egyptian banking data spanning 10 cities over 2015–2024 reveals significant implementation risks requiring strategic management. The study identifies critical concerns including potential socioeconomic polarization, with urban centers like Cairo showing pronounced high-skilled employment gains (+1.99 percentage points) while rural areas like Aswan exhibit contrasting patterns (−0.35 percentage points). Function-specific vulnerabilities emerge in highly automated domains, particularly loan processing where middle-skilled employment declined by 5.21 percentage points. The longitudinal fixed-effects analysis strengthens confidence in these findings, though future research should explore instrumental variable approaches to address potential endogeneity concerns. Beyond quantitative employment shifts, the study reveals broader implications for career mobility patterns, wage structure evolution, and social cohesion within Egypt's transforming financial sector. These findings necessitate a multi-faceted response strategy combining phased AI implementation, region-specific workforce development programs, and proactive policy interventions. Stakeholders must coordinate responses: policymakers should establish early warning systems for occupational displacement while expanding digital infrastructure investment; financial institutions should conduct comprehensive impact assessments in vulnerable functions while creating clear reskilling pathways; and educational institutions should develop specialized transition programs targeting at-risk middle-skilled occupations..

References

1. Ibrahim, A., Almasria, N.A., Alhatabat, Z.A., Ershaid, D.J.A., Aldboush, H.H.: Transforming financial services with artificial intelligence and machine learning: insights into opportunities and emerging trends. In: Utilizing AI and Machine Learning in Financial Analysis, pp. 129–148. IGI Global Scientific Publishing (2025). https://doi.org/10.4018/979-8-3693-8507-4.ch008
2. AI-Powered Career Navigation Platform: Predicting Automation Risk and Guiding Professionals to Optimal Occupation Transitions with a Focus on Generation X in Portugal - ProQuest. [Online]. Available: https://www.proquest.com/openview/708bee0aee04a14c1c9705e2d509077b/1?cbl=2026366&diss=y&pq-origsite=gscholar. Accessed 26 Apr 2025
3. Tariq, M.U.: The Role of AI in Skilling, Upskilling, and Reskilling the Workforce. In: Integrating Generative AI in Education to Achieve Sustainable Development Goals, pp. 408–420. IGI Global Scientific Publishing (2024). https://doi.org/10.4018/979-8-3693-2440-0.ch023
4. The Impact of Artificial Intelligence Applications on Financial Services Quality and Financial Performance: Evidence from the Egyptian Bank Sector by Akram Ramadan Ali :: SSRN. [Online]. Available: https://papers.ssrn.com/sol3/papers.cfm?abstract_id=5094438. Accessed 26 Apr 2025
5. Fintech: Digital Transformation in Banking and Financial Services. Custom Software Development Company. [Online]. Available: https://maddevs.io/blog/digital-transformation-in-banking-and-financial-services/. Accessed 14 Mar 2024
6. Vidhya, A.: What Is Principal Component Analysis (PCA)? Analytics Vidhya. [Online]. Available: https://www.analyticsvidhya.com/blog/2016/03/pca-practical-guide-principal-component-analysis-python/. Accessed 17 Dec 2024
7. Marukatat, S.: Tutorial on PCA and approximate PCA and approximate kernel PCA. Artif. Intell. Rev. **56**(6), 5445–5477 (2023). https://doi.org/10.1007/s10462-022-10297-z
8. Zounemat-Kermani, M.: Principal Component Analysis (PCA) for estimating chlorophyll concentration using forward and generalized regression neural networks. Appl. Artif. Intell. **28**(1), 16–29 (2014). https://doi.org/10.1080/08839514.2014.862771
9. Alita, D., Putra, A.D., Darwis, D.: Analysis of classic assumption test and multiple linear regression coefficient test for employee structural office recommendation. Indonesian J. Comput. Cybern. Syst. **15**(3), 295 (2021). https://doi.org/10.22146/ijccs.65586
10. Chang, V., Baudier, P., Zhang, H., Xu, Q., Zhang, J., Arami, M.: How Blockchain can impact financial services – the overview, challenges and recommendations from expert interviewees. Technol. Forecast. Soc. Change. **158**, 120166 (2020). https://doi.org/10.1016/j.techfore.2020.120166
11. AI Powered Credit Scoring and Fraud Detection Models for Financial Technology Applications | Research Briefs on Information and Communication Technology Evolution. [Online]. Available: https://rebicte.org/index.php/rebicte/article/view/203. Accessed 26 Apr 2025.

12. CEEOL - Article Detail. [Online]. Available: https://www.ceeol.com/search/article-detail?id=1060089. Accessed 26 Apr 2025.
13. Central Agency for Public Mobilization and Statistics (CAPMAS), "Annual Bulletin of Employment, Wages, and Working Hours Statistics. Cairo: CAPMAS, 2015–2024. Available at: https://www.capmas.gov.eg/Pages/Publications.aspx?page_id=5104
14. Central Bank of Egypt: Financial Stability Report: Digital Transformation in the Egyptian Banking Sector. Cairo: Central Bank of Egypt (2022). Available at: https://www.cbe.org.eg/en/pages/publications.aspx
15. Information and Decision Support Center (IDSC): Regional Development Indicators in Egypt: Digital Infrastructure and Labor Market Dynamics. Egyptian Cabinet IDSC, Cairo (2021) Available at: https://www.idsc.gov.eg/IDSC/publication/View.aspx

MSIO-DLD Algorithm: Multi-objective Optimization for Health Science Short Video Recommendation

Shuang Geng[1], Nan Yang[1], Yanghui Li[1], Rui Wang[1], and Xusheng Wu[2(✉)]

[1] College of Management, Shenzhen University, Shenzhen, Guangdong 518000, China
gs@szu.edu.cn, wangrui2023@email.szu.edu.cn

[2] Shenzhen Health Development Research and Data Management Center, Shenzhen, Guangdong 518000, China
drniuben@gmail.com, jysgwuxusheng@163.com

Abstract. This paper addresses the multi-objective optimization problem in health science popularization short video recommendation by proposing a Multi-Stage Iterative Optimization Algorithm based on Disease Label Detection Mutation Strategy (MSIO-DLD). The algorithm introduces a multi-stage optimization framework that iteratively optimizes the diversity, professionalism, and accuracy of the recommendation list at each stage, aiming to achieve a comprehensive balance among these three core objectives. During the diversity optimization phase, MSIO-DLD incorporates a mutation strategy based on disease label detection, effectively mitigating the over-concentration of recommended content within a single category. Simultaneously, cosine similarity is employed to control the deviation between the mutated recommendation list and the original one. Experimental results demonstrate that MSIO-DLD outperforms seven representative baseline algorithms across multiple evaluation metrics, particularly excelling in professionalism and accuracy. Furthermore, MSIO-DLD exhibits superior performance in key indicators such as Hypervolume and Inverted Generational Distance, validating its effectiveness and robustness in addressing multi-objective optimization problems. Future research will focus on integrating domain knowledge from the medical field to further refine the recommendation process and explore more innovative strategies to enhance algorithm performance.

Keywords: recommendation system · health science · optimization algorithm · MSIO-DLD

1 Introduction

Recommendation systems (RS) aim to predict users' future interests by mining their historical behavioral data to model latent preferences. As public awareness of healthy living continues to grow, the recommendation of short videos focused on health science popularization has become a novel application area for recommendation systems. This approach seeks to enhance public health literacy by intelligently delivering scientific,

X. Xie et al. (Eds.): ICSS 2025, LNOR, pp. 158–172, 2026.
https://doi.org/10.1007/978-3-032-13116-4_13

authoritative, and easily understandable health information [1]. Conventional recommendation systems generally prioritize the enhancement of recommendation accuracy as their main goal [2, 3]. However, recent studies have indicated that solely improving the accuracy of recommendation lists, although it can enhance the alignment between the recommended content and users' interests to a certain extent, does not significantly improve users' overall satisfaction [4]. Hence, incorporating supplementary evaluation metrics is essential to deliver more varied and tailored recommendations to users.

In contrast to single-objective optimization, multi-objective optimization methods enable the concurrent improvement and coordination of various aspects, including user preferences, content quality, topic diversity, and the professionalism of videos [5]. In response, an increasing number of studies have incorporated multi-objective optimization algorithms to develop more comprehensive video recommendation models. Although multi-objective recommendation systems based on content-based filtering, collaborative filtering, hybrid approaches, and deep learning have been extensively studied and applied across various fields, Substantial obstacles persist in meeting the intricate demands associated with recommending health science short videos [6]. To begin with, the fundamental attributes of health science short videos are rooted in their expert knowledge and credibility, yet traditional methods struggle to directly assess the professional quality of video content, leading to insufficient guarantees regarding the scientific rigor and authority of the recommended videos [7]. Second, variation strategies in traditional multi-objective recommendation algorithms often involve randomly altering recommended videos under a given variation probability [8]. Although this approach may partially improve the diversity of recommended content, it often fails to avoid suggesting videos related to the same disease category as those meant to introduce variety, which may lead to user fatigue due to repetitive exposure. Moreover, traditional multi-objective recommendation algorithms, which perform iterative optimization across multiple targets simultaneously, inevitably suffer from incomplete convergence for some objectives, ultimately compromising the overall quality of the recommendation list [9].

To address the aforementioned challenges, this paper proposes a multi-stage iterative optimization algorithm based on a disease label detection variation strategy (MSIO-DLD). The goal is to comprehensively balance three core objectives—diversity, professionalism and accuracy—in the recommendation of health science popularization short video. The algorithm introduces a variation strategy based on short video disease labels to enhance the diversity of recommended lists along the dimension of disease types, thereby preventing content from being overly concentrated in a single category. Furthermore, by adopting a multi-stage optimization framework that optimizes each objective individually, the method enables more targeted optimization compared to traditional simultaneous multi-objective optimization approaches. Additionally, this paper conducts a systematic evaluation of the proposed framework in real-world recommendation scenarios. The experimental findings indicate that the proposed method outperforms alternatives on various evaluation indicators, achieving an effective balance among multiple objectives and converging toward a representative Pareto-optimal front.

2 Literature Review

2.1 Multi Objective Recommendation System

In traditional recommendation systems, the optimization process typically focuses on a single objective, such as maximizing the accuracy of the recommendation list or improving the ranking quality of recommended items [8]. Nonetheless, emerging research emphasizes that incorporating novelty and diversity into recommendation outcomes is vital for improving the overall effectiveness of recommendation systems [10]. In other words, relying solely on accuracy metrics is insufficient to comprehensively reflect the effectiveness of a Recommendation system, as highly accurate recommendations may not adequately satisfy users' diverse and personalized interest needs. Given the potential intrinsic conflicts among multiple optimization objectives—such as accuracy and diversity—traditional single-objective or weighted-sum approaches often struggle to effectively address such complex optimization tasks [11]. In this context, heuristic optimization techniques, including Genetic Algorithms and Particle Swarm Optimization, have been extensively utilized for multi-objective recommendation challenges owing to their effectiveness in managing nonlinear and high-dimensional problem spaces, and complex search spaces, as well as their ability to effectively balance conflicting objectives [12]. In recent years, researchers have proposed various heuristic-based multi-objective recommendation algorithms, achieving significant progress both theoretically and in practical applications. For instance, reference [13] implemented personalized movie recommendations through content analysis combined with user preference modeling, leveraging a genetic algorithm to simultaneously optimize the accuracy, diversity, and novelty of the recommended items. Another study introduced a multi-objective ant colony optimization algorithm based on an archive strategy to simultaneously account for factors such as attraction popularity, travel time, cost, and tourist preferences, thereby generating precise points-of-interest recommendations and efficient itinerary planning solutions for travelers [14].

2.2 Multi Objective Video Recommendation System

With the rapid advancement of internet technologies, short-form video platforms now serve as significant mediums for accessing information, engaging with entertainment content, and distributing knowledge. However, given the massive volume of short video content, providing users with personalized and high-quality recommendations has emerged as a critical challenge. Reference [15] considered multiple modalities of short videos and leveraged the correlations among these modalities to assess the similarity between videos. They further dynamically adjusted recommendation results based on user click behaviors to reflect users' varying levels of interest in different video segments. One study introduced a dynamic framework for multi-objective optimization, framing the sequential recommendation problem as a dynamic task and addressing it via an interactive evolutionary method [16]. A study developed a decomposition-based system that enables the collaborative recommendation of both popular and long-tail videos [12]. Another study introduced a health science short video recommendation system grounded in multi-objective optimization. Through the development of a project graph search mechanism and the refinement of encoding strategies, heuristic optimization techniques were applied to derive a Pareto-optimal solution set, facilitating a dynamic trade-off among multiple objectives [17]. In summary, multi-objective Recommendation systems show great promise in the short video domain, particularly in the important area of health science short video recommendation. Through continuous optimization of recommendation algorithms and models, it is possible to deliver more personalized and high-quality recommendation services, meet users' needs for health knowledge, and promote the development of health science communication.

3 Preliminaries

3.1 Multi-objective Optimization

Multi-objective optimization challenges typically encompass several competing objectives, which renders it unfeasible to optimize all targets simultaneously [18]. Consequently, the aim shifts to identifying a range of trade-off solutions, collectively termed the Pareto Set (PS), with their associated objective outcomes forming the Pareto Front (PF) [19]. These types of challenges fall under the category of multi-objective optimization problems, which are generally formulated as follows:

$$\max\ F(x) = (f_1(x), \ldots, f_m(x)) \tag{1}$$

where m denotes the number of objectives and x represents the decision vector.

3.2 Objective Function

This research conceptualizes the recommendation of health science short videos as a multi-objective optimization problem, characterized by three core evaluation criteria: diversity, professionalism, and accuracy. In this study, we redefined the calculation method of diversity and professionalism according to the research questions. Diversity is evaluated based on the proportion of different disease tags among the videos in the recommendation list, thereby measuring the richness of the recommended health science content. Professionalism is assessed by leveraging the associated hospital level, physician rank, and the videos' performance metrics (likes, favorites, shares, and views) to evaluate the professional quality of the recommended health science videos. Accuracy is calculated using the average of user ratings for the recommended short videos, aiming to assess the degree of alignment between the recommended videos and users' preferences [20].

Diversity: In particular, this study measures diversity by assessing the distribution ratio of various disease categories included within the recommendation list:

$$\text{Diversity}(L) = \frac{|\ s_L\ |}{|\ z_L\ |} \tag{2}$$

where $z_L$denotes the list of disease labels associated with the videos in the recommendation list L, allowing for repetitions, while s_L represents the list obtained by removing duplicate disease labels from z_L.

Professionalism: For a recommended list of health science popularization short videos of length |L|, this research introduces an integrated evaluation framework aimed at measuring the overall quality and level of professionalism within the recommendation list:

$$\text{Professionalism} = \frac{1}{|\mathrm{L}|}\left(\sum_{\mathrm{i}=1}^{|\mathrm{L}|}\left(\mathrm{q}_{\mathrm{i}}^{\text{hospital}} + \mathrm{q}_{\mathrm{i}}^{\text{doctor}} + \mathrm{q}_{\mathrm{i}}^{\text{video}}\right)\right) \tag{3}$$

where $\mathrm{q}_{\mathrm{i}}^{\text{hospital}}$ denotes the hospital level, $\mathrm{q}_{\mathrm{i}}^{\text{doctor}}$ represents the physician rank, and $\mathrm{q}_{\mathrm{i}}^{\text{video}}$ refers to the video's performance metrics. All these scores have been normalized.

Accuracy: This study evaluates the predictive accuracy of the recommendation system by calculating the mean user ratings of the items presented in the recommendation list. The accuracy metric is formally expressed by the following equation:

$$\text{Accuracy}(\mathrm{L}) = \frac{1}{|\mathrm{L}|}\left(\sum\nolimits_{\mathrm{i}=1}^{|\mathrm{L}|} \mathrm{r}_{\mathrm{ui}}\right) \tag{4}$$

Here, |L| refers to the length of the recommendation list, while r_{ui} denotes the estimated rating of user ufor item i. For items lacking explicit user ratings, this study utilizes the Graph Transformer Recommendation Model (GFormer) to generate predicted values r_{ui}. By combining generative self-supervised learning with a graph transformer structure, GFormer substantially improves both the accuracy and robustness of the prediction process [21].

4 The Proposed Content Search Scheme

4.1 Overall Framework

In this research, each solution represents a recommendation list comprising several items drawn from the item pool, with its quality assessed through fitness scores reflecting diversity, professionalism, and accuracy. The final recommendation outcomes consist of those solutions situated on the optimal Pareto front within the defined search space.

As illustrated in the overall algorithm framework (Fig. 1), the first stage focuses on ensuring the diversity of solutions. By employing a mutation strategy based on disease label detection proposed in this study, We are able to significantly improve the variety within the recommendation list, thus better fulfilling the heterogeneous preferences of users in the domain of health science short video content. Only after the diversity objective has reached convergence does the process proceed to the second stage. In the following stage, the algorithm iteratively optimizes for professionalism to ensure the recommended health science videos are of high professional quality and authority. Upon the convergence of professionalism, the final stage involves iterating on accuracy to guarantee that the overall recommendation list aligns with users' personalized preferences for health science short videos.

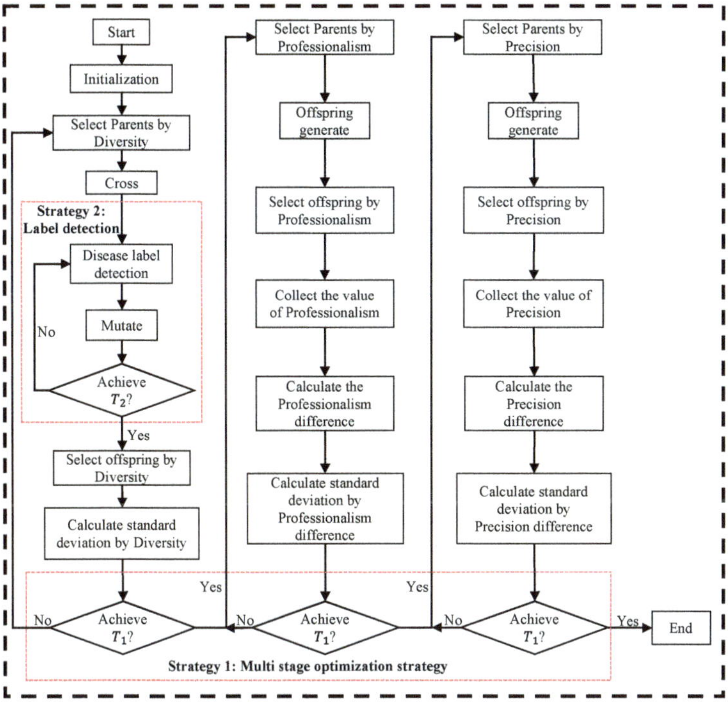

Fig. 1. Overall framework diagram of MSIO-DLD algorithm

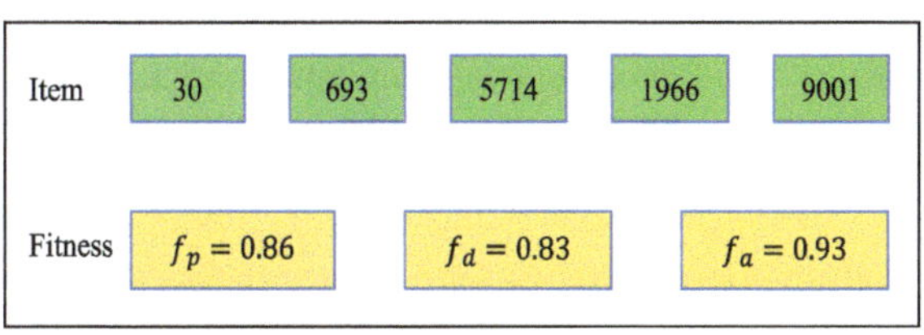

Fig. 2. The coding of individual

4.2 Encoding

In this study, the solution encoding consists of two primary components: the pool of potential items along with the evaluation scores corresponding to each objective. The above figure demonstrates an example of encoding for an individual. Every recommended item is represented by a distinct identifier, and the length of the recommendation list depends on the number of items, which is set based on the particular needs of the recommendation system. The fitness values include diversity evaluation f_d, professionalism evaluation f_p, and accuracy evaluation f_a, which serve to evaluate the diversity, professionalism, and accuracy of the recommendation results, respectively (Fig. 2).

4.3 Multi Stage Optimization Strategy

In recommending health science short videos, the inherent conflicts among multiple objectives create considerable complexity in simultaneously optimizing these goals. To overcome this difficulty, the present study introduces a multi-phase optimization approach, separately addressing diversity, professionalism, and accuracy at different stages

During the initial phase, a mutation strategy informed by disease label detection is implemented to maintain the diversity of the recommendation list. The second stage focuses on iteratively enhancing the professionalism of the recommendation list to guarantee its authority and reliability. In the third stage, the optimization targets the accuracy of the recommendation list to ensure that it aligns with users' preferences. As the optimization proceeds through these sequential stages, the overall population gradually achieves a more desirable distribution, and the state of the population tends toward stability. Consequently, in each stage, the maximum or minimum values of the fitness functions associated with each objective also gradually stabilize. Based on this concept, this study defines a metric L_s to represent the current population's degree of iterative stability:

$$L_s = f_i^{max} - f_i^{min} \tag{5}$$

where $f_i^{\max/\min}$ represents the highest and lowest values of the corresponding objective function in each iteration. Based on this metric, the study first utilizes an archive to store the difference L_s between the highest and lowest values of the objective functions for each generation of the population. Subsequently, the standard deviation of these differences over a selected number of generations g is calculated and compared against a predefined threshold T_1 to determine whether the current population satisfies the stopping condition for the given stage of iteration. Specifically, when the standard deviation T_1 falls below the threshold, it reflects that the population has stabilized in relation to the present objective, at which point the algorithm proceeds to the next optimization stage; otherwise, the optimization continues within the current stage. It should be noted that not all data in the archive are used for this calculation. Based on empirical experience, this study only considers the most recent 30 generations when calculating the standard deviation, with the number of generations g thus set to 30.

4.4 Mutation Strategy Based on Disease Label Detection

In traditional multi-objective recommendation algorithms, mutation strategies typically involve randomly altering recommended items when a certain mutation probability is met. Although this approach can enhance the diversity of the recommendation list to some extent, it inevitably leads to the selection of items of similar types for users. Moreover, random mutations often result in significant deviations from the original recommendation list, thereby substantially reducing recommendation accuracy. To address these issues, this study proposes two targeted improvements. As illustrated in Fig. 3, the first step involves collecting the disease label data for all videos in the recommendation list based on the disease labels associated with each health science short video. The

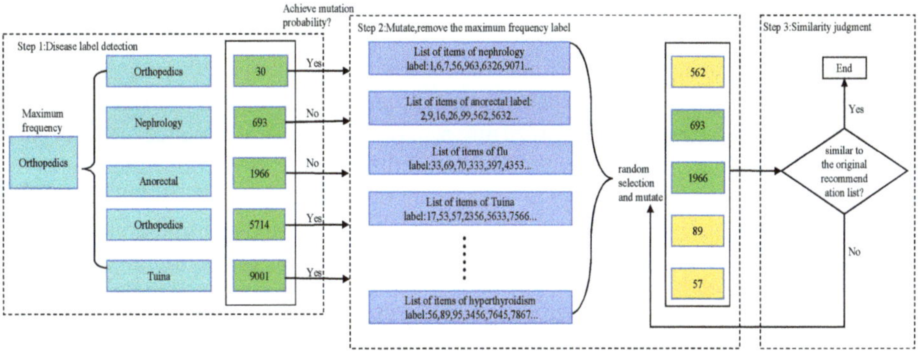

Fig. 3. The mutation strategy based on disease label detection

most frequently repeated disease type within the recommendation list is then identified. Mutation operations are subsequently conducted only among videos excluding the most frequently repeated disease type, thereby avoiding excessive recommendations of videos with identical disease labels.

To further ensure that the mutated recommendation list does not substantially deviate from the original list and evaluate the similarity between the mutated and original recommendation lists, cosine similarity is applied. A mutation is considered successful only if the cosine similarity exceeds a predefined threshold T_2; otherwise, the mutation process is repeated to ensure considerable similarity is retained between the lists.

5 Experiment

5.1 Dataset

To assess the effectiveness of the proposed algorithm, we utilized a dataset provided by a collaborating consultation platform. The dataset comprises 306,573 ratings made by 101,254 users on 11,624 health science short videos. In addition, it includes information on 1,369 disease categories, as well as metadata for each video such as the number of likes, number of favorites, the affiliated hospital level, and the physician rank associated with the video.

5.2 Benchmark

The effectiveness of MSIO-DLD is evaluated by comparing it with seven established baseline algorithms. The first three are traditional algorithms commonly used in multi-objective recommendation tasks, namely NSGA2 [22], SPEA2 [23], and MOEA/D [24]. The remaining four are newly proposed multi-objective recommendation algorithms in recent years, specifically ASDNSGA2 [25], RPD_NSGA2 [26], NSGA2_SDR [27], and MOEA/D-TS [28].

5.3 Parameter Settings

In the experimental setup, the algorithm was run for a maximum of 200 iterations with a population size of 100 individuals. The crossover and mutation probabilities were configured at 0.9 and 0.2, respectively. To ensure reliable average performance metrics, each algorithm was executed independently 30 times. Furthermore, the parameter settings for the baseline algorithms were adopted from their respective reference publications. In the proposed MSIO-DLD algorithm, the threshold used to determine the end of each optimization stage was T_1, set to 5.0E-03, and the threshold for the mutation strategy based on disease label detection was T_2, set to 0.8.

5.4 Results

As shown in Table 1, this study conducts a performance comparison between the proposed MSIO-DLD algorithm and seven baseline algorithms, covering recommendation list lengths ranging from 6 to 20, with a detailed comparative analysis based on three objective function values. To present the performance differences among the algorithms more intuitively, the top-performing results in the table are emphasized using bold font, and the data are further visualized in Figs. 4 and 5.

Experimental evidence demonstrates that the MSIO-DLD algorithm consistently achieves superior objective function values across most recommendation list lengths compared to the other seven algorithms. This phenomenon suggests that the MSIO-DLD algorithm demonstrates significant superiority in the context of health science short video recommendation, particularly in terms of professionalism and accuracy. The root of this advantage lies in the three-stage iterative framework employed by MSIO-DLD, which enables more refined control over the optimization process and, through multiple rounds of iterative refinement, further enhances the algorithm's performance with respect to professionalism and accuracy. Additional comparisons show that algorithms utilizing the fast non-dominated sorting genetic algorithm (NSGA-II) tend to achieve superior performance compared to those that do not incorporate NSGA-II. This result underscores the strong capability of the fast non-dominated sorting method to effectively manage multi-objective recommendation tasks by achieving a more balanced trade-off among conflicting objectives.

However, the MSIO-DLD algorithm shows relatively weaker performance in terms of diversity. Specifically, for certain recommendation list lengths, the diversity objective in the first optimization stage did not fully converge due to a limited number of iterations (only 30 rounds). Nevertheless, in the vast majority of cases, MSIO-DLD still achieves optimal levels of diversity. These results suggest that although there remains room for improvement in some aspects, the overall performance of the MSIO-DLD algorithm remains superior to that of other baseline algorithms in multi-objective optimization tasks.

In addition, this study also collected and calculated the hypervolume (HV) of each algorithm, an important criterion for assessing effectiveness in multi-objective optimization tasks [29]. The final HV performance for each algorithm on the dataset was derived by computing the average HV values across multiple executions. Table 2 summarizes the mean HV values along with their respective standard deviations for all algorithms

Table 1. Comparison of objective values for all algorithms with recommendation list lengths ranging from 6–20

List Length	Evaluation metrics		NSGA2	SPEA2	MOEAD	ASDNSGA2	RPD_NSGA2	NSGA2_SDR	MOEA/D-TS	MSIO-DLD
	Diversity	Mean	7.5645E-01	7.4475E-01	7.3240E-01	8.1315E-01	8.4053E-01	8.3464E-01	8.4307E-01	8.6615E-01
K=6	Professionalism	Mean	7.5434E-01	8.0368E-01	7.3671E-01	7.9950E-01	8.1065E-01	8.2614E-01	8.0576E-01	8.7109E-01
	Accuracy	Mean	5.9277E-01	5.8161E-01	5.7552E-01	6.1634E-01	6.9525E-01	6.3927E-01	6.2111E-01	7.4970E-01
	Diversity	Mean	7.2424E-01	7.2187E-01	7.1599E-01	8.3871E-01	8.3024E-01	8.3358E-01	8.5334E-01	8.4443E-01
K=8	Professionalism	Mean	8.1537E-01	8.1828E-01	8.0644E-01	8.5615E-01	8.6131E-01	8.5629E-01	8.4072E-01	8.9456E-01
	Accuracy	Mean	6.5859E-01	6.2971E-01	6.3195E-01	7.2929E-01	7.1613E-01	7.3538E-01	7.2708E-01	7.9883E-01
	Diversity	Mean	7.9178E-01	7.1964E-01	7.1344E-01	8.2006E-01	8.2184E-01	8.5895E-01	8.3753E-01	8.4528E-01
K=10	Professionalism	Mean	8.0704E-01	8.3709E-01	7.9584E-01	8.2994E-01	8.4918E-01	8.5469E-01	8.4978E-01	8.9145E-01
	Accuracy	Mean	6.6728E-01	6.5407E-01	6.1772E-01	7.3363E-01	7.3109E-01	7.4492E-01	7.2268E-01	8.2952E-01
	Diversity	Mean	7.9202E-01	7.9461E-01	7.7318E-01	8.0574E-01	8.0058E-01	8.2362E-01	8.2973E-01	8.5598E-01
K=12	Professionalism	Mean	8.2201E-01	8.5466E-01	8.3381E-01	8.8997E-01	8.9801E-01	8.9565E-01	8.8066E-01	9.1310E-01
	Accuracy	Mean	6.9001E-01	6.7140E-01	6.6853E-01	7.1894E-01	7.2726E-01	7.3088E-01	7.3678E-01	7.6751E-01
	Diversity	Mean	7.8115E-01	8.0321E-01	7.7244E-01	8.2465E-01	8.1873E-01	8.2370E-01	8.2973E-01	8.5682E-01
K=14	Professionalism	Mean	8.3777E-01	8.4456E-01	8.3889E-01	8.4366E-01	8.5948E-01	8.5628E-01	8.6135E-01	9.1957E-01
	Accuracy	Mean	7.1685E-01	7.2385E-01	6.9370E-01	7.5284E-01	7.2948E-01	7.2954E-01	7.4887E-01	7.8703E-01
	Diversity	Mean	7.8265E-01	7.3596E-01	7.8300E-01	8.1538E-01	8.2094E-01	8.6913E-01	8.2973E-01	8.5122E-01
K=16	Professionalism	Mean	8.0231E-01	8.1873E-01	7.9311E-01	8.3540E-01	8.4792E-01	8.5132E-01	8.4135E-01	8.9189E-01
	Accuracy	Mean	6.8443E-01	7.2329E-01	6.5113E-01	7.2337E-01	7.4090E-01	7.3099E-01	7.6887E-01	8.0581E-01
	Diversity	Mean	8.0621E-01	7.7138E-01	7.9703E-01	8.2829E-01	8.2240E-01	8.3152E-01	8.3085E-01	8.5961E-01
K=18	Professionalism	Mean	8.2721E-01	7.6752E-01	8.0953E-01	8.4414E-01	8.5323E-01	8.6577E-01	8.6338E-01	9.0116E-01
	Accuracy	Mean	6.4790E-01	6.3748E-01	6.0246E-01	7.2975E-01	7.3207E-01	7.1808E-01	7.2098E-01	7.5658E-01
	Diversity	Mean	7.5644E-01	7.1085E-01	7.4620E-01	8.2553E-01	8.4020E-01	8.3965E-01	8.5441E-01	8.4431E-01
K=20	Professionalism	Mean	8.2175E-01	7.7550E-01	8.0706E-01	8.4708E-01	8.5502E-01	8.4018E-01	8.5954E-01	8.9184E-01
	Accuracy	Mean	6.5236E-01	6.8752E-01	6.3073E-01	7.2308E-01	7.1095E-01	7.0636E-01	7.2098E-01	7.6932E-01

Fig. 4. Comparison of objective values for algorithms with recommended lists of 6–12

evaluated on the collaborative consultation platform dataset. The standard deviation provides insight into the stability and consistency of algorithm performance, where a lower standard deviation indicates greater stability across multiple runs, while a higher standard deviation may suggest fluctuations in performance.

As illustrated in Table 2, the proposed MSIO-DLD algorithm consistently attains the highest HV values across the majority of cases, demonstrating its ability to identify solutions nearer to the true Pareto front while also achieving wider coverage in the objective space, thus more effectively balancing multiple objectives. Moreover, the relatively low standard deviation demonstrates that MSIO-DLD can effectively optimize multiple objectives while maintaining high performance stability, further highlighting its

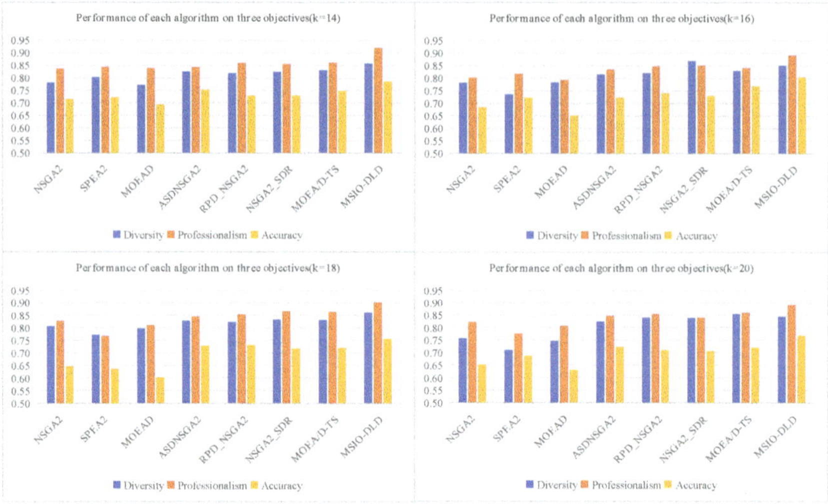

Fig. 5. Comparison of objective values for algorithms with recommended lists of 14–20

significant advantages over other algorithms in addressing multi-objective optimization challenges.

Table 2. HV values of all algorithms in the recommendation list with a length of 6–20

List Length	Evaluation metrics	NSGA2	SPEA2	MOEAD	ASDNSGA2	RPD_NSGA2	NSGA2_SDR	MOEAD-TS	MSIO-DLD
K=6	HV_Mean	6.3321E-01	6.7564E-01	1.0872E-01	6.1824E-01	8.4455E-01	8.8170E-01	6.1027E-01	9.6780E-01
	HV_Std	1.2004E-01	1.0872E-01	5.8535E-02	1.2313E-01	1.0202E-01	7.3889E-02	1.0313E-01	1.2136E-02
K=8	HV_Mean	6.6191E-01	6.5580E-01	8.7479E-01	5.8212E-01	8.4544E-01	8.6115E-01	6.2013E-01	9.5732E-01
	HV_Std	9.0608E-02	1.0704E-01	5.3048E-02	9.8448E-02	1.0449E-01	8.9321E-02	1.0616E-01	1.2776E-02
K=10	HV_Mean	6.9683E-01	6.5299E-01	8.6798E-01	6.0520E-01	8.8598E-01	9.0046E-01	6.8880E-01	9.6685E-01
	HV_Std	7.5985E-02	9.5636E-02	4.9992E-02	5.7188E-02	5.6958E-02	5.4734E-02	1.1456E-01	1.3641E-02
K=12	HV_Mean	6.4542E-01	6.9380E-01	8.1295E-01	5.8656E-01	8.3357E-01	8.3731E-01	6.8602E-01	9.6393E-01
	HV_Std	7.3200E-02	9.0298E-02	9.2507E-02	5.5303E-02	8.4234E-02	7.3679E-02	9.6977E-02	1.6283E-02
K=14	HV_Mean	6.3588E-01	6.4681E-01	7.5491E-01	5.7583E-01	8.4416E-01	8.9861E-01	6.8853E-01	8.8215E-01
	HV_Std	7.5633E-02	7.4414E-02	8.8076E-02	7.3102E-02	3.9059E-02	6.3548E-02	9.5876E-02	1.5015E-02
K=16	HV_Mean	5.7866E-01	6.1319E-01	8.0215E-01	5.4457E-01	8.5348E-01	8.4755E-01	6.8753E-01	8.9391E-01
	HV_Std	6.6469E-02	6.8582E-02	6.7539E-02	8.3948E-02	1.1169E-02	3.1287E-02	9.6033E-02	1.2845E-02
K=18	HV_Mean	5.8590E-01	6.0449E-01	7.6653E-01	5.3420E-01	8.3691E-01	8.3365E-01	6.8060E-01	8.8298E-01
	HV_Std	6.2321E-02	7.3537E-02	9.6045E-02	6.0515E-02	6.5819E-02	6.2658E-02	1.0086E-01	8.9606E-03
K=20	HV_Mean	5.3481E-01	5.7208E-01	7.1442E-01	5.2754E-01	8.2523E-01	8.1674E-01	6.8010E-01	8.9698E-01
	HV_Std	6.4420E-02	6.2209E-02	9.0670E-02	6.4470E-02	7.4433E-02	5.8028E-02	9.9727E-02	8.6361E-03

Table 3 summarizes the mean values and standard deviations of the inverted generational distance (IGD) metric for all tested algorithms. IGD serves as a comprehensive measure that evaluates the discrepancy between the approximated Pareto front generated by an algorithm and the true Pareto front, considering both convergence and diversity aspects [30]. Specifically, IGD calculates the average Euclidean distance from each point on the true Pareto front to its nearest counterpart on the approximated front. Smaller IGD values indicate better algorithmic performance, signifying that the solution set is closer to the true Pareto front and more uniformly spread. According to the results in Table 3, the MSIO-DLD algorithm achieves the lowest IGD values in most cases, demonstrating its effectiveness in balancing convergence and diversity. This further validates the proposed MSIO-DLD method's proficiency in multi-objective optimization, particularly in closely approximating the true Pareto front while maintaining an even distribution of solutions.

Table 3. IGD values of all algorithms in the recommendation list with a length of 6–20

List Length	Evaluation metrics	NSGA2	SPEA2	MOEAD	ASDNSGA2	RPD_NSGA2	NSGA2_SDR	MOEAD-TS	MSIO-DLD
K=6	IGD_Mean	2.6633E-01	2.2620E-01	1.0412E-01	2.7552E-01	1.2050E-01	1.0226E-01	2.7911E-01	6.6607E-02
	IGD_Std	1.1901E-01	9.7561E-02	3.7917E-02	1.2727E-01	7.8737E-02	4.9543E-02	1.0757E-01	1.7430E-03
K=8	IGD_Mean	2.4884E-01	2.4488E-01	2.3213E-01	2.6173E-01	2.3767E-01	2.3512E-01	2.5292E-01	5.4513E-02
	IGD_Std	2.3927E-02	1.2223E-02	9.6917E-03	3.2965E-02	1.3881E-02	1.2433E-02	2.4649E-02	2.7456E-03
K=10	IGD_Mean	2.1249E-01	2.5848E-01	9.1552E-02	2.9759E-01	6.7791E-02	6.9169E-02	2.2606E-01	5.1102E-02
	IGD_Std	7.8693E-02	1.0365E-01	4.1511E-02	6.3356E-02	4.9607E-02	4.1827E-02	1.2134E-01	2.6985E-03
K=12	IGD_Mean	2.5008E-01	2.2751E-01	1.4363E-01	3.0552E-01	1.2400E-01	1.2627E-01	2.2778E-01	5.3538E-02
	IGD_Std	7.2775E-02	8.7114E-02	7.2265E-02	6.1087E-02	6.8121E-02	6.0475E-02	9.7069E-02	3.6985E-02
K=14	IGD_Mean	2.4084E-01	2.3217E-01	1.7305E-01	2.8772E-01	1.1264E-01	1.2499E-01	2.0672E-01	7.4271E-02
	IGD_Std	6.8001E-02	7.1746E-02	7.0350E-02	7.6344E-02	1.9870E-02	4.1300E-02	8.8488E-02	2.9909E-02
K=16	IGD_Mean	2.9896E-01	2.7281E-01	1.5928E-01	3.2701E-01	1.2700E-01	1.3477E-01	2.2891E-01	7.2756E-02
	IGD_Std	6.1206E-02	5.8367E-02	4.2180E-02	7.9286E-02	1.9175E-02	1.6967E-02	7.7606E-02	3.9194E-02
K=18	IGD_Mean	2.0866E-01	2.0798E-01	2.0875E-01	2.2246E-01	2.0946E-01	2.0976E-01	2.1093E-01	2.1752E-01
	IGD_Std	2.8315E-02	2.8933E-02	1.8597E-02	3.6503E-02	1.7271E-02	1.8968E-02	3.1591E-02	4.2523E-03
K=20	IGD_Mean	3.4369E-01	3.1617E-01	2.2382E-01	3.5645E-01	1.4307E-01	1.4872E-01	2.4330E-01	6.6589E-02
	IGD_Std	5.7777E-02	5.3731E-02	7.0946E-02	5.7487E-02	5.4764E-02	4.0849E-02	8.3892E-02	3.3259E-03

To further validate the performance of the MSIO-DLD algorithm, this study employed the sign test method to conduct a comparative analysis between the MSIO-DLD algorithm and other baseline algorithms, focusing on the top 25 solutions in the solution space. The sign test is a non-parametric statistical method used to assess the superiority of algorithms across multiple independent experiments [31]. At a significance level of 0.05, the MSIO-DLD algorithm must outperform the corresponding solutions of the compared algorithms on at least 18 out of 25 cases to demonstrate statistical significance. This criterion is based on the principles of hypothesis testing, ensuring that the results possess a considerable degree of certainty.

The sign test results are presented in Table 4, where "win" denotes the number of solutions where the MSIO-DLD algorithm outperforms the corresponding solutions of the compared algorithm, and "lose" indicates the number of solutions where it underperforms. Wins equal to or greater than 18 have been highlighted in bold in the table. The statistical results show that, in most cases, the number of wins for the MSIO-DLD algorithm is significantly greater than the number of losses, demonstrating that its performance significantly surpasses that of the baseline algorithms. Even in instances where the threshold of 18 wins was not reached, the number of wins consistently exceeded 15, further corroborating the outstanding performance of the MSIO-DLD algorithm in the task of health science short video recommendation.

Table 4. The sign text of MSIO-DLD

	K=6						K=8						K=10						K=12					
	Diversity		Professionalism		Accuracy		Diversity		Professionalism		Accuracy		Diversity		Professionalism		Accuracy		Diversity		Professionalism		Accuracy	
	Wins(+)	Loses(-)	Wins(+)	Loses(-)	Wins(+)	Loses(-)	Wins(+)	Loses(-)	Wins(+)	Loses(-)	Wins(+)	Loses(-)	Wins(+)	Loses(-)	Wins(+)	Loses(-)	Wins(+)	Loses(-)	Wins(+)	Loses(-)	Wins(+)	Loses(-)	Wins(+)	Loses(-)
NSGA2	21	4	20	5	20	5	21	4	22	3	20	5	20	5	23	2	21	4	21	4	21	4	20	5
SPEA2	22	3	20	5	21	4	22	3	22	3	21	4	22	3	20	5	20	5	20	5	20	5	21	4
MOEA/D	24	1	24	1	20	5	23	2	21	4	23	2	20	5	23	2	21	4	21	4	22	3	23	2
ASDNSGA2	20	5	19	6	19	6	18	7	19	6	19	6	19	6	19	6	18	7	20	5	19	6	20	5
RPD_NSGA2	23	2	22	3	20	5	17	8	19	6	16	9	18	7	21	4	20	5	17	8	20	5	21	4
NSGA2_SDR	17	8	20	5	22	3	17	8	22	3	19	6	16	9	20	5	22	3	20	5	20	5	21	4
MOEA/D_TS	20	5	19	6	21	4	15	10	19	6	20	5	20	5	20	5	22	3	18	7	19	6	20	5
	K=14						K=16						K=18						K=20					
	Diversity		Professionalism		Accuracy		Diversity		Professionalism		Accuracy		Diversity		Professionalism		Accuracy		Diversity		Professionalism		Accuracy	
	Wins(+)	Loses(-)	Wins(+)	Loses(-)	Wins(+)	Loses(-)	Wins(+)	Loses(-)	Wins(+)	Loses(-)	Wins(+)	Loses(-)	Wins(+)	Loses(-)	Wins(+)	Loses(-)	Wins(+)	Loses(-)	Wins(+)	Loses(-)	Wins(+)	Loses(-)	Wins(+)	Loses(-)
NSGA2	19	6	23	2	20	5	22	3	21	4	23	2	22	3	21	4	21	4	20	5	19	6	20	5
SPEA2	22	3	21	4	21	4	20	5	23	2	22	3	20	5	22	3	21	4	23	2	22	3	20	5
MOEA/D	19	6	23	2	23	2	20	5	20	5	21	4	23	2	21	4	22	3	22	3	21	4	21	4
ASDNSGA2	17	8	16	9	16	9	21	4	20	5	20	5	21	4	19	6	22	3	22	3	21	4	21	4
RPD_NSGA2	17	8	19	6	20	5	20	5	20	5	21	4	16	9	20	5	19	6	17	8	20	5	19	6
NSGA2_SDR	17	8	19	6	19	6	17	8	21	4	22	3	16	9	17	8	19	6	17	8	19	6	18	7
MOEA/D_TS	19	6	20	5	22	3	19	6	22	3	22	3	18	7	19	6	20	5	15	10	18	7	19	6

6 Conclusion

This research presents a multi-stage iterative optimization algorithm (MSIO-DLD) that incorporates a disease label detection variation strategy to tackle the multi-objective optimization challenge within health science short video recommendations. Through a multi-phase optimization framework, the algorithm sequentially enhances the diversity, professionalism, and accuracy of the recommendation list, striving to effectively balance these three key objectives and produce high-quality recommendations. Additionally, the incorporation of a disease label-based mutation strategy during the diversity optimization stage further enhances the diversity of the recommendation list. In the experimental section, the MSIO-DLD algorithm was comprehensively evaluated using a dataset provided by a collaborative consultation platform and was benchmarked against seven representative baseline algorithms. The results demonstrate that MSIO-DLD consistently exhibits superior performance across multiple evaluation metrics, particularly excelling in professionalism and accuracy.

Finally, although this study focuses on short video recommendation within the medical domain, it relies solely on user-item interactions for recommendation and does not incorporate domain-specific medical knowledge. In future work, we plan to integrate relevant medical knowledge to guide the optimization process and explore more innovative optimization strategies. These initiatives seek to improve the performance of the MSIO-DLD algorithm further and offer more robust support for advancing health science popularization short video recommendation systems.

References

1. Tian, Y., Liu, S., Zhang, D.: Clearness qualitative comparative analysis of the spread of TikTok health science knowledge popularization accounts. Digit. Health. **9**, 20552076231219116 (2023)
2. Joorabloo, N., Jalili, M., Ren, Y.: Improved recommender systems by denoising ratings in highly sparse datasets through individual rating confidence. Inf. Sci. **601**, 242–254 (2022)
3. Pan, Y., Huo, Y., Tang, J., Zeng, Y., Chen, B.: Exploiting relational tag expansion for dynamic user profile in a tag-aware ranking recommender system. Inf. Sci. **545**, 448–464 (2021)

4. Tian, Y., et al.: Evolutionary large-scale multi-objective optimization: a survey. ACM Computing Surveys (CSUR). **54**(8), 1–34 (2021)
5. Zhang, L., Liu, S., Yang, H., Chen, Z., Zhao, H.: MOREM: an evolutionary multitasking optimization algorithm for multi-objective recommendations. Inf. Sci. **678**, 120921 (2024)
6. Zaizi, F.E., Qassimi, S., Rakrak, S.: Multi-objective optimization with recommender systems: a systematic review. Inf. Syst. **117**, 102233 (2023)
7. Tan, Y., Geng, S., Chen, L., Wu, L.: How doctor image features engage health science short video viewers? Investigating the age and gender bias. Ind. Manag. Data Syst. **123**(9), 2319–2348 (2023)
8. Zheng, Y., Wang, D.X.: A survey of recommender systems with multi-objective optimization. Neurocomputing. **474**, 141–153 (2022)
9. Zuo, Y., Gong, M., Zeng, J., Ma, L., Jiao, L.: Personalized recommendation based on evolutionary multi-objective optimization. IEEE Comput. Intell. Mag. **10**(1), 52–62 (2015)
10. Castells, P., Hurley, N., Vargas, S.: Novelty and diversity in recommender systems. In: Recommender Systems Handbook, pp. 603–646. Springer, New York (2021)
11. Roijers, D.M., Vamplew, P., Whiteson, S., Dazeley, R.: A survey of multi-objective sequential decision-making. J. Artif. Intell. Res. **48**, 67–113 (2013)
12. Wang, S., Gong, M., Li, H., Yang, J.: Multi-objective optimization for long tail recommendation. Knowl.-Based Syst. **104**, 145–155 (2016)
13. Almeida, M.S., Britto, A.: MOEA-RS: a content-based recommendation system supported by a multi-objective evolutionary algorithm. In: Artificial Intelligence and Soft Computing: 19th International Conference, pp. 265–276. Springer (2020)
14. Zhang, J., Li, Y.: Application of improved multi-objective evolutionary algorithm in intelligent tourism interest point recommendation and itinerary planning. Informatica. **49**(7) (2025)
15. Mei, T., Yang, B., Hua, X.S., Li, S.: Contextual video recommendation by multimodal relevance and user feedback. ACM Trans. Inf. Syst. **29**(2), 1–24 (2011)
16. Zhou, W., et al.: Dynamic multi-objective optimization framework with interactive evolution for sequential recommendation. IEEE Trans. Emerging Top. Comput. Intell. **7**(4), 1228–1241 (2023)
17. Geng, S., Fu, C., Wang, R., Qian, Y., Li, Y.: Multi-criteria health science short video recommendation inspired by heuristic optimization. In: Wuhan International Conference on E-Business, pp. 372–381. Springer (2024)
18. Deb, K., Sindhya, K., Hakanen, J.: Multi-objective optimization. In: Decision Sciences, pp. 161–200. CRC Press (2016)
19. Burachik, R.S., Kaya, C.Y., Rizvi, M.M.: Algorithms for generating pareto fronts of multi-objective integer and mixed-integer programming problems. Eng. Optim. **54**(8), 1413–1425 (2022)
20. Geng, S., He, X., Liang, G., Niu, B., Liu, S., He, Y.: Accuracy-diversity optimization in personalized recommender system via trajectory reinforcement based bacterial colony optimization. Inf. Process. Manag. **60**(2), 103205 (2023)
21. Li, C., Xia, L., Ren, X., Ye, Y., Xu, Y., Huang, C.: Graph transformer for recommendation. In: Proceedings of the 46th International ACM SIGIR Conference on Research and Development in Information Retrieval, pp. 1680–1689 (2023)
22. Deb, K., Agrawal, S., Pratap, A., Meyarivan, T.: a fast elitist non-dominated sorting genetic algorithm for multi-objective optimization: NSGA-II. In: International Conference on Parallel Problem Solving from Nature, pp. 849–858. Springer, Berlin/Heidelberg (2000)
23. Zitzler, E., Laumanns, M., Thiele, L.: SPEA2: improving the strength Pareto evolutionary algorithm. TIK Rep. **103** (2001)
24. Zhang, Q., Li, H.: MOEA/D: a multiobjective evolutionary algorithm based on decomposition. IEEE Trans. Evol. Comput. **11**(6), 712–731 (2007)

25. Deng, W., Zhang, X., Zhou, Y., Liu, Y., Zhou, X., Chen, H., Zhao, H.: An enhanced fast non-dominated solution sorting genetic algorithm for multi-objective problems. Inf. Sci. **585**, 441–453 (2022)
26. Elarbi, M., Bechikh, S., Gupta, A., Said, L.B., Ong, Y.S.: A new decomposition-based NSGA-II for many-objective optimization. IEEE Trans. Syst. Man Cybern. Syst. **48**(7), 1191–1210 (2017)
27. Tian, Y., Cheng, R., Zhang, X., Su, Y., Jin, Y.: A strengthened dominance relation considering convergence and diversity for evolutionary many-objective optimization. IEEE Trans. Evol. Comput. **23**(2), 331–345 (2018)
28. Cao, J., Zhang, J., Zhao, F., Chen, Z.: A two-stage evolutionary strategy based MOEA/D to multi-objective problems. Expert Syst. Appl. **185**, 115654 (2021)
29. Shang, K., Ishibuchi, H., He, L., Pang, L.M.: A survey on the hypervolume indicator in evolutionary multiobjective optimization. IEEE Trans. Evol. Comput. **25**(1), 1–20 (2020)
30. Santos, T., Xavier, S.: A convergence indicator for multi-objective optimisation algorithms. TEMA (São Carlos). **19**, 437–448 (2018)
31. Derrac, J., García, S., Molina, D., Herrera, F.: A practical tutorial on the use of nonparametric statistical tests as a methodology for comparing evolutionary and swarm intelligence algorithms. Swarm ad Evol. Comput. **1**(1), 3–18 (2011)

Health Literacy in Online Health Platforms: A Markov Chain Analysis of User Behavioral Transitions

Shuang Geng[1], Huimei Tao[1], Jingyu Chen[1], Ben Niu[1], and Xusheng Wu[2](✉)

[1] College of Management, Shenzhen University, Shenzhen 518000, China
gs@szu.edu.cn

[2] Shenzhen Health Development Research and Data Management Center, Shenzhen, China
jysgwuxusheng@163.com

Abstract. User engagement behavior and health check purchasing decisions are crucial for e-health service platforms (e-HSPs) to enhance service design and improve profitability. However, the relationship between user engagement behaviors and the factors motivating users to make health check purchasing decisions remains unclear. In this study, we employ a Markov chain modeling approach to understand the underlying mechanisms of user behaviors and their transition states on an e-health service platform. We tracked 9,903 users with 38,603 behavioral activities who had experience purchasing health-check products on an e-health service platform offering various health services. We analyzed the dynamics of user behaviors by modeling the transition probabilities across different participation states. Our analysis identifies three distinct behavioral transition patterns: one-way transitions, cyclical transitions, and self-cyclical transitions. Additionally, we identified three e-health literacy development paths: health knowledge learning, self-health management, and support exchange. Among these paths, the support exchange path has the highest transition probability leading to health check purchasing decisions. Users who actively search for health information are more likely to make health check purchases than those who passively receive health information. Our study offers valuable practical implications for the design and management of online health service platforms.

Keywords: E-health literacy · Health check-up purchasing · User engagement behavior · Markov chain analysis

1 Introduction

E-health service platforms (e-HSPs) play a pivotal role in providing an environment for medical users to access health information [1]. Information systems researchers have increasingly focused on examining the relationship between users' e-health literacy and health-related behaviors within e-health service platforms (e-HSPs). These behaviors encompass disease management [2, 3], medication adherence [4], disease prevention [5, 6], and other activities aimed at maintaining health [7]. For patients, seeking and

X. Xie et al. (Eds.): ICSS 2025, LNOR, pp. 173–185, 2026.
https://doi.org/10.1007/978-3-032-13116-4_14

using information about medical issues, self-care, and disease prevention enhances their understanding of risk factors and preventive strategies [5, 8]. Health literacy pertains to an individual's capacity to actively seek and utilize health information [9]. A definition provided by the U.S. Institute of Medicine for health literacy is "the ability of individuals to access, process, and comprehend fundamental health information or services to make appropriate health-related decisions" [10, 11]. With the proliferation of online healthcare services, e-health literacy emerges as an important patient informational skill, referring to "the capacity to seek, find, comprehend, and appraise health information from electronic sources and subsequently apply the knowledge gained to addressing or tackling health problems" [12].

Health checks, a key practice in health promotion, prevention, and management, allow users to identify health issues, prevent diseases, and enhance health awareness [13, 14]. Especially for the past few years, the advent of COVID-19 has spurred a surge in demand for preventive medical services, particularly health check products. As of 2021, the number of health checks in China stands at approximately 488 million, with a market size reaching 189 billion CNY, which indicates a gradual transition in people's health consciousness from passive medical treatment to proactive health management. Several studies found that eHealth literacy is closely related to people's engagement in health examinations and preventive health behaviors [14–16]. Most of these studies use self-reported questionnaires to evaluate individuals' eHealth literacy levels [15, 16]. However, the users' eHealth literacy-related behaviors on e-health service platforms and their connections with users' health check purchasing decisions remain unexplored. Therefore, we focus on the transition process between users' eHealth literacy-related behaviors on the service platforms, as well as their influence on health check product purchasing.

User engagement behaviors within e-HSPs include behaviors such as health knowledge-seeking [2, 4], social support exchange [2, 17], health information consumption and production [18], as well as doctor-patient communication [4, 17]. Active user involvement in health-related actions holds promise in bolstering health literacy, attitudes, and motivation for disease screening [19]. This can positively impact health outcomes and enhance the quality of care experienced [2, 5, 17]. Building on these findings, we further argue that users' health-related decisions on platforms are not solely shaped by isolated behaviors but are also influenced by a series of ongoing engagement patterns. This suggests that the synergistic effects of multiple user engagement behaviors may contribute to users' decisions to purchase health check products. Therefore, we adopt a Markov chain model-based approach to analyze the dynamic changes of user behaviors via modeling the transition probabilities between distinct participation states, thus understanding the underlying mechanisms of user behaviors and their collective impact on health check product purchasing. Overall, we aim to contribute to the eHealth literacy research by focusing on the following research questions:

1. What are the users' e-health literacy related behavioral patterns in e-HSPs?
2. How do different e-health literacy-related behavioral patterns relate to users' health check purchase decisions?

To address these research questions, we first propose a conceptual e-health behavior pathway model elucidating the relations between users' e-health literacy-related behaviors, including stages of health needs cognition, health information participation and evaluation, and health decision making. Drawing upon this model, we analyze the user behavioral data by applying Markov chains to compute patient participation state transition probabilities. Our findings suggest that there are three behavioral transition patterns: one-way transitions, cycled transitions, and self-cycled transitions. Additionally, we identified three e-health literacy development paths: health knowledge learning, self-health management, and support exchange. Among these paths, the support exchange path has the highest transition probability leading to health check purchasing decisions. Users who actively search for health information are more likely to make health check purchases than those who passively receive health information. The insights from this study can be helpful for platform managers to improve the service provision design to enhance user engagement and achieve platform business success.

2 E-Health Literacy

Norman and Skinner's seminal conceptualization characterizes e-health literacy as "the ability to seek, find, understand, and evaluate health information from electronic sources and to apply the knowledge gained to address health problems" [12]. A few studies have been working on e-health literacy measurement due to the rapid development of online health service platforms. According to the Digital Health Literacy Tool developed by Van Der Vaart et al. (2017), they measure e-health literacy in terms of skills related to health 1.0 and health 2.0 aspects of personal search and use and interaction with online health information [20]. After that, the conceptual framework of e-health literacy proposed by Liu et al. (2021) considers the skills related to health 3.0, suggesting that individuals can enhance their e-health literacy by creating their health data and using that data to communicate, respond, and share [18]. So far, existing literature can be categorized into four major domains: theoretical or conceptual frameworks (defining the dynamic attributes and social embeddedness of health literacy), development and validation of measurement tools (enabling cross-cultural assessment and real-time tracking), analysis of dynamic processes and mechanisms (revealing mediating pathways and moderating effects of health literacy), and intervention strategies and systemic optimization (designing multi-level solutions to address structural barriers). These four research streams are closely related. For example, theoretical models guide the development of measurement tools, and dynamic mechanism analyses provide targets for precision interventions. Notably, most of these studies rely on cross-sectional data and questionnaires for data analysis, leaving a gap in understanding the dynamic evolution of health literacy and the interrelationships between different health literacy-related behaviors.

2.1 E-Health Literacy and Preventive Health Behaviors

Preventive health behaviors, such as disease screening and health checks, are effective tools for individuals to promptly identify potential disease risk factors and health issues, which also equip healthcare providers with effective information for diagnosis and treatment strategies [13, 14, 26]. Prior studies have demonstrated that individual health literacy positively influences the adoption of preventive health behaviors. Unlike users with limited e-health literacy, those with high levels of e-health literacy demonstrate adeptness in searching for, engaging with, and utilizing to learn about health-related information and knowledge such as disease symptoms, healthy lifestyles, and prevention methods [21]. People are more inclined to practice preventive health behaviors when they have higher health self-perception [6, 7, 13, 15, 16, 22]. Individuals who were more concerned about their health and regularly engagement in proactive health information acquisition usually exhibited better e-health literacy and had an active preventive health behavioral pattern [23]. Empirically, Kim et al. (2023) 's meta-analysis points to the pivotal role of e-health literacy as a critical mediating factor within the pathway whereby health-related information prompts health behavioral modifications such as disease prevention and management [7].

2.2 E-Health Behavior Pathway Model

Drawing on the consumer decision-making process model and health literacy path model [24, 25], we propose a e-Health Behavior Pathway model (eHBP, Fig. 1) presuming that that users (including patients and all other individuals interested in health issues) shall go through a series of stages before health product consumption decision-making, beginning with health needs cognition, to health information participation and evaluation, and at last making health decisions. The eHBP model comprises three user behavioral stages and accounts for the unique health product consumption decision-making process of medical users, who may transit from one stage to the next, skipping certain stages, or transit between different behavioral states within the same stage [30]. The stages definitions are built on Kasl and Cobb's (1966) seminal definitions: preventive health behaviors refer to "activities undertaken by individuals perceiving themselves as healthy to prevent disease or detect it in an asymptomatic stage," while illness behaviors are "activities initiated by individuals feeling ill to define their health status or seek remedies." Over online healthcare platforms, preventive behaviors can be subdivided into two types: 1) Health awareness-building behaviors: including knowledge acquisition, patient community interactions, and health information sharing [17, 18]; 2) Disease prevention actions: such as health screenings to detect asymptomatic conditions [14, 26]. Users' engagement in the cognition and participation stages (e.g., browsing, liking, posting) reflects the gradual accumulation of health awareness and knowledge, which inherently signifies the development of e-health literacy. Preventive consumption decisions (e.g., purchasing screenings) at the health decision-making stage are perceived as the outcome of e-health literacy development. Meanwhile, Illness behaviors on e-HSPs primarily manifest as online consultations, where users describe symptoms to physicians and request diagnoses [4, 17]. These behaviors correspond to treatment-related consumption decisions (e.g., purchasing consultation services) in the decision-making stage.

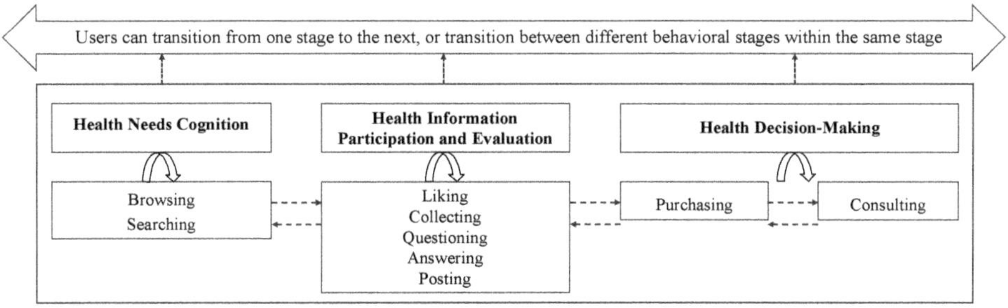

Fig. 1. E-Health Behavior Pathway model (eHBP)

3 Research Design

3.1 Data Collection

Our data were sourced from an e-health service platform offering integrated services, including appointment booking, health consultations, prescriptions, and health checks, with over three million monthly active users and 720,000 registered physicians. Anonymous and deidentified data were collected under user privacy agreements. Inclusion criteria covered users who have health-check purchasing behavior between January 1, 2022, and May 31, 2023. The exclusion criteria removed users who purchased health checks due to employment requirements or driver's license applications. Behavioral data (e.g., reading, liking, commenting, Q&A) during this period were retrieved. After excluding invalid/duplicate entries, 38,603 valid behavioral records from 9,903 users were obtained.

3.2 Behavioral State

The behavioral states of the users in this study were defined according to their behavior records, with a total of 14 behavioral states, detailed in Table 1.

Table 1. Detailed Information on User Behavioral Status

Behavioral states	Description
Browse article	Read a health science promotion article.
Browse video	View a health science promotion video.
Like article	Like a health science promotion article.
Like video	Like a health science promotion video.
Collect article	Collect a health science promotion article.
Collect video	Collect a health science promotion video.
Article comment	Comment health science promotion articles.
Video comment	Comment for health science promotion videos.

(*continued*)

Table 1. *(continued)*

Behavioral states	Description
Post note	Post health notes, record health status, treatment experience, set health goals, etc.
Note comment	Comment for health notes.
Question	Ask questions to obtain information in patient Q&A communities.
Answer	Answering questions to provide information in patient Q&A communities.
Consult	Communicate with healthcare providers to seek health advice.
Health check purchasing	Make purchasing decisions for health check products.

3.3 Data Analysis

We utilize a first-order Markov chain to model the sequential behaviors of users. Markov chains are a widely adopted tool for analyzing user trajectories, including users' continuous engagement within health information platforms [27, 28]. In this study, the developed model comprises a discrete state space S and a transition matrix P. The discrete state space S is mathematically defined as: $S = \{S_1, S_2, \ldots, S_i, \ldots, S_n\}$, consisting of n distinct behavioral states and n = 14. S_i indicates that the user is in the ith behavioral state, where $S_i \in$ {Browse Article, Browse Video, Like Article, Like Video, Collect Article, Collect Video, Article Comment, Video Comment, Post Note, Note Comment, Question, Answer, Consult, Health Check Purchase}

The transition matrix P is mathematically defined as Eq. (1), computed using formula (2):

$$P = \begin{bmatrix} p_{11} & p_{12} & \cdots & p_{1j} \\ p_{21} & p_{22} & \cdots & p_{2j} \\ \vdots & \vdots & \ddots & \vdots \\ p_{i1} & p_{i2} & \cdots & p_{ij} \end{bmatrix} \tag{1}$$

$$p_{ij} = \frac{n_{ij}}{\sum_{k=1}^{n} n_{ik}} (i = 1, \ldots, n; j = 1, \ldots, n) \tag{2}$$

where n_{ij} denotes the transition frequency from state S_i to state S_j, S_i denotes the transition probability of a user moving from state S_i to state S_j, and the transition probability is estimated by the transition frequency n_{ij} between behavioral states. To ensure the Markov property of random variable sequence, we constructed a $X^2\chi^2$ statistic testing [29]. Given a significance level of $\alpha = 0.05$, a random sequence that follows a distribution with $(m-1)^2(m-1)^2$ degrees of freedom exhibits Markov properties. The result $\chi^2 = 25726.478$ is greater than the critical value $\chi_\alpha^2((m-1)^2) = \chi_\alpha^2(169)$, indicating that the user's state transition sequence possesses Markov properties.

Then, the constructed first-order Markov chain is used to analyze the user behavior transition matrix. This method identifies transition patterns with high probability, summarizes user e-health literacy development paths, and explores its association with health check purchasing decisions. We use a heat map to visualize the behavior transition matrix (P) and reveal user behavior transition rules. To ensure the readability of the results, the critical values representing frequent transition probabilities are set as follows: (1) In the same column of the transition matrix, analyze the grids with transition probabilities>0.15. (2) On the main diagonal of the transition matrix, analyze the grids with a transition probability>0.4. These thresholds were chosen because the majority of transition probabilities fall below these values, making probabilities above 0.15 (for column-wise transitions) or 0.4 (for diagonal self-transitions) particularly indicative of significant and representative behavioral transitions. Using these criteria, we combine the transition law and draw the main path of user behavior participation. The analyses are conducted using R (version 4.3) and Python (version 3.0).

4 Results

4.1 User Behavior Transition Pattern

We created a heat map to demonstrate the Markov chain transition matrix in Fig. 2. This matrix analyzes 28,700 behavioral transitions involved in the participation behaviors of 9,903 users. The numbers in the grids of the heatmap indicate the probability of transitioning from one behavioral state to another, while the color signifies the magnitude of these probabilities. The transition probabilities in each row of the heatmap sum to 1. As shown in Fig. 2, users who engage in Q&A community questioning have the highest probability (0.462) to transition to health check purchasing. These users ask about disease symptoms, medication details, disease-related diagnosis and treatment processes, and appropriate health check products for their health conditions. Users who answer the questions in Q&A communities have the second-highest transition probability (0.370), followed by those participating in online consultations (0.296), reading articles (0.290), and posting notes (0.248). The results suggest that users with active engagement behaviors are more likely to buy health check products.

In addition, Fig. 2 shows that there are three behavioral patterns. The transition between the same behavior is named as self-transition (S_iS_i). Behaviors with a high probability of self-transition include health check purchase (0.626), consultation (0.618), browsing video (0.6), browsing article (0.463), and note comment (0.495). The transition between different behaviors occur can be divided into two types: one-way transition ($S_i \rightarrow S_j$) and cycled transitions (S_iS_j). The probability of a one-way transition from one behavioral state to another is high for the following pairs: questioning (0.167) or answering (0.222) → consultation, article comments → browsing articles (0.222), article comments → liking articles (0.194), collecting articles → liking articles (0.160), video comment → browsing video (0.179) and liking video (0.179). The probability of cycled transitions between the same behavioral states is high for the following behaviors: consultation ↔ health check purchase (0.296, 0.219), posting note ↔ commenting on notes (0.212, 0.347), collecting video ↔ liking video (0.187, 0.392), liking video ↔ liking article (0.164, 0.265).

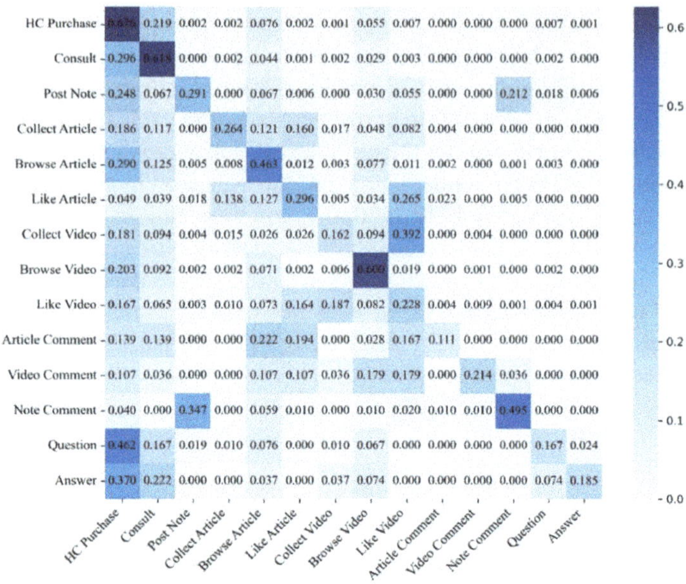

Fig. 2. The engagement behavior transition Matrix of health check users

4.2 User E-Health Literacy Behavioral Pathways

We identify three types of e-health literacy behavioral pathways with a high probability of occurrence and distinct behavioral characteristics.

We name the first path as a health knowledge learning path. In this path, users learn and interact with popular health science promotion articles and videos to develop their e-health literacy. These users are likely to transit between reading articles states (0.463) and between watching videos states (0.6), forming a continuous learning process. A high probability for cycled transition was found for liking videos $\leftrightarrow$ collecting videos (0.187, 0.392), liking articles $\leftrightarrow$ collecting articles (0.138, 0.160), and liking videos $\leftrightarrow$ liking articles (0.164, 0.265), suggesting higher co-occurrence for "liking" and "collecting" behavior. Users' engagement in interesting videos and articles further reinforces their engagement in similar content. In addition, the high probability of one-way transition for article comment $\rightarrow$ browsing articles (0.222), video comment $\rightarrow$ browsing videos (0.179), and video comment $\rightarrow$ liking videos (0.179) suggests that after acquiring knowledge, users may shift from passive receivers of information to active information seekers.

We name the second path as the self-management path, which is characterized by users who predominantly post health notes to record their health conditions and health-related goals. Users with behavior on this path also interact with related notes by posting comments to disseminate health information. Notably, there is a high probability (0.495) for self-cycled transition within users' posting note states. Maintaining a health profile by regularly writing notes enables users to track their health issues effectively. Additionally, there was a high probability for cycled transition (0.212, 0.347) between the posting note state and note comment state. This transition occurs when users are willing to exchange health information by communicating with other users after receiving messages. The probability for the transition from posting health notes to health check purchasing is 0.248, indicating that the note posting probably reflects the patients' concern about their health conditions.

We refer to the third pathway as the 'providing and obtaining support' pathway. The transition from Q&A community responses to health check purchasing occurs with a probability of 0.370, while the transition from Q&A community questions to health check purchasing has a probability of 0.462. These patients discuss topics relevant to disease prevention, diagnostic procedures, and medical checkup programs in the Q&A community, indicating that they have certain interested health topics to discuss with others. Results also show that the probability of users' transitioning from answering in Q&A communities to online consulting is 0.222, and the probability for transitioning from asking questions to online consulting is 0.167. This may imply that users may fail to find satisfactory answers for their queries in the Q&A communities, and therefore, they turn to online consulting to seek professional advice and guidance.

Furthermore, it is found that among the three behavior paths, the providing and obtaining support path has a higher probability for transitioning to health check purchasing. In the health product purchasing decision-making stage, there is a high probability of self-transition (0.626) for health check purchases. Online consulting also has a high probability of self-transition (0.618), which may occur when users consult different doctors several times over a period of time or engage in continuous counselling for specific diseases. There exists cycled transitions between online consulting and the purchase of health checks (probability of 0.296 and 0.219), suggesting that users tend to consult doctors before making a health check purchasing decision and may seek further guidance after the check-ups.

5 Discussion

5.1 Principal Findings

This study constructs a e-Health Behavior Pathway model (eHBP) comprising of three user behavior stages and accounts for the unique health product consumption decision-making process of medical users, who may transit from one stage to the next, skipping certain stage, or transit between different behavioral states within the same stage [30]. Drawing on the proposed eHBP model, a Markov chain analysis is conducted to explore associations between e-health literacy behavior pathways and health check purchasing. Our analysis reveals three behavioral transition patterns (self-cycled transition, one-way transition, and cycled transition) and three literacy development pathways (knowledge acquisition, self-management, and support exchange). Results suggest that users' information-seeking behavior (e.g., article browsing) and information-exchange behavior (e.g., community interactions) have higher transition probability to health check purchasing behavior. This partially confirms the prior findings that passive users ("lurkers") exhibit higher activity frequency but shorter information retention and limited social support [31, 32], leading to decision-making distrust [33]. In contrast, active users leverage peer feedback to enhance health self-efficacy [34], forming an "information-literacy-decision" virtuous cycle. Our results also suggest that the support exchange pathway demonstrates higher transition probability than other paths ($p < 0.01$), highlighting the peer support's role in decision-making [35].

5.2 Further Discussion on Behavioral Patterns

Beyond the principal findings, a deeper exploration into the identified behavioral patterns reveals nuanced insights into user engagement on e-health platforms. The prevalence of self-transition behaviors, particularly for activities like health check purchasing (0.626) and consultation (0.618), suggests a strong habit formation or ongoing need for these services. For platform managers, this highlights the importance of fostering positive initial experiences to encourage sustained engagement. For instance, simplifying the health check purchasing process or providing consistent, high-quality consultation services can reinforce these self-transitions.

The one-way transitions, such as questioning or answering leading to consultation, indicate a clear progression in user needs. Users often begin with information-seeking behaviors in Q&A communities and, if their needs are not fully met, transition to more direct, professional consultation. This underscores the critical role of Q&A forums as a preliminary support mechanism. Platforms should ensure these forums are well-moderated and provide accurate, timely information to potentially address user queries at an earlier stage, or seamlessly guide them to consultation services when necessary. This also suggests that users who actively engage in information exchange are more likely to seek professional advice, indicating a higher level of health literacy and proactive health management.

The cyclical transitions, particularly between consultation and health check purchasing, reveal an iterative decision-making process. Users may consult doctors before purchasing health checks and then seek further guidance after receiving results. This highlights the interconnectedness of different health behaviors and the need for integrated service pathways on e-health platforms. Designing seamless transitions between consultation, health check scheduling, and post-check-up follow-ups can significantly enhance user experience and adherence to health recommendations. This also supports the idea that health decisions are often not singular events but rather a series of interconnected actions influenced by evolving information and needs [36].

Furthermore, the observation that the support exchange path has the highest transition probability to health check purchasing emphasizes the power of peer influence and social learning in health decision-making. Users who actively engage in Q&A communities, both asking and answering, are more likely to proceed with health checks. This is consistent with social cognitive theory, which posits that individuals learn from observing others and through social interactions [37]. Platforms can leverage this by promoting active community engagement to foster a supportive environment that encourages health-promoting behaviors. The trust built within these communities can significantly impact health-related decisions, as users often value peer experiences and recommendations [38].

5.3 Practical Implications

This study provides a framework for modeling users' health literacy development and preventive and therapeutic health decision-making. For patients, the behavioral pathway of e-health literacy reveals a close connection between their health decisions and health information engagement behaviors, highlighting that actively engaging with health information is crucial for participating in healthcare and making informed health choices. At the patient level, implementing health education and promotion programs can guide the development of targeted health literacy interventions. By recognizing the different stages of health literacy activities, platform managers and healthcare policymakers can adopt knowledge-focused strategies to support patients' understanding of their health behaviors. For platform managers, they can design diverse channels for health information access and evaluation, and encourage user participation in health information sharing and interaction through incentive mechanisms to enhance health literacy. Furthermore, the credibility of the e-health service platform itself plays a crucial role in shaping user behavior. A platform with high service credibility, characterized by accurate information, professional and responsive services, and transparent user feedback mechanisms, is more likely to foster user trust. This trust, in turn, can significantly influence users' willingness to engage in various health-related behaviors, from seeking information to purchasing health checks.

5.4 Limitations and Future Directions

Despite the contributions of this study, there are some limitations. First, our sampling excluded non-platform users, limiting the findings applicable to online health platforms. Second, the simplicity of first-order Markov chains highlights the need for higher-order models to comprehensively decode behavioral pathways. Third, data access constraints omitted critical variables like education level, which may offer novel insights. We hope this study can energize more in-depth studies on eHealth literacy.

Acknowledgments. Disclosure Statement No potential conflict of interest was reported by the author(s).

Data Availability Statement The data that support the findings of this study are not available due to a confidentiality agreement.

Additional Information This research was funded by the National Natural Science Foundation of China (72471154), General Project of Ministry of Education Foundation on Humanities and Social Sciences (24YJC630049), and Guangdong Philosophy and Social Science Planning Project (GD24YTS01).

References

1. Jiang, X., Xie, H., Tang, R., Du, Y., Li, T.: Characteristics of online health care services from China's largest online medical platform: cross-sectional survey study. J. Med. Internet Res. **23**(4), e25817 (2021)

2. Zhou, J., Wang, C.: Improving cancer survivors' e-health literacy via online health communities (OHCs): a social support perspective. J. Cancer Surviv. **14**, 244–252 (2020)
3. Tandon, A., Dhir, A., Islam, N.: Mobile health interventions for cancer care and support: the next level of digitalization in healthcare? IEEE Trans. Eng. Manag. **71**, 6173–6189 (2023)
4. Lu, X., Zhang, R.: Association between eHealth literacy in online health communities and patient adherence: cross-sectional questionnaire study. J. Med. Internet Res. **23**(9), e14908 (2021)
5. Lee, H.Y., Jin, S.W., Henning-Smith, C., Lee, J., Lee, J.: Role of health literacy in health-related information-seeking behavior online: cross-sectional study. J. Med. Internet Res. **23**(1), e14088 (2021)
6. Choi, H.S., Lee, J.E.: Factors affecting COVID-19 preventive behaviors of young adults based on eHealth literacy and the health belief model: a cross-sectional study. Inquiry. **60**, 00469580231159487 (2023)
7. Kim, K., Shin, S., Kim, S., Lee, E.: The relation between eHealth literacy and health-related behaviors: systematic review and meta-analysis. J. Med. Internet Res. **25**, e40778 (2023)
8. Thapa, D.K., Visentin, D.C., Kornhaber, R., West, S., Cleary, M.: The influence of online health information on health decisions: a systematic review. Patient Educ. Couns. **104**(4), 770–784 (2021)
9. Champlin, S., Mackert, M., Glowacki, E.M., Donovan, E.E.: Toward a better understanding of patient health literacy: a focus on the skills patients need to find health information. Qual. Health Res. **27**(8), 1160–1176 (2017)
10. Kindig, D.A., Panzer, A.M., Nielsen-Bohlman, L. (eds.): Health Literacy: a Prescription to End Confusion. National Academies Press (2004)
11. Berkman, N.D., Davis, T.C., McCormack, L.: Health literacy: what is it? J. Health Commun. **15**(S2), 9–19 (2010)
12. Norman, C.D., Skinner, H.A.: eHealth literacy: essential skills for consumer health in a networked world. J. Med. Internet Res. **8**(2), e506 (2006)
13. Labeit, A., Peinemann, F., Baker, R.: Utilisation of preventative health check-ups in the UK: findings from individual-level repeated cross-sectional data from 1992 to 2008. BMJ Open. **3**(12), e003387 (2013)
14. Morimoto, A., et al: Relationship between communicative and critical health literacy, health information sources, and participation in health checkups among middle-aged Japanese community residents. Prev. Med. **161**, 107112 (2022)
15. Mitsutake, S., Shibata, A., Ishii, K., Oka, K.: Association of eHealth literacy with colorectal cancer knowledge and screening practice among internet users in Japan. J. Med. Internet Res. **14**(6), e153 (2012)
16. Bao, X., Chen, D., Shi, L., Xia, Y., Shi, Z.: The relationship between COVID-19-related prevention cognition and healthy lifestyle behaviors among university students: mediated by e-health literacy and self-efficacy. J. Affect. Disord. **309**, 236–241 (2022)
17. Chen, S., Guo, X., Wu, T., Ju, X.: Exploring the influence of doctor--patient social ties and knowledge ties on patient selection. Internet Res. **32**(1), 219–240 (2022)
18. Liu, H.X., Chow, B.C., Liang, W., Hassel, H., Huang, Y.W.: Measuring a broad spectrum of eHealth skills in the web 3.0 context using an eHealth literacy scale: development and validation study. J. Med. Internet Res. **23**(9), e31627 (2021)
19. Jin, S.W., Lee, Y., Dia, D.A.: Analyzing paths from online health information seeking to colorectal cancer screening using health literacy skills frame and cognitive mediation model. Patient Educ. Couns. **102**(3), 416–423 (2019)
20. Van Der Vaart, R., Drossaert, C.: Development of the digital health literacy instrument: measuring a broad spectrum of health 1.0 and health 2.0 skills. J. Med. Internet Res. **19**(1), e27 (2017)

21. Mitsutake, S., Shibata, A., Ishii, K., Oka, K.: Associations of eHealth literacy with health behavior among adult internet users. J. Med. Internet Res. **18**(7), e192 (2016)
22. Neter, E., Brainin, E.: Association between health literacy, e-health literacy, and health outcomes among patients with long-term conditions. Eur. Psychol. **24**(1) (2019)
23. Yang, S.C., Luo, Y.F., Chiang, C.H.: The associations among individual factors, eHealth literacy, and health-promoting lifestyles among college students. J. Med. Internet Res. **19**(1), e15 (2017)
24. Engel, J.F., Kollat, D.T., Roger, D.: Consumer Behavior. Blackwell (1973)
25. Edwards, M., Wood, F., Davies, M., Edwards, A.: The development of health literacy in patients with a long-term health condition: the health literacy pathway model. BMC Public Health. **12**, 1–15 (2012)
26. Hills, O., Shah, D.: Online health information seeking, medical care beliefs and timeliness of medical check-ups among African Americans. Patient Educ. Couns. **103**(12), 2468–2476 (2020)
27. Lim, S., Yim, D., Khuntia, J., Tanniru, M.: A continuous-time Markov chain model-based business analytics approach for estimating patient transition states in online health Infomediary. Decis. Sci. **51**(1), 181–208 (2020)
28. Wu, D., Huyan, X., She, Y., Hu, J., Duan, H., Deng, N.: Exploring and characterizing patient multi-behavior engagement trails and patient behavior preference patterns in pathway-based m-health hypertension self-management: analysis of use data. JMIR Mhealth Uhealth. **10**(2), e33189 (2022)
29. Zhang, Y.F., Zhang, Q.F., Yu, R.H.: Markov property of Markov chains and its test. In: 2010 International Conference on Machine Learning and Cybernetics, vol. 4, pp. 1864–1867. IEEE (2010)
30. Karimi, S., Holland, C.P., Papamichail, K.N.: The impact of consumer archetypes on online purchase decision-making processes and outcomes: a behavioural process perspective. J. Bus. Res. **91**, 71–82 (2018)
31. James, T.L., Calderon, E.D.V., Bélanger, F., Lowry, P.B.: The mediating role of group dynamics in sharing received social support from active and passive use in online health communities. Inf. Manag. **59**(3), 103606 (2022)
32. Mo, P.K., Coulson, N.S.: Empowering processes in online support groups among people living with HIV/AIDS: a comparative analysis of 'lurkers' and 'posters'. Comput. Human Behav. **26**(5), 1183–1193 (2010)
33. Zhang, S., Bantum, E.O.C., Owen, J., Bakken, S., Elhadad, N.: Online cancer communities as informatics intervention for social support: conceptualization, characterization, and impact. J. Am. Med. Inform. Assoc. **24**(2), 451–459 (2017)
34. Deng, Z., Liu, S.: Understanding consumer health information-seeking behavior from the perspective of the risk perception attitude framework and social support in mobile social media websites. Int. J. Med. Inform. **105**, 98–109 (2017)
35. Rupert, D.J., et al: Peer-generated health information: the role of online communities in patient and caregiver health decisions. J. Health Commun. **21**(11), 1187–1197 (2016)
36. Prochaska, J.O., DiClemente, C.C.: Stages and processes of self-change of smoking: toward an integrative model of change. J. Consult. Clin. Psychol. **51**(3), 390–395 (1983)
37. Bandura, A.: Social Foundations of Thought and Action: A Social Cognitive View. Prentice-Hall, Inc. (1986)
38. Fan, H., Lederman, R.: Online health communities: how do community members build the trust required to adopt information and form close relationships? Eur. J. Inf. Syst. **27**(1), 62–89 (2018)

Clustering Analysis of Cecal Microbiota Dynamics in *Eimeria Maxima*-Infected Chickens

Sina Aghakhani[1], Mohammad Fili[1(✉)], Guiping Hu[1], Guolong Zhang[2], and Lizhi Wang[1]

[1] School of Industrial Engineering and Management, Oklahoma State University, Stillwater, OK 74078, USA
{sina.aghakhani,mohammad.fili,guiping.hu,lizhi.wang}@okstate.edu
[2] Department of Animal and Food Sciences, Oklahoma State University, Stillwater, OK 74078, USA
glenn.zhang@okstate.edu

Abstract. Understanding how intestinal microbiota responds to Eimeria maxima infection is vital for advancing microbiome-based strategies against coccidiosis. This study analyzed temporal changes in the cecal microbial community of chickens infected with E. maxima, utilizing hierarchical clustering based on cosine distance and variance ratio criterion (VRC) scores. The cecal digesta samples were collected from infected and mock-infected broiler chickens at 3, 5, 7, 14, and 21 days post-infection. After filtering amplicon sequence variants (ASVs) and normalizing abundances, optimal clustering structures were determined. Results indicated a distinct clustering pattern between infected and mock-infected groups, highlighting bacterial groups associated with infection stages. This study provides a computational perspective on the dynamic restructuring of the intestinal microbiota community following coccidiosis and emphasizes the potential for microbiome-based intervention strategies.

Keywords: Coccidiosis · Microbiota dynamics · Hierarchical clustering · Eimeria maxima · Subcommunity detection

1 Introduction

Coccidiosis, a protozoan disease caused by multiple species of the genus *Eimeria*, is one of the most economically devastating illnesses affecting the global poultry industry, with estimated annual losses exceeding $13 billion [1]. *Eimeria maxima*, among the seven recognized *Eimeria* species infecting chickens, is a primary agent of this disease, characterized by the disruption of small intestinal epithelial cells, leading to impaired nutrient absorption, growth retardation, and vulnerability to secondary infections such as necrotic enteritis [1–3]. A recent dose-response study [4] demonstrated that higher Eimeria maxima challenge levels magnify both microbiota disturbance and necrotic-enteritis risk, underscoring the need to intervene at the peak of dysbiosis. The life cycle of *E. maxima* (EM) involves both asexual and sexual reproductive stages, occurring in

X. Xie et al. (Eds.): ICSS 2025, LNOR, pp. 186–196, 2026.
https://doi.org/10.1007/978-3-032-13116-4_15

the epithelial cells of the jejunum and ileum over 6–7 days post-infection. Complementary work [5] reported that microbial shifts in the ileum lag roughly two days behind those in the jejunum, highlighting the importance of sampling multiple gut regions. During the acute phase (5–7 days post-infection), rapid intracellular replication leads to extensive mucosal damage and inflammation, which manifests as diarrhea, intestinal lesions, and weight loss [1, 6, 7]. Upon completion of the cycle, the host may begin to recover, although oocyst shedding and low-grade inflammation may persist, potentially leading to reinfection under certain housing conditions [1, 7]. The gastrointestinal microbiota, composed predominantly of bacteria, plays essential roles in maintaining intestinal homeostasis, regulating immune responses, fermenting undigested feed components, and synthesizing short-chain fatty acids (SCFAs) [1, 8, 9]. In healthy chickens, lactic acid bacteria such as *Lactobacillus, Ligilactobacillus*, and *Limosilactobacillus* dominate the upper gastrointestinal tract, while SCFA-producing anaerobes such as *Faecalibacterium* and *Romboutsia* prevail in the cecum [1, 8, 9]. Recent studies have demonstrated that *Eimeria*-induced intestinal damage is closely associated with alterations in the gut microbiota. Disruption of this balance by EM infection leads to dysbiosis, characterized by the suppression of beneficial anaerobes and the enrichment of aerotolerant or facultative anaerobes such as *Escherichia, Staphylococcus*, and *Enterococcus* [1]. These changes are believed to stem from the pro-inflammatory intestinal environment during the acute phase, which favors the expansion of bacteria capable of surviving oxidative stress while inhibiting obligate anaerobes that contribute to SCFA production [1].

Liu et al. provided a comprehensive study of the microbiota response to EM infection, capturing dynamic changes in both the ileum and cecum over a 14-day period. They observed minimal changes at 3 dpi, followed by significant dysbiosis during the acute phase (5–7 days post-infection), and partial recovery by 14 dpi [1]. However, while this work quantified diversity indices and genus-level shifts, there is a need for higher-resolution analyses to characterize the temporal restructuring of specific ASVs and to understand how community composition fluctuates across infection stages.

This study builds upon the foundational insights of Liu et al. by employing unsupervised hierarchical clustering and distance-based validation to analyze microbial patterns in healthy versus EM-infection chickens. We aim to quantify the structural complexity of the cecal microbiota during EM infection and reveal trajectories that may inform therapeutic strategies for mitigating coccidiosis.

The rest of this paper is organized as follows: In Sect. 2, we describe the experimental design, data preprocessing, and clustering methodology employed to analyze the temporal restructuring of the cecal microbiota. Section 3 presents the clustering results and the comparative co-clustering analysis between EM-infected and Mock groups. Finally, Sect. 4 concludes the study by summarizing the key findings and outlining directions for future research.

2 Method

In this section, we describe the methodology used to analyze the temporal dynamics of the cecal microbiota in both EM-infected (EM) and mock-infected (Mock) broiler chickens. The analysis involved data preprocessing, hierarchical clustering of ASVs based on temporal abundance patterns, and optimal cluster selection using the variance ratio criterion (VRC) to compare microbial restructuring between the two groups.

2.1 Experimental Design and Data Sources

In this study, we utilized 16S rRNA gene sequencing data collected from cecal samples of broiler chickens subjected to experimental EM infection [10]. The chickens were divided into two groups: an EM group and a mock-infected control group (Mock). Sample collection occurred at 3, 5, 7, 14, and 21 days post-infection (DPI), covering the early infection stage, the peak of infection and intestinal damage, and the recovery phase. This longitudinal design allowed for a comprehensive analysis of the temporal dynamics of microbiota restructuring. The input datasets consisted of: (i) feature tables with ASV counts for cecum and ileum samples, and (ii) a metadata file detailing batch, treatment, and dpi for each sample.

2.2 Data Preprocessing

To ensure high-quality downstream analyses, several preprocessing steps were applied to the raw ASV count data:

- *Sample Selection*: Samples were filtered based on treatment type. Specifically, for each treatment group (i.e., EM and Mock), the corresponding cecal samples were retained.
- *Normalization*: Since sequencing depths can vary widely across samples, raw ASV counts were normalized to relative abundances. Each sample's ASV counts were divided by the total counts in that sample and then multiplied by 100, resulting in a matrix of abundance proportions.
- *Abundance Aggregation*: Following normalization and filtering, data were aggregated using sample median: For each ASV, and at each DPI point, the median abundance across all replicates was calculated. Median aggregation reduces the impact of outliers and sample-specific anomalies, providing a more robust representation of microbiota dynamics at each time point.
- *Selection of Common ASVs*: To enable consistent inter-group comparisons, we identified and retained only those ASVs that were shared across treatment groups. Specifically, 146 ASVs were common across both EM and Mock groups. By focusing on these 146 shared ASVs, we captured approximately 98.37% of the total microbial abundance in the EM group and 99.39% in the Mock group. This strategy ensured that our clustering and downstream analyses were based on a representative subset of the microbiota, preserving the major biological signals.

2.3 Clustering Analysis Framework

To investigate temporal microbiota restructuring, hierarchical clustering was performed on the processed datasets. Cosine distance was chosen as the dissimilarity metric between ASVs. Unlike Euclidean distance, cosine distance is robust to magnitude differences and focuses on the similarity (relative proportion) between abundance trajectories. This property is desirable for microbiome data where absolute abundances may vary but relative compositional trends are biologically meaningful.

Pairwise cosine-based distances were computed between ASVs using Eq. (1):

$$d(x, y) = 1 - \frac{\sum_i x_i y_i}{\sqrt{\sum_i x_i{}^2}\sqrt{\sum_i y_i{}^2}} \tag{1}$$

where, x and y represent two ASV abundance vectors across time points.

We used Ward's minimum variance method as the linkage strategy for hierarchical clustering. This method minimizes the total within-cluster variance, leading to compact and well-separated clusters. Specifically, at each step, the algorithm merges the pair of clusters that results in the smallest increase in the total within-cluster variance. The resulting linkage matrix captures the hierarchical relationships between ASVs and forms the basis for dendrogram visualization and cluster assignment.

2.4 Optimal Clusters Choice

Identifying the optimal number of clusters (k) is critical for meaningful interpretation. We evaluated clustering quality across a range of cluster numbers (k = 2 to 140) using the VRC metric shown in **Eq. (2)**:

$$VRC = \frac{Between\ cluster\ dispersion/(k-1)}{Within\ cluster\ dispersion/(n-k)} \tag{2}$$

where n is the number of ASVs.

Higher VRC scores indicate better-defined and more separated clusters. The optimal k was chosen either as the value corresponding to the maximum VRC score or as the largest k within 10% of the maximum VRC value (epsilon-band strategy).

3 Results

In this section, we present the results of clustering and comparative co-clustering analyses performed on the cecal microbiome data from EM and Mock groups. First, we determined the optimal number of microbial clusters for each group using the VRC and examined the structural patterns of ASV temporal abundance within these clusters. Then, we compared the clustering outcomes between groups to assess the extent to which infection with EM influenced the stability and reorganization of microbial associations over time.

3.1 Clustering Results

We analyzed the VRC score across a wide range of cluster sizes ($k = 2$ to 140) for both EM-infected and Mock samples (see Fig. 1). The optimal number of clusters for the EM group was found to be $k = 119$, where the VRC score peaked sharply, reaching a value slightly above 5.0. This peak was preceded by a steep upward trend starting at $k \approx 100$, indicating the emergence of distinct substructures in the EM microbiota.

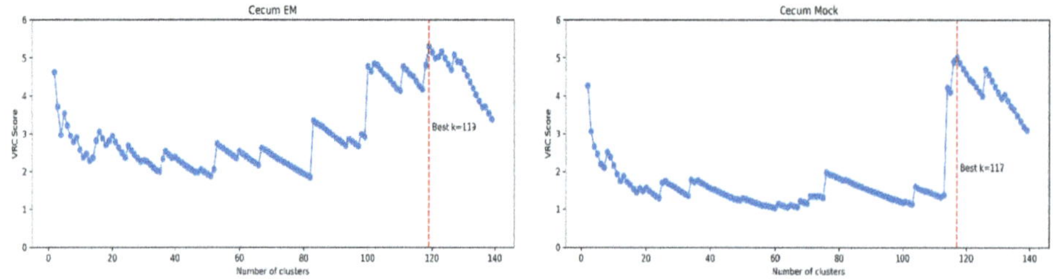

Fig. 1. VRC score-based determination of optimal cluster number for cecal microbiome samples.

In contrast, the VRC score curve for the Mock group exhibited a flatter overall profile. While the optimal k was close to that of the EM group ($k = 117$), the maximum VRC score remained lower (just under 5.0), and the progression toward the peak was more gradual and less variable. This may indicate a more stable and homogeneous microbial community in the absence of infection, with a more gradual emergence of distinct clustering patterns as the number of clusters (k) increases. To gain deeper insights into how EM infection influences the temporal behavior of individual microbial taxa, we visualized the normalized abundance trajectories of ASV clusters over time for both EM-infected and Mock groups. Each plot in Figs. 2 and 3 represents a distinct cluster of ASVs that share similar longitudinal patterns across multiple days (3, 5, 7, 14, and 21 DPI). These clusters were generated through hierarchical clustering using cosine-based distance to capture trajectory shape similarities and Ward linkage to ensure compact, variance-minimizing groupings. The optimal number of clusters was determined through the VRC analysis. Figures 2 and 3 specifically illustrate the normalized abundance profiles of clusters containing at least three ASVs (i.e., multi-member clusters). In the EM group, a total of 10 multi-member clusters (comprising three or more members), 12 two-member clusters, and 97 singleton clusters were identified. Similarly, in the Mock group, 8 multi-member clusters, 21 two-member clusters, and 88 singleton clusters were identified. The clustering results demonstrate that ASVs grouped together exhibited highly synchronized temporal abundance patterns, underscoring the biological coherence captured by the clustering strategy.

3.2 Comparative Analysis of Sub-communities

To assess the extent to which microbial community structures were preserved or altered following EM infection, we performed a comparative analysis between the EM- and mock-infected groups, which is summarized in Table 1.

A comparative analysis revealed that EM infection led to a substantial restructuring of microbial community relationships. Out of all unique ASV pairs evaluated (i.e., all combinations of two ASVs co-assigned to the same cluster), only one pair remained together in both EM-infected and Mock groups: "Oscillospiraceae_F194" and "Anaerotignum_F170". This may suggest that infection disrupts pre-existing bacterial associations. Furthermore, 44 pairs are clustered together exclusively under the EM condition, while 51 pairs are clustered together exclusively under the Mock condition. These distinct relationships are illustrated in Figs. 4 and 5.

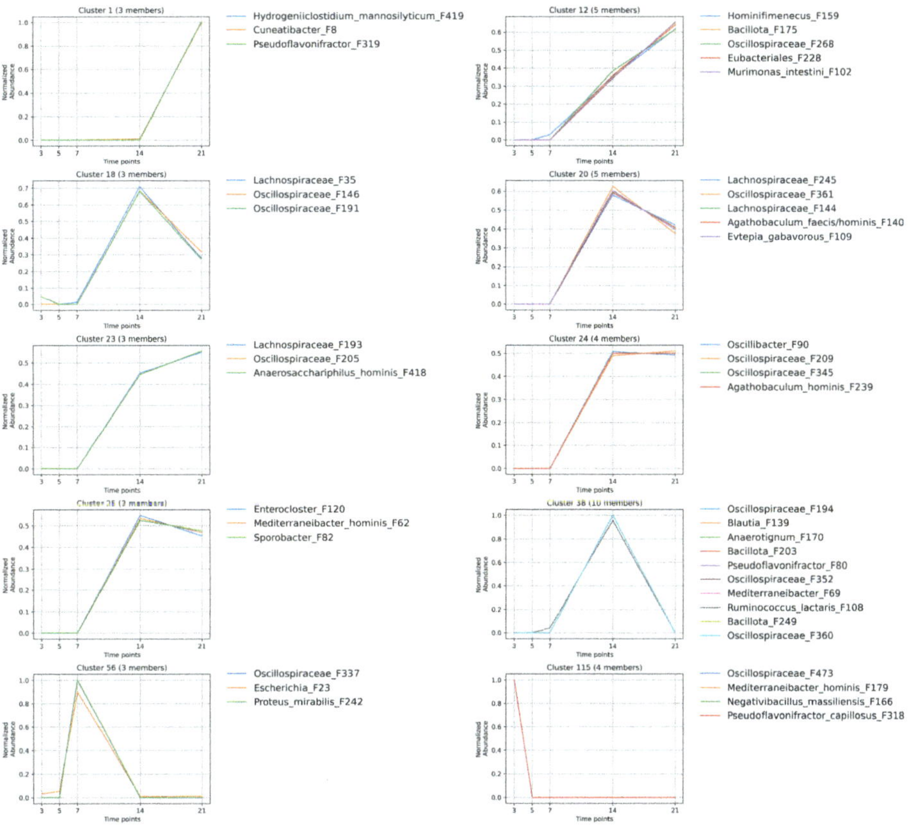

Fig. 2. Normalized abundance trajectories for clusters of ASVs in cecal microbiome of EM-infected chickens. Only clusters comprising three or more ASVs are presented.

From Fig. 2, for EM-infected chickens, we can see that most of the clusters containing three or more ASVs begin with their lines at 3 dpi close to zero and show little movement up to 7 dpi. One notable exception is Cluster 115: its members start high at 3 dpi, plunge to zero by 5 dpi, and stay flat through 21 dpi. In contrast, every other multi-member cluster—apart from Clusters 1, 56, and 115—undergoes a sharp rise between 7 dpi and 14 dpi. Cluster 56 rises a bit earlier, peaking between 5 dpi and 7 dpi before tapering off toward 14 dpi, whereas Cluster 1 delays its surge until 14 dpi. After day 14, the trajectories diverge: some clusters keep climbing, others decline at varying speeds. These distinct yet internally consistent patterns illustrate both the complex, time-dependent behavior of individual ASVs and the ability of the clustering method to group ASVs whose behavior is close over time.

The story for Mock-infected chickens is almost different. Clusters 7, 8 and 9 sit near baseline at 3 dpi, begin a gentle rise by 7 dpi, and keep climbing through 21 dpi. Cluster 2 stays flat for the first 14 days, then shows a late but pronounced lift. Clusters 76 and 77 take the opposite path: they start high on day 3 and decline sharply thereafter, stabilizing at low abundance. Cluster 97 is mixed: two ASVs peak at 7 dpi before falling quickly, while

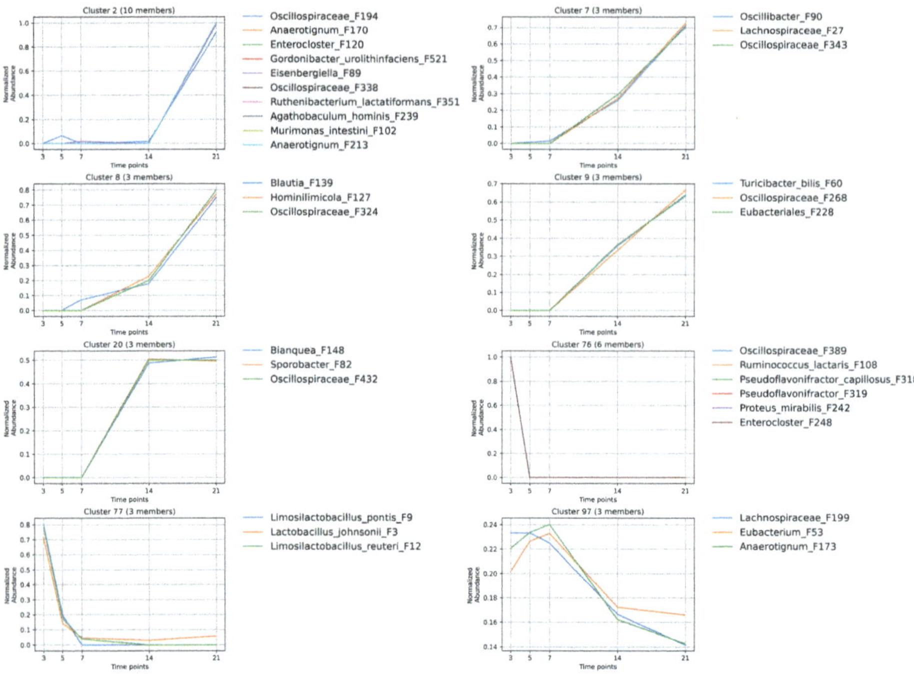

Fig. 3. Normalized abundance trajectories for clusters of ASVs in cecal microbiome of Mock-infected chickens. Only clusters comprising three or more ASVs are presented.

Lachnospiraceae_F199 holds steady until 5 dpi and then drops to baseline. Although the clusters follow different overall trajectories, the ASVs inside each cluster move in unison, confirming that the clustering method groups taxa with genuinely similar time-courses.

Table 1. Co-clustering comparison between the EM- and mock-infected chickens

Category	Number of ASV Pairs	Description
Present in Both	1	Pairs of ASVs* assigned to the same cluster in both EM and Mock groups
Only Present in EM	44	Pairs clustered together in the EM group but not in the Mock group
Only Present in Mock	51	Pairs clustered together in the Mock group but not in the EM group

* Pairs of ASVs refer to any combination of ASVs like (*a*, *b*). In this context, (*a*, *b*) and (*b*, *a*) are considered the same.

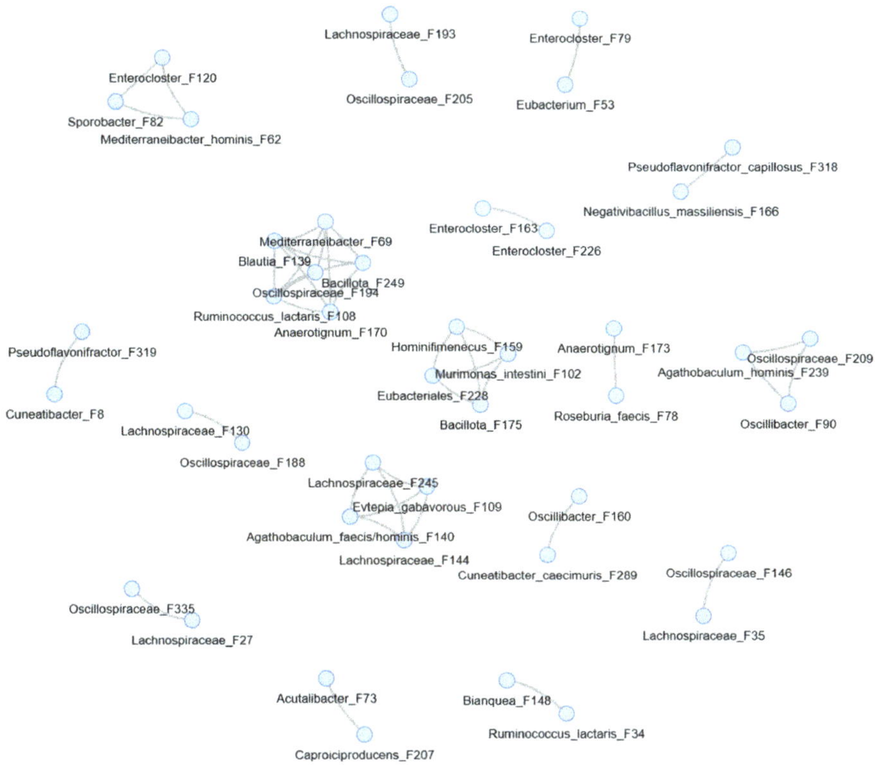

Fig. 4. Pairs of ASVs identified only in the cecal microbiome of EM-infected chickens.

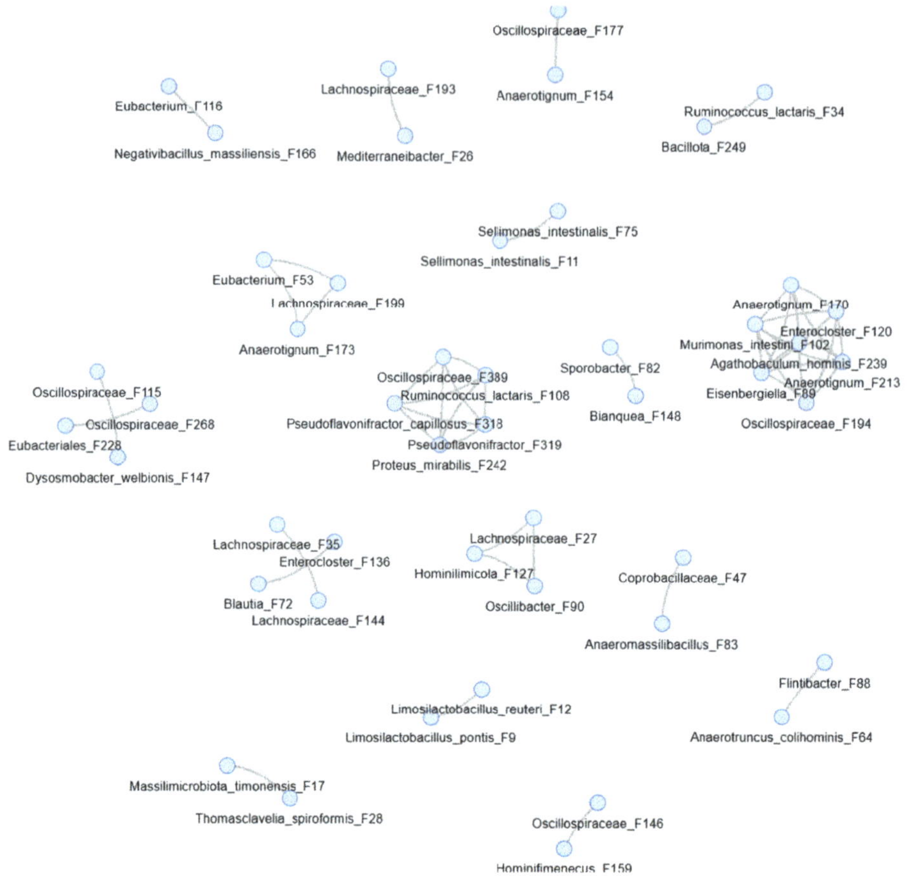

Fig. 5. Pairs of ASVs identified only in the cecal microbiome of mock-infected chickens.

4 Conclusion

In this study, we investigated the temporal restructuring of the cecal microbiota in broiler chickens following EM infection using a combination of hierarchical clustering and comparative analysis. Through longitudinal 16S rRNA gene sequencing data, we captured the dynamic changes in microbial communities over five key time points post-infection. Using the VRC, we determined the optimal number of microbial clusters for both EM-infected and Mock groups, revealing notable differences in clustering behavior.

Temporal clustering analysis showed that ASVs within each cluster exhibited highly synchronized abundance patterns across time points, validating the effectiveness of our clustering framework. Furthermore, the comparative analysis revealed that infection with EM led to substantial disruption of pre-existing microbial associations. Out of all ASV pairs examined, only one pair remained together in both EM and Mock groups, whereas 44 pairs co-clustered exclusively under infection and 51 pairs exclusively under control conditions. These findings may suggest that EM infection not only disrupts the microbial equilibrium but also promotes the formation of new, context-specific bacterial interactions.

Tracking the rise and fall of key bacterial clusters across 3, 5, 7, 14, and 21 days post-infection enables farm managers to anticipate the onset of dysbiosis before flock performance declines. Because ASVs within the same cluster rise and fall together, monitoring can focus on one low-cost marker from each cluster. Collecting samples at these fixed time points—3, 5, 7, 14, and 21 days post-challenge—creates a concise set of benchmarks that staff can use to compare successive flocks. By spotting departures from these established patterns early, producers can anticipate gut-health issues sooner, target follow-up tests more narrowly, and trim both diagnostic expense and unwarranted interventions.

While this study provides important insights into infection-driven microbial restructuring, there are natural opportunities for further refinement. First, integrating complementary biological datasets such as host immune responses or metabolite profiles could enhance the interpretation of observed microbial patterns. Second, exploring clustering frameworks that consider phylogenetic relationships among ASVs, or different distance metrics could offer a broader view of microbial dynamics. Expanding the analysis to include additional time points and other gut segments could further enrich the understanding of microbiota responses during infection and recovery.

Acknowledgments. This study was supported by a National Science Foundation grant (1830478), USDA National Institute of Food and Agriculture grants (2022-67016-37208 and 2024-67016-42415), the Ralph F. and Leila W. Boulware Endowment Fund, and Oklahoma Agricultural Experiment Station Project H-3112.

References

1. Liu, J., et al.: Dynamic response of the intestinal microbiome to Eimeria maxima-induced coccidiosis in chickens. Microbiol. Spectr. **12**, e00823–e00824 (2024)
2. Mesa-Pineda, C., Navarro-Ruíz, J.L., López-Osorio, S., Chaparro-Gutiérrez, J.J., Gómez-Osorio, L.M.: Chicken coccidiosis: from the parasite lifecycle to control of the disease. Front. Vet. Sci. **8**, 787653 (2021)
3. Lee, Y., Lu, M., Lillehoj, H.S.: Coccidiosis: recent progress in host immunity and alternatives to antibiotic strategies. Vaccine. **10**, 215 (2022)
4. Goo, D., et al.: Effects of Eimeria maxima infection doses on growth performance and gut health in dual-infection model of necrotic enteritis in broiler chickens. Front. Physiol. **14**, 1269398 (2023)
5. Miska, K.B., Campos, P.M., Cloft, S.E., Jenkins, M.C., Proszkowiec-Weglarz, M.: Temporal changes in jejunal and ileal microbiota of broiler chickens with clinical coccidiosis (Eimeria maxima). Animals. **14**, 2976 (2024)
6. Dubey, J., Jenkins, M.: Re-evaluation of the life cycle of Eimeria maxima Tyzzer, 1929 in chickens (Gallus domesticus). Parasitology. **145**, 1051–1058 (2018)
7. Al-Badri, R., Barta, J.R.: The kinetics of oocyst shedding and sporulation in two immunologically distinct strains of Eimeria maxima, GS and M6. Parasitol. Res. **111**, 1947–1952 (2012)
8. Maki, J.J., Klima, C.L., Sylte, M.J., Looft, T.: The microbial pecking order: utilization of intestinal microbiota for poultry health. Microorganisms. **7**, 376 (2019)

9. Yadav, S., Jha, R.: Strategies to modulate the intestinal microbiota and their effects on nutrient utilization, performance, and health of poultry. J. Anim. Sci. Biotechnol. **10**, 1–11 (2019)
10. Guo, J., et al.: Is intestinal microbiota fully restored after chickens have recovered from coccidiosis? Pathogens. **14**, 81 (2025)

Enhancing Logistics Management Education for Non-STEM Students in the AI Era: Evaluating the Role of Industry Experts in Logistics Education

Toshikuni Sato(✉), Kazuo Kikuchi, and Ippei Machida

School of Commerce, Meiji University, Chiyoda, Tokyo 101-8301, Japan
{tsato,kazukiku,cyouy}@meiji.ac.jp

Abstract. The worldwide shortage of logistics workers has emerged as a pressing issue. Especially in Japan, which has not accepted large numbers of immigrants, the implementation of the 2024 Working Style Reform Act, coupled with the country's acute aging population, is anticipated to cause significant disruptions to logistics services, creating a complex and challenging environment for the Japanese logistics industry. In response to escalating labor shortages, several Japanese logistics companies are increasingly leveraging information and digital technologies. Global trends also indicate the increasing adoption of AI and digital transformation within the logistics sector. Given this context, logistics education and research are relatively straightforward to develop with engineering students due to their technical background; however, this study focuses on the educational needs of Non-STEM students to expand the talent pool capable of contributing to the industry. This study introduced a lecture series featuring industry experts and investigated its educational impact on students' attitudes towards logistics knowledge and employment. The results indicate that the lectures deepened the students' logistics knowledge; however, a notable gap exists between students' interest in logistics and their motivation to work in the logistics industry.

Keywords: logistics and supply chain education · Non-STEM students · guest lectures · labor shortages

1 Introduction

Japan is currently experiencing a period of population decline [1]. In particular, labor shortages have become a major issue in Japanese industry. There are concerns that labor shortages have continued in the logistics industry. Trucks account for over 90% of the transportation volume measured by tonnage, resulting in a long-term shortage of truck drivers. The background to this is the long working hours and low wages of truck drivers [2]. In addition, the Japanese government applied the Working Style Reform Act to the logistics industry in 2024. This act restricts the upper limit of working hours for truck drivers. This has created the "2024 issue," which means that goods can no longer be transported as quickly and smoothly as before in Japanese supply chain [3]. In other words, the logistics industry can no longer meet the high demand for logistics services.

X. Xie et al. (Eds.): ICSS 2025, LNOR, pp. 197–209, 2026.
https://doi.org/10.1007/978-3-032-13116-4_16

The truck driver shortage problem is a global issue, including in the United States, EU, and Japan [4, 5], but several companies are trying to solve the shortage of logistics workers using information and digital technologies. For example, in Japan, NEXT Logistics Japan uses quantum computers to calculate the optimal product combination of cargo from different shippers (e.g., weight, destination, volume, and temperature controls) to improve truck loading efficiency. Consequently, productivity increases, reducing the burden on truck drivers and improving their salaries [6, 7]. CBcloud Co., Ltd. directly matches truck drivers and shippers on a digital platform, responding to temporary cargo demands and improving the income of truck drivers [8]. Daifuku Co., Ltd. also uses AR simulations to design distribution centers and is attempting to automate warehouses by introducing robots and sensor technologies [9].

The application of computer and information technologies to address logistics challenges, alongside the growing integration of AI and digital transformation's advancement, is projected to become increasingly widespread. Not only is there a shortage of people who can develop and use digital technology, but there is also a shortage of management personnel who can apply technology in the context of logistics businesses or supply chains [10]. Consequently, contemporary international research on university logistics education primarily focuses on educational methods that incorporate simulations and games, specifically designed for science and engineering students to understand logistics and supply chains [11–14]. However, traditional instructor-led pedagogies remain the predominant approach in humanities and social sciences at Japanese universities. This necessitates exploring pedagogical strategies that can adapt to evolving technological landscapes. Therefore, a significant challenge lies in motivating Non-STEM (science, technology, engineering and mathematics) students to enhance their logistical problem-solving competencies and prepare for careers in the logistics industry.

This study specifically focuses on the logistics education of Non-STEM students in Japan who typically lack formal technical training in computer science and engineering. This study's primary objective is to elucidate these students' current perceptions and attitudes towards logistics issues and discuss the issues of logistics education in response to Japan's 2024 logistics problems. Specifically, this study investigates the impact of a guest lectures series, designated as "The Art of Logistics" (detailed in Sect. 3.1), on these students.

2 Literature Review

Previous research on logistics and supply chain education has consistently highlighted the gap between university or graduate school education and industry needs, resulting in a discrepancy between the capabilities required by industry and those possessed by students. Consequently, various efforts have been made to bridge this research gap.

Various pedagogical approaches have been used to enhance logistics education. An analysis of 112 graduate courses from 61 schools worldwide revealed that 15.13% utilized case assignments to enhance market-related skills, 37.82% employed research projects to improve research capabilities, while 23.53% incorporated both [10]. For example, Erasmus students in the Netherlands were reported to analyze real-world corporate cases in teams to improve their understanding of logistics activities [15]. Simulation games have also been utilized, such as those designed to help practitioners and graduate students understand the Forrester effect and inventory fluctuations [16]. Furthermore, field trips have been found to build knowledge and enhance attitudes among Austrian higher education students [17].

In addition to these methods, collaborative research projects with companies have been emphasized as valuable tools. For instance, one study highlighted importance of "research projects with a company," in which students (undergraduate and graduate) from Mexican and American universities collaborated to develop competencies through proposals to reduce CO_2 emissions from retail companies' delivery vehicles [18]. Guest lecturers have also been recognized for their utility in providing industry insights into supply chain programs offered in undergraduate, graduate, and executive education [19]. However, faculties also faced challenges in finding suitable guest lecturers, suggesting that they should be considered as complementary to traditional lecture methods rather than a primary approach [19].

In the context of evolving educational strategies, game- or simulation-based learning is considered in engineering student education, potentially assisting in closing the gap between industrial practices and academic education. These educational methods have positive impacts not only on the acquisition of knowledge and skills, but also on students' attitudes, such as motivation [11, 12], engagement [12, 13], self-directed learning [11, 12], and reflection [12]. While educational methods using games or simulations are also being discussed within the logistics business and management [20–22], previous studies have primarily investigated engineering or industrial engineering students [11, 13, 14], with few cases involving business students [12]. Therefore, we focus on the lack of discussion and research on logistics education, particularly for Non-STEM students, including humanities and social sciences students.

3 Methodology

3.1 The Lectures on "The Art of Logistics"

Meiji University's School of Commerce offers lectures based on traditional teaching methods, such as logistics and transportation theories, for Non-STEM students. In addition, some seminar classes are conducted with companies, but not many. Therefore, Non-STEM students have few opportunities to learn about the current state of logistics. This study's lecture subject was "the Art of Logistics." This lecture is funded by corporate donations, unlike regular university lectures. "The Art of Logistics" lecture is characterized by the fact that longstanding practitioners in logistics management are invited to give all the lectures, with the aim of understanding the overall picture of the logistics industry. Twelve guest lecturers took the stage each time and discussed their company's logistics strategy, how to respond to the "2024 issue," and the use of digital technology in each company. Guest lecturers can also ask students questions during lectures, making them somewhat interactive. Students from all faculties at Meiji University can take this course, but they were mainly economics and business students. Through this lecture, students can gain a deeper understanding of the on-site situation surrounding logistics, such as the 2024 issue and the use of information and digital technology. Therefore, the aim of this course was to enhance logistics management education for Non-STEM students in the age of AI.

3.2 Data Collection

Data were collected from students enrolled in this lecture, which was conducted from April to the end of July 2024. Participation in this study was voluntary and data were collected only from students who provided informed consent for their data to be used for research purposes. Data collection consisted of administering questionnaires at two time points: before the lecture and after its completion. To ensure anonymity while matching pre- and post-lecture responses, each participant was assigned a unique, randomly generated identification (ID) number. The ID number was used to link the pre- and post-lecture questionnaires and create a paired dataset. Ultimately, the dataset was comprised of 67 students. The details of all questionnaires, which were measured using a 7-point Likert scale (1 = strongly disagree to 7 = strongly agree), with some exceptions, are provided in the next section.

4 Results and Discussions

4.1 Respondent Demographics

Figure 1 shows a stacked bar chart of the respondents according to gender, grade, and faculty. The respondents were male (58.2%) and female (41.8%), with 77.6% being 3rd-grade students and 22.4% being 4th-grade students. Although the class was open to students from all faculty at Meiji University, not all of them majored in science or engineering. More than half the enrolled students were from the top three faculties: Commerce, Arts and Letters, and Political Science and Economics. This highlights that the lectures garnered interest from students beyond the traditional boundaries of logistics, business, and management.

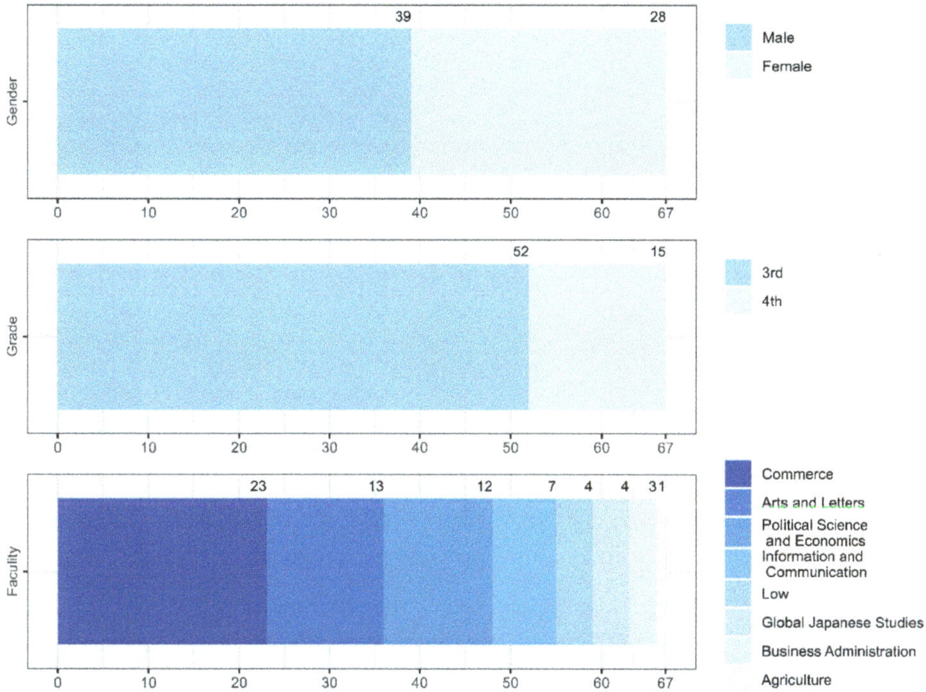

Fig. 1. Percentage of respondents by demographic (Note: the numbers indicate the count of respondents)

4.2 Overall Assessment of the Lecture

To evaluate the lectures' overall effectiveness, we illustrate the distribution of student post-satisfaction and changes in awareness of Japan's 2024 issue in Figs. 2 and 3, respectively. Figure 2 indicates that all students reported a satisfaction level of "somewhat satisfied" or higher, thereby demonstrating the absence of dissatisfaction among the students. Notably, the most frequent answer was "very satisfied," with 34 students (approximately half of the respondents) selecting this option.

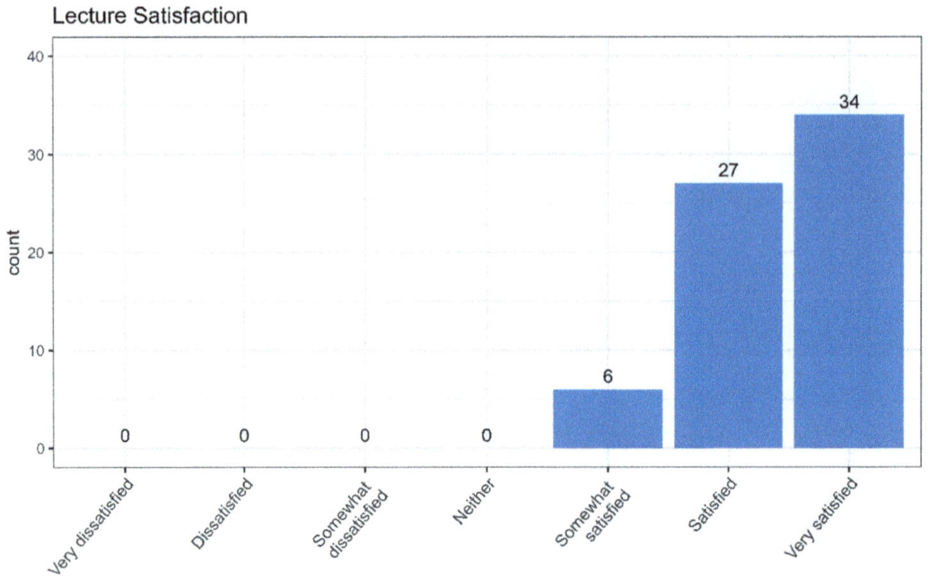

Fig. 2. Student satisfaction after the lecture

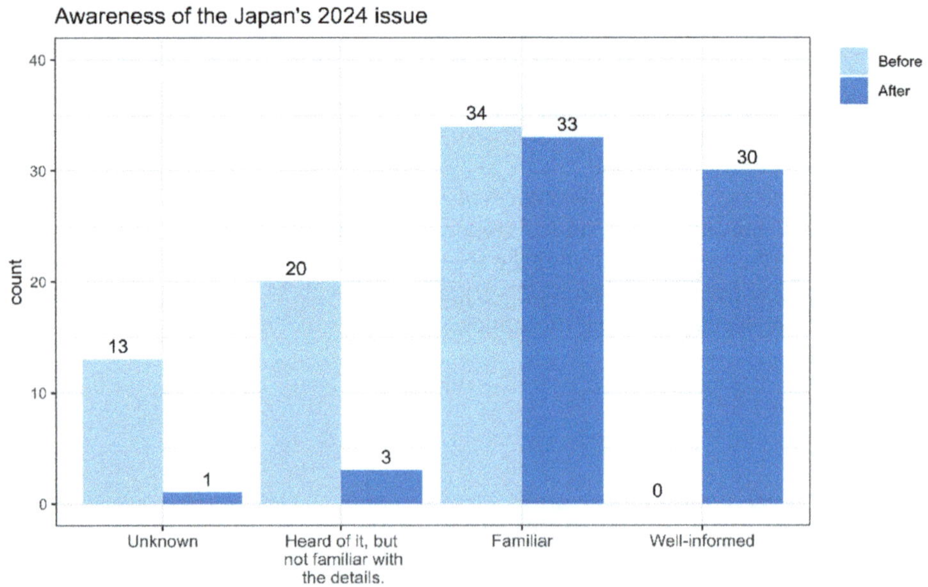

Fig. 3. Pre- and post- comparison of student awareness of the Japan's 2024 issue

Figure 3 provides evidence of the deepened awareness of Japan's 2024 issue among students through the lecture, as shown by the rightward shift in the distribution from preto post lecture. Prior to the lecture, no students reported being "well-informed" regarding the Japan's 2024 issue. However, after the lecture, this response category increased to 30 students. Including respondents who answered "Familiar," the overall awareness of Japan's 2024 issue among the students reached approximately 94%.

4.3 Attitude Towards Logistics Knowledge

To explore the lectures' detailed educational effectiveness, we conducted paired t-tests for a comparative analysis of students' subjective attitudes before and after the lectures using five questionnaire items related to logistics knowledge. Figure 4 shows boxplots of the pre- and post-attitude scores for each questionnaire item. In each box, the bold black horizontal line indicates the median position, and the dotted red point indicates the mean position. The numerical mean value is displayed to the lower left of the red point. The whisker is set to a 1.5 interquartile range; therefore, black dots exceeding this whisker range indicate outliers. The upper numerical value on the bar indicates the p-value obtained from paired t-tests.

Figure 4 demonstrates statistically significant differences in the following four attitude scores after the lecture: "I know about logistics" ($t(66) = 3.46$, $p < .01$), "I am confident in my knowledge of logistics" ($t(66) = 3.57$, $p < .01$), "(Before the lecture,) I expect to gain knowledge about logistics business through this lecture / (After the lecture,) I gained knowledge about logistics business through this lecture" ($t(66) = 3.61$, $p < .01$), "I can imagine the jobs in the logistics industry other than transporting goods by truck" ($t(66) = 5.60$, $p < .01$). These results imply that lectures had a positive impact on students' attitudes towards logistics knowledge. Although the difference in "I am interested in logistics." was not statistically significant, the mean score was high (5.49) before the lecture, and the median score increased from 5 to 6 after the lecture. However, most students who attended the lecture had a high initial interest in logistics, suggesting a potential bias in the results owing to a sample that may not adequately represent students with low motivation.

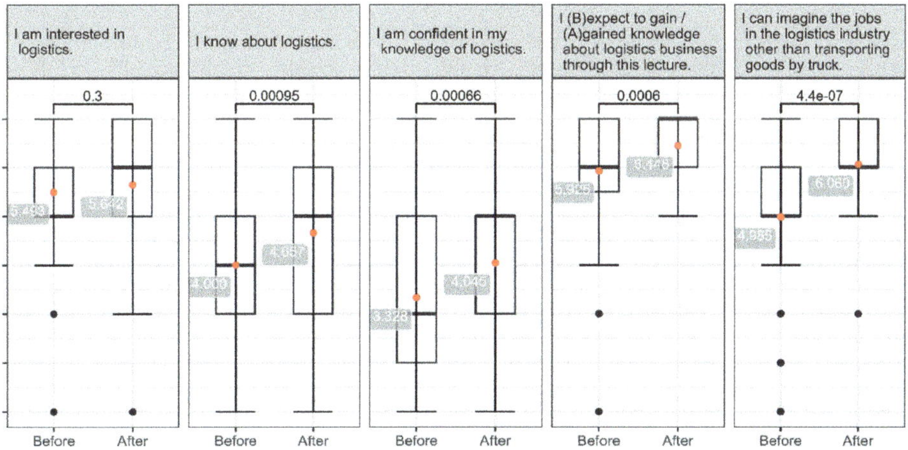

Fig. 4. Pre- and post- comparison of student attitude towards logistics knowledge

4.4 Attitude Towards Working in the Logistics Industry

The preceding section demonstrates that students maintained a high initial interest in logistics and showed positive changes in their attitudes towards logistics knowledge. Next, we examined students' attitudes towards working in the logistics industry. Similar to the investigation of attitudes towards logistics knowledge, we conducted paired t-tests to comparatively analyze students' subjective attitudes before and after the lecture using five questionnaire items related to working in the logistics industry. The results are shown in Fig. 5, and the interpretation follows the principles described for Fig. 4 in the previous section.

According to Fig. 5, the attitude score showed a statistically significant increase in "I think that the logistics industry is an essential industry." ($t(66) = 3.16$, $p < .01$). However, a statistically significant decrease was observed in "I would consider logistics companies a top priority in my job search." ($t(66) = -2.63$, $p < .05$). Additionally, there were no significant changes in the other two attitude scores: "I consider logistics companies when looking for employment." and "I think the logistics companies pay well." These score distributions tend to skew positively. Furthermore, the number of future careers in the logistics industry will not increase. Therefore, the lecture played an important role in enhancing students' understanding of the significance of the logistics industry, but it did not effectively increase their motivation to work in the field.

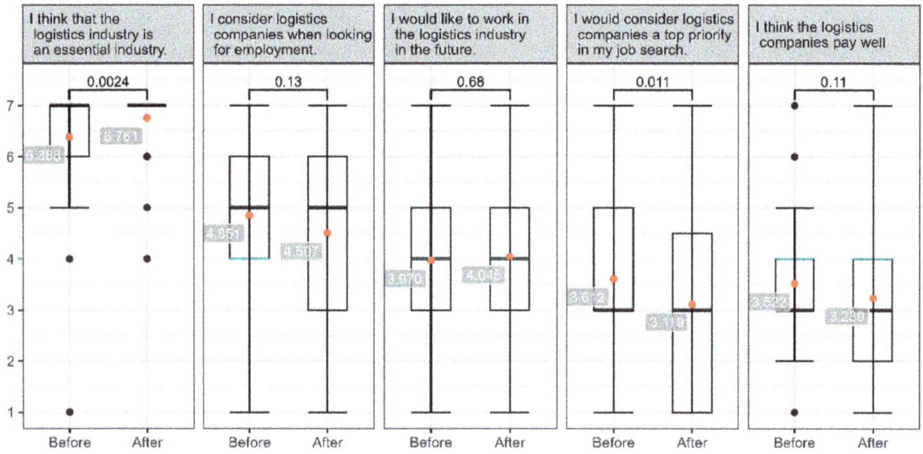

Fig. 5. Pre- and post- comparison of student attitude towards working in logistics industry

4.5 Images of the Logistics Industry

To better understand students' specific perceptions of the logistics industry, we conducted paired t-tests on changes in scores across 14 image-related questionnaire items. Figure 6 shows boxplots illustrating the difference in students' perceptions before and after the lecture. The bold black and red dots indicate the median and mean difference scores, respectively. The range represented by the two red lines in each boxplot shows the 95% confidence interval (95% CI), and a statistically significant difference is indicated when this range does not include the zero point. It is marked by the red dotted horizontal line in Fig. 6. The 12 items on the left side, ranging from "Hard" to "No idea," represent negative images, while the remaining two items, "Advanced" and "International," represent positive images. A positive skew in the negative image scores or negative skew in the positive image scores suggests an improvement in the respective images through the lecture.

Although the positive perceptions of "Advanced" and "International" improved after the lecture, the lecture failed to significantly mitigate several negative perceptions, such as "Hard," "Dirty," "Difficult," "Low salary," and "Many companies do not care about their employees," as shown in Fig. 6. Conversely, we found that other negative perceptions, namely "Gloom," "Outdated," "Unskilled labor," "Long working hours," and "Few opportunities for women to show of ability," were significantly improved. Figure 7 presents the detailed results of the improved perception scores. As explained in previous sections, this interpretation follows the principles shown in Figs. 4 and 5. Figure 7 indicates the lecture significantly succeeded in reducing the negative perceptions of "Gloom" ($t(66) = -4.34$, $p < .01$), "Outdated" ($t(66) = -3.87$, $p < .01$), "Unskilled labor" ($t(66) = -3.45$, $p < .01$) "Long working hours" ($t(66) = -2.99$, $p < .05$) and "Few opportunities for women to show of ability" ($t(66) = -3.81$, $p < .01$). However, the two negative images of "Long working hours" and "Few opportunities for women to show of ability" still maintained high negative levels, as their post-lecture means remained above the midpoint score.

These results highlight points for reflection on both our educational method and the future of the Japanese logistics industry. While the lecture appeared to have a positive educational impact on students, as discussed in previous sections, it may have inadvertently lessened the motivation to seek employment in the logistics industry by providing them with a deeper understanding of its reality. This indicates a serious issue for the Japanese logistics industry. Despite growing efforts to promote AI and digital transformation in the industry—and the expansion of logistics education among engineering students specializing in these areas—these Non-STEM students continue to avoid careers in logistics after learning about the industry's realities. Given that a serious labor shortage is expected to persist in Japan, facilitating the transition of highly interested Non-STEM students to appropriate career paths within the logistics industry represents a significant challenge. Despite the limitations of our educational method, it demonstrably changed students' attitudes towards logistics knowledge and their awareness of the current situation in the Japanese industry. We recommend continuous cooperation between industry professionals and academic institutions to overcome this challenge.

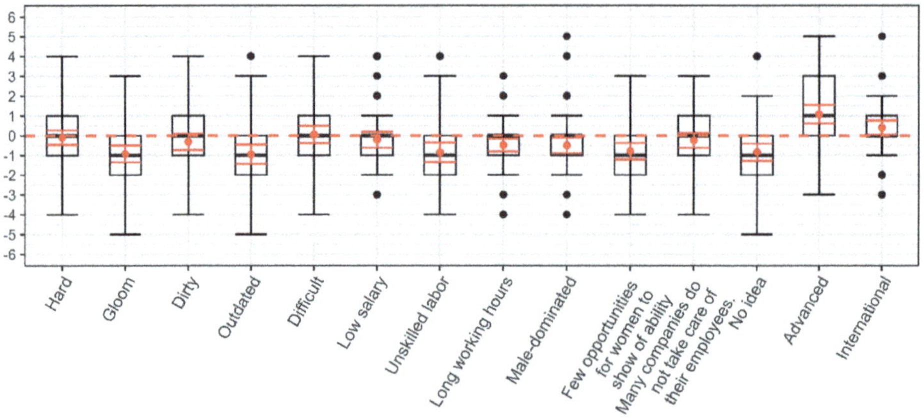

Fig. 6. Pre- and post- comparison of students' images of the logistic industry

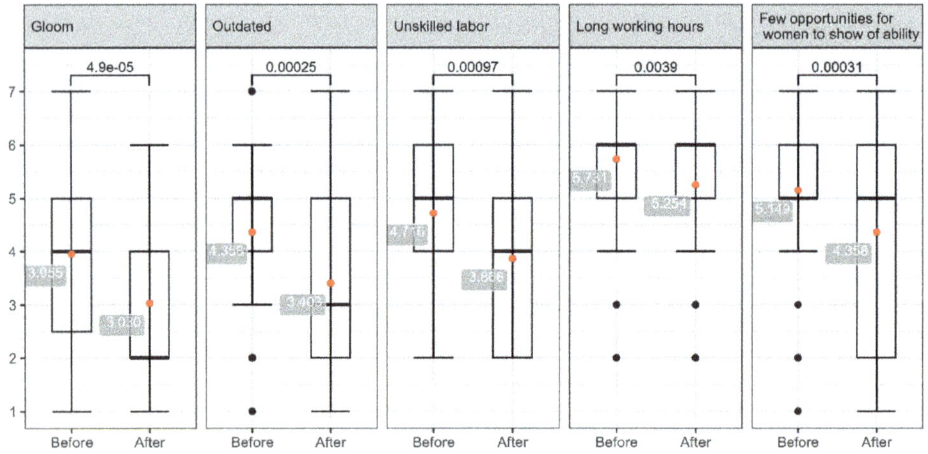

Fig. 7. Selected negative perceptions towards logistic industry

5 Conclusions

Currently, promoting AI utilization and digital transformation is intensifying across all sectors, and the logistics industry is no exception. In student education specializing in computer science and engineering, this social context is reflected in curricula centered on computer simulations essential for logistics. Logistics has also played a significant role in marketing and business, such as the "Place" component of the 4Ps of marketing. Nevertheless, educational approaches for Non-STEM students remain underexplored, particularly for those who lack access to practical logistics training through specialized software or digital platforms that link academic knowledge with real-world applications.

This study aimed to enhance Non-STEM students' comprehension of the logistics industry and foster their career interests through a dedicated lecture series. Subsequently, we evaluated the educational impact of these interventions. While the lecture series successfully maintained the students' high interest in logistics and strongly emphasized the importance of the logistics industry, which is currently facing a severe labor shortage, it may have discouraged students from pursuing careers in that industry. Considering, however, that prior research in logistics education has extensively explored students' knowledge of and attitudes towards learning [e.g., 11–14, 17], this study's unique contribution lies in its investigation of attitudes towards employment in the logistics industry, thereby revealing a significant gap in motivation between academic engagement and professional aspirations.

This study had three main limitations. First, the sample's representativeness was limited; therefore, future research should consider university education and its evaluation on a nationwide scale in Japan. As mentioned in Sect. 4.3, most respondents initially had a strong interest in logistics, making the effectiveness of the lecture series on students with less interest unclear. However, this implies that even students with a strong interest in logistics did not have their job motivation altered. The lecture series design should be revised to address this point. Second, the questionnaire items used to measure educational effectiveness were simple. To contribute theoretical perspectives, future research should consider creating questionnaires based on established constructs from prior research in the education field [e.g., 11–13] and investigate the acquisition of more specific knowledge to demonstrate educational effectiveness in a more detailed manner. Third, because this study is a preliminary step in exploring the Non-STEM student situation in logistics education, our empirical analysis was limited mainly to descriptive analysis. To estimate more valid educational effects, our analytical framework should be extended with established statistical methods, such as the mixed effect model, in the educational field [e.g., 23–25]. In addition to these limitations, the continuous labor shortage in the logistics industry is particularly severe in Japan. It is also important to continue addressing the educational challenge of how to connect not only computer science and engineering students, but also those with a strong interest in the logistics industry, to relevant employment opportunities.

References

1. Ministry of Health, Labour and Welfare of Japan: VITAL STATISTICS OF JAPAN 2019. https://www.mhlw.go.jp/english/database/db-hw/dl/81-1a2en.pdf (2022).
2. Ministry of Land, Infrastructure, Transport and Tourism of Japan: Summary of the White Paper on Land, Infrastructure, Transport and Tourism in Japan, 2023. https://www.mlit.go.jp/en/statistics/content/001717675.pdf (2023).
3. JIJI PRESS LTD.: Japan Logistics Biz Braced for "2024 Issue" of Driver Shortages. https://sp.m.jiji.com/english/show/25491 (2023).
4. International Road Transport Union (IRU): Worse than you thought: truck driver shortages and demographics. https://www.iru.org/news-resources/newsroom/worse-you-thought-truck-driver-shortages-and-demographics (2024).
5. American Trucking Associations (ATA): FreightWaves gets it wrong on driver shortage. https://www.trucking.org/news-insights/freightwaves-gets-it-wrong-driver-shortage (2023).
6. Sumitomo Corporation: Sumitomo Corporation Invests in NEXT Logistics Japan. http://www.sumitomocorp.com/en/jp/news/topics/2024/group/20240910 (2024).
7. NEXT Logistics Japan: https://www.next-logistics-jp.jp/. Last accessed 2025/03/23.
8. CBcloud: https://cb-cloud.com/en. Last accessed 2025/03/23.
9. DAIFUKU: https://www.daifuku.com/. Last accessed 2025/03/23.

10. Lutz, H., Birou, L., Walden, J.: Survey of graduate supply chain courses: content, coverage and gaps. Supply Chain Manag. Int. J. **27**, 625–636 (2021). https://doi.org/10.1108/SCM-12-2020-0637
11. Pacheco, E., Palma-Mendoza, J.: Using serious games in logistics education. In: 2021 the 2nd International Conference on Industrial Engineering and Industrial Management, pp. 51–55. Association for Computing Machinery, New York, NY, USA (2021). https://doi.org/10.1145/3447432.3447438
12. Pacheco-Velazquez, E.: Effects of the use of simulators and an online platform in logistics education. Int. J. Interact. Des. Manuf. **16**, 439–457 (2022). https://doi.org/10.1007/s12008-021-00791-z
13. Pacheco-Velázquez, E.A., Rodés-Paragarino, V., Ramírez-Echeverri, S.A.: Student's experience in using simulators for the development of complex skills in logistics education: assessing enjoyment of learning. Smart Learn. Environ. **12**, 18 (2025). https://doi.org/10.1186/s40561-025-00372-7
14. Ramírez-Montoya, M.S., Rodés-Paragarino, V., Pacheco-Velazquez, E., Ramírez-Etcheverry, S.: Transforming logistics education by a virtual logistics simulation generator: UX pilot study. In: 2024 IEEE Gaming, Entertainment, and Media Conference (GEM), pp. 1–6 (2024). https://doi.org/10.1109/GEM61861.2024.10585758
15. van Hoek, R.I.: Logistics education: achieving market and research driven skill development. Int. J. Phys. Distrib. Logist. Manage. **31**, 505–519 (2001). https://doi.org/10.1108/EUM0000000005906
16. Holweg, M., Bicheno, J.: Supply chain simulation – a tool for education, enhancement and Endeavour. Int. J. Prod. Econ. **78**, 163–175 (2002). https://doi.org/10.1016/S0925-5273(00)00171-7
17. Putz, L.-M., Treiblmaier, H., Pfoser, S.: Field trips for sustainable transport education: impact on knowledge, attitude and behavioral intention. Int. J. Logist. Manag. **29**, 1424–1450 (2018). https://doi.org/10.1108/IJLM-05-2017-0138
18. Gámez-Pérez, K.M., Sarmiento, A.M., Garcia-Reyes, H., Velázquez-Martínez, J.C.: An international university-industry collaboration model to develop supply chain competences. SCM. **25**, 475–487 (2020). https://doi.org/10.1108/SCM-08-2019-0317
19. van Hoek, R.I., Godsell, J., Harrison, A.: Embedding "insights from industry" in supply chain programmes: the role of guest lecturers. Supply Chain Manag. Int. J. **16**, 142–147 (2011). https://doi.org/10.1108/13598541111115383
20. Rodriguez-Garcia, E.E., et al.: A systematic literature review on the use of simulators to train undergraduate students in logistics. In: 2024 Portland International Conference on Management of Engineering and Technology (PICMET), pp. 1–7 (2024). https://doi.org/10.23919/PICMET64035.2024.10653134
21. Schmuck, R.: Education and training of manufacturing and supply chain processes using business simulation games. Procedia Manuf. **55**, 555–562 (2021). https://doi.org/10.1016/j.promfg.2021.10.076
22. Miščević, G., Hadžić, A.P., Tijan, E.: Applications and software for teaching logistics and logistics management. In: 2018 41st International Convention on Information and Communication Technology, Electronics and Microelectronics (MIPRO), pp. 0666–0670 (2018). https://doi.org/10.23919/MIPRO.2018.8400125
23. Fielding, A., Yang, M.: Generalized linear mixed models for ordered responses in complex multilevel structures: effects beneath the school or College in Education. J. R. Stat. Soc. A. Stat. Soc. **168**, 159–183 (2005). https://doi.org/10.1111/j.1467-985X.2004.00342.x

24. Gordon, K.R.: How mixed-effects modeling can advance our understanding of learning and memory and improve clinical and educational practice. J. Speech Lang. Hear. Res. **62**, 507–524 (2019). https://doi.org/10.1044/2018_JSLHR-L-ASTM-18-0240
25. Leckie, G., Parker, R., Goldstein, H., Tilling, K.: Mixed-effects location scale models for joint modeling school value-added effects on the mean and variance of student achievement. J. Educ. Behav. Stat. **49**, 879–911 (2024). https://doi.org/10.3102/10769986231210808

Mathematical Foundations of AI-Augmented Leadership: The NOVA Framework for Multi-agent AI Optimization

Tuhin Chattopadhyay(✉)

Jagdish Sheth School of Management (JAGSoM), Bengaluru, India
tuhin.c@jagsom.edu.in

Abstract. Executive decision-making is increasingly challenged by volatility, uncertainty, and cognitive limitations. While AI has transformed business operations, its application in strategic leadership remains constrained by trust deficits, cognitive biases, and lack of counterfactual reasoning. This study introduces a mathematically grounded AI-Augmented Leadership framework, operationalized through the NOVA (Neuroscience-Oriented Virtual Agents) and DAMA (Decision-Augmenting Multi-Agent AI) models. The framework enhances executive decision-making via Bayesian bias compensation, trust calibration using the Perceived Explainability Index (PEI), and counterfactual strategy exploration. Simulation-driven validation, employing Monte Carlo methods, reinforcement learning, ANOVA, and regression analysis, demonstrates a 62.3% reduction in cognitive bias, 18–26% improvement in decision accuracy, and a rise in AI adoption likelihood from 20% to 90% with improved explainability. A case study in mergers and acquisitions confirms the model's practical relevance, recommending Partial Stake Acquisition (51%) as the optimal strategy. The findings underscore the strategic imperative of integrating explainable, trust-calibrated AI into executive workflows. By bridging human cognition with multi-agent AI optimization, this research positions AI-Augmented Leadership as a transformative model for future-ready organizations.

Keywords: AI-Augmented Leadership · Multi-Agent AI · Trust Calibration · Counterfactual Reasoning · Executive Decision-Making

1 Introduction

Executive decision-making today faces unprecedented complexity, driven by economic volatility, geopolitical flux, and digital disruption. While Artificial Intelligence (AI) has transformed operational analytics, its integration into strategic leadership remains limited due to cognitive biases, trust deficits, and an inability to model counterfactual scenarios. Executives, bounded by limited cognition and heuristic reasoning, often hesitate to rely on AI due to interpretability challenges and misalignment with strategic intuition.

This paper introduces AI-Augmented Leadership—a mathematically grounded framework that positions AI as a cognitive partner in executive decision-making. At

X. Xie et al. (Eds.): ICSS 2025, LNOR, pp. 210–222, 2026.
https://doi.org/10.1007/978-3-032-13116-4_17

its core are the NOVA (Neuroscience-Oriented Virtual Agents) Framework and the DAMA (Decision-Augmenting Multi-Agent AI) Model (Fig. 1). Together, they integrate Bayesian bias compensation, reinforcement learning, and trust calibration via the Perceived Explainability Index (PEI) to dynamically refine decisions and generate counterfactual strategies.

Unlike traditional static or rule-based AI systems, NOVA + DAMA supports real-time, explainable, and adaptable decision augmentation. The paper formulates AI-Augmented Leadership as a constrained optimization problem that bridges algorithmic intelligence and executive cognition. It is guided by the following research questions:

i. How can Multi-Agent AI be structured to align with executive cognitive processes?
ii. What mechanisms ensure trust calibration in AI-driven executive decision-making?
iii. Can AI-generated counterfactual strategies redefine strategic decision-making beyond human intuition?
iv. How does the NOVA Framework compare against traditional AI-driven decision systems in terms of accuracy and adaptability?

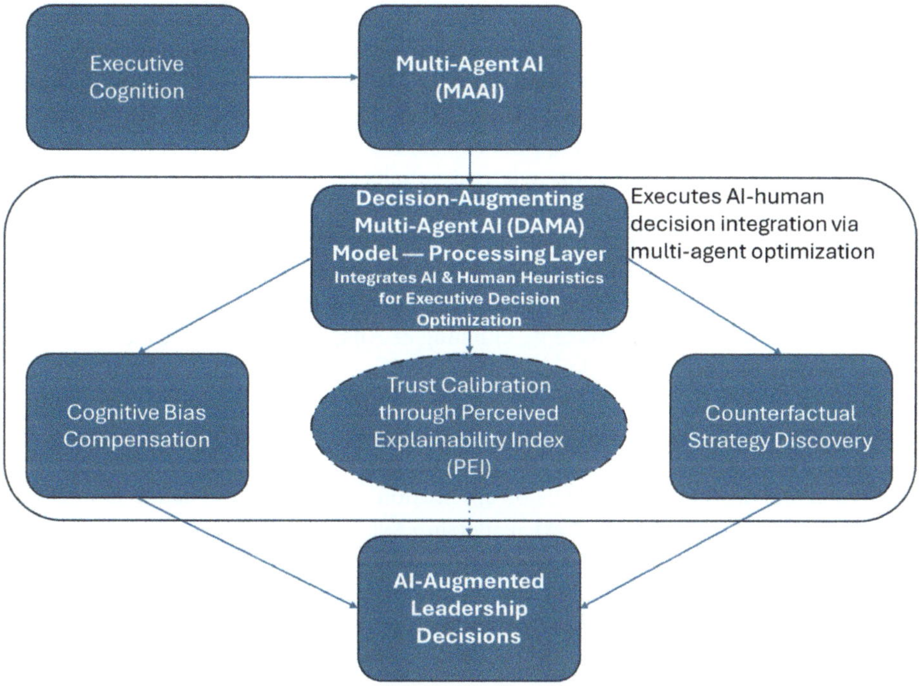

Fig. 1. The NOVA Framework – Cognitive Architecture for AI-Augmented Decision-Making

Contributions include a maturity model of AI adoption in leadership, a formal decision utility model, and empirical validation via Monte Carlo simulations. This research aims to establish AI not as a replacement, but as a strategic co-pilot for human leadership in uncertain environments. This research is particularly relevant for strategic decision-making in high-stakes domains such as mergers and acquisitions, crisis management, policy formulation, and digital transformation initiatives.

2 Literature Review

The integration of Artificial Intelligence into executive decision-making has advanced through multi-agent models, reinforcement learning, and trust-based explainability frameworks, yet critical gaps remain in aligning these approaches with strategic leadership cognition.

Multi-Agent AI (MAAI) frameworks, particularly in reinforcement learning, enable distributed decision-making where AI agents interact dynamically. While hierarchical planning (Russell & Norvig, 2021) and game-theoretic modeling (Silver et al., 2016) offer strategic foresight, they often lack customization for executive heuristics, explainability, and counterfactual exploration.

Cognitive bias modeling has introduced Bayesian calibration and bias-weighted optimization (Tversky & Kahneman, 1974), but few systems dynamically adapt bias correction across iterative decision cycles. This study addresses that limitation through an evolving Bayesian trust mechanism embedded in the NOVA+DAMA framework.

Reinforcement learning, including Markov Decision Processes (Sutton & Barto, 2018) and Multi-Agent RL (Lowe et al., 2017), has shown promise in operational automation, though its empirical application in executive leadership remains limited. This research extends its utility through trust-weighted learning loops.

Trust calibration and explainability—measured by the Perceived Explainability Index (PEI)—are pivotal for AI adoption (Doshi-Velez & Kim, 2017), yet prior models lacked mathematical integration of trust into adoption dynamics. This paper formally embeds PEI within decision optimization. This aligns with recent work on explainable AI in executive decision systems (e.g., Gunning & Aha, 2019; Rai, 2020), emphasizing the role of transparency in AI adoption.

As most studies fail to simulate AI adoption likelihood dynamically over iterative decision cycles, this study incorporates computational validation through Monte Carlo simulation and regression modeling, bridging the gap between theoretical AI leadership models and real-world strategic decision applications. Table 1 shared below summarizes the key research gaps in current AI leadership systems and outlines how this study addresses them through the NOVA + DAMA framework.

Table 1. Identified Research Gaps & How This Study Bridges Them

Research Gap	How This Study Addresses It
Multi-agent AI lacks application in leadership decision-making.	The study develops a structured AI-human synergy model for executive strategy.
Cognitive biases are modeled but not embedded in AI decision-making.	This research integrates Bayesian cognitive bias correction into AI leadership models.

(continued)

Table 1. *(continued)*

Research Gap	How This Study Addresses It
Reinforcement Learning (RL) is applied in AI automation but not in leadership augmentation.	The study adapts RL-based learning to refine executive decision policies.
AI trust calibration lacks formal mathematical representation.	A trust-weighted optimization function using PEI is introduced.
AI leadership augmentation models lack mathematical validation.	Monte Carlo simulations validate AI-human trust dynamics and decision accuracy.

3 Methodology

This study adopts a computational-first methodology to validate the effectiveness of the proposed AI-Augmented Leadership framework. Unlike conventional empirical research reliant on real-world datasets, we simulate executive decision-making using Monte Carlo methods, Bayesian inference, and Multi-Agent Reinforcement Learning (MARL). The goal is to evaluate the NOVA + DAMA model's capacity to enhance leadership decisions by mitigating cognitive bias, calibrating trust, and enabling counterfactual reasoning.

3.1 Research Design

Unlike traditional decision-support systems that rely on static rules or linear optimization, this study proposes a novel hybrid simulation methodology that integrates Bayesian inference, trust calibration modeling (via PEI), and counterfactual analysis within a multi-agent reinforcement learning loop. This combination is purpose-built for cognitive alignment with executive leadership heuristics — a methodological departure from prior work.

The methodology follows three core phases:

- Mathematical Formulation: Executive decisions are modeled as a constrained optimization problem integrating AI insights, human heuristics, and cognitive distortions.
- Simulation Implementation: Monte Carlo simulations, Bayesian updates, and reinforcement learning are used to operationalize the NOVA + DAMA system.
- Performance Evaluation: Outcomes are benchmarked against traditional AI and human-only decision-making models based on bias reduction, decision accuracy, and AI adoption likelihood.

3.2 NOVA + DAMA Simulation Framework

The simulation environment models interactions between AI agents and human executives across strategic decision scenarios. The NOVA Framework mimics executive cognition, while DAMA serves as the computational engine responsible for bias mitigation, trust calibration, and counterfactual discovery.

Mathematical Representation:

Let $A = \{A_1, A_2, \ldots, A_n\}$ be a set of AI agents, where each agent A_i specializes in one aspect of decision augmentation:

1. Bias Mitigation: Adjusts for overconfidence, anchoring, and availability biases.
2. Counterfactual Strategy Discovery: Evaluates alternative strategic outcomes.
3. Trust Calibration: Measures and optimizes AI explainability for human adoption.

The optimal executive decision is formulated as:

$$D^{*} = {}_{w1}A + {}_{w2}H - {}_{w3}B(D) + C(A, H) \cdot T(A) \tag{1}$$

where:

A = AI-generated insights, H = Human heuristics, $B(D)$ = Cognitive bias distortion, C (A, H) = Counterfactual strategy discovery, $T(A)$ = Trust calibration function, and

- w_1 (AI-Generated Insights Weight): Represents the relative influence of AI-driven recommendations on the decision-making process. A higher value of w_1 indicates greater reliance on AI's predictive analytics, data-driven risk assessment, and counterfactual strategy simulations.
- w_2 (Human Heuristics Weight): Captures the impact of executive intuition, experience, and domain expertise in decision-making. A higher w_2 reflects stronger reliance on human judgment, even when AI provides alternative insights.
- w_3 (Bias Compensation Factor): Accounts for the influence of cognitive biases (e.g., overconfidence, anchoring, loss aversion) on executive decisions. The weight w_3 is calibrated to minimize decision distortions by applying AI-driven bias correction techniques.

The assignment of decision weights (w_1, w_2, w_3) follows empirical findings from Kahneman (2011) on cognitive bias mitigation, Brynjolfsson & McAfee (2017) on AI trust calibration, and Sutton & Barto (2018) on reinforcement learning. These weights were refined through Monte Carlo sensitivity testing to ensure robustness across multiple decision scenarios.

Expected Decision Utility:

$$E(D) = \sum_{t=1}^{T} P(A_t \mid H_t) U(D_t) \tag{2}$$

where $P(A_t|H_t)$ represents the probability of AI adoption based on human-AI trust dynamics.

This framework ensures that AI recommendations are continuously refined based on past decision feedback.

3.3 Simulation Design and Parameters

Ten thousand Monte Carlo iterations were conducted to simulate three scenarios:

1. Human-only decision-making
2. Traditional AI decision support
3. AI-Augmented Leadership via NOVA + DAMA

The simulation parameters and assumptions, outlined in Table 2, are derived from prior literature and refined to replicate real-world executive decision-making environments. Key simulation parameters are as follows:

Table 2: Simulation parameters and assumptions

Parameter	Description	Initial Value/Range	Reference
Perceived Explainability Index (PEI)	Measures trust in AI, starting low and improving with explainability.	0.45 → 0.88 (logistic growth)	Brynjolfsson & McAfee (2017)
AI Adoption Likelihood	Probability of executives adopting AI in decision-making.	20% → 90% over iterations	Rogers (2003), Russell & Norvig (2021)
Human-Only Bias Levels	Cognitive biases distorting decision-making (e.g., overconfidence, anchoring).	32%–45%	Kahneman (2011)
AI-Adjusted Bias Reduction	Expected reduction in cognitive bias due to AI assistance.	60%–65%	Silver et al. (2016)
Monte Carlo Simulations	Generates probability distributions for AI-human decision interactions.	10,000 iterations	Metropolis & Ulam (1949)
Reinforcement Learning Model	AI adapts decision strategies based on human feedback.	Q-Learning, PPO	Sutton & Barto (2018)
Trust Calibration Mechanism	Bayesian updates adjusting AI transparency and executive trust over time.	Modeled via MDPs	Doshi-Velez & Kim (2017), Puterman (1994)
Bias Compensation Framework	AI agents dynamically counteract known executive biases.	Bayesian Inference	Kahneman (2011), Russell & Norvig (2021)

At each iteration, cognitive bias is sampled from a probability distribution, AI recommendations are applied or rejected based on evolving PEI, and key performance indicators—decision accuracy, bias reduction, and trust—are logged.

3.4 Statistical Validation

To confirm the framework's robustness, two statistical techniques are employed:

- ANOVA tests whether AI-Augmented Leadership leads to significantly higher decision accuracy compared to other models.
  - *Null hypothesis (H_0)*: No difference in performance between models.

  - *Alternative hypothesis (H_1)*: NOVA + DAMA outperforms traditional and human-only systems.

- Multiple Regression identifies the predictive strength of trust (PEI), bias reduction, and decision accuracy on AI adoption:

$$AI_{adoption} = \beta_0 + \beta_1\, PEI + \beta_2\, Accuracy + \beta_3\, BiasReduction + \varepsilon \tag{3}$$

where:

- $AI_{adoption}$ represents the likelihood of executives adopting AI recommendations.
- *PEI* denotes the Perceived Explainability Index, capturing the role of AI transparency in trust formation.
- *Accuracy* represents the degree of decision improvement due to AI augmentation.
- *BiasReduction* quantifies the extent to which AI reduces decision distortions.
- ε is the residual error term.

Statistical significance ($p < 0.05$) is used to confirm the impact of each predictor, and model fit is evaluated using R^2.

3.5 Case Application: Mergers and Acquisitions (M&A)

To ground the simulation in real-world strategy, the model is applied to a high-stakes M&A decision involving four acquisition options. Bayesian probabilistic modeling, counterfactual scoring, and trust-adjusted decision paths are used to evaluate optimality. Simulation results confirm that Partial Stake Acquisition (51%) is the most robust strategic option, with the final decision score (D*) exceeding the threshold of 1.0 for AI recommendation adoption.

4 Results and Analysis

This section presents the computational validation of the AI-Augmented Leadership framework using a combination of Monte Carlo simulations, multi-agent reinforcement learning, Bayesian bias compensation, and statistical modeling. These techniques allow for controlled experimentation that evaluates how the NOVA + DAMA system improves decision accuracy, reduces cognitive bias, and increases AI adoption through trust calibration.

4.1 Simulation Design and Setup

The simulation environment replicates high-stakes executive decision contexts, including investment planning, market entry, and M&A strategy. Three decision models are evaluated:

1. Human-only judgment,
2. Traditional AI decision support,
3. AI-Augmented Leadership using NOVA + DAMA.

Each iteration incorporates:

- Bias modeling: Decision distortion due to anchoring, overconfidence, and heuristic shortcuts.
- Trust modeling: AI transparency captured through the Perceived Explainability Index (PEI).
- Counterfactual analysis: Alternative strategic paths evaluated using Bayesian logic.

Simulation parameters (e.g., PEI range from 0.45 to 0.88, AI adoption from 20% to 90%, baseline bias at 32–45%) were informed by prior studies and fine-tuned through 10,000 Monte Carlo iterations.

4.2 Comparative Performance Evaluation

AI-Augmented Leadership demonstrated a 62.3% reduction in executive cognitive biases through reinforcement learning and Bayesian calibration. Human-only models showed significant susceptibility to distortion, while traditional AI systems lacked adaptability to executive intuition. NOVA + DAMA emerged as the only model that dynamically evolved over decision cycles to mitigate bias while integrating human heuristics.

Decision accuracy improved by 18.4%–26.1%, confirming AI's role as a strategic optimizer. These results underscore the superiority of NOVA + DAMA in producing stable, risk-aware outcomes across scenarios.

ANOVA (Analysis of Variance) is applied to determine whether statistically significant differences exist between human-only decision-making, traditional AI support, and AI-Augmented Leadership (NOVA + DAMA Framework). The hypothesis tested is:

$$H_0 : \mu_{\text{Human}} = \mu_{\text{Traditional AI}} = \mu_{\text{AI-Augmented Leadership}} \tag{4}$$

where μ represents the mean decision accuracy for each category. If the p-value is below a predefined significance threshold ($p < 0.05$), the null hypothesis is rejected, indicating a meaningful performance variation among decision-making models. Table 3 presents the results of the ANOVA test, indicating statistically significant performance variation across decision models.

Table 3. ANOVA Results: AI-Augmented Leadership vs. Alternative Models

Source of Variation	SS (Sum of Squares)	df (Degrees of Freedom)	MS (Mean Square)	F-Value	p-Value
Between Groups	1218.3	2	609.15	24.78	p < 0.001
Within Groups	2834.1	147	19.28		
Total	4052.4	149			

Interpretation & Strategic Implications

- The F-Value (24.78) confirms that AI-Augmented Leadership exhibits a significantly different decision accuracy compared to traditional models. These finding highlights AI's potential to systematically enhance strategic foresight in executive decision-making.
- p-value ($p < 0.001$) indicates that AI-Augmented Leadership significantly outperforms both human-only and traditional AI-based decision-making.
- Post-hoc Tukey's HSD test further substantiates that AI-Augmented Leadership is the most statistically robust decision model among the three.

4.3 Regression Analysis: Predictors of AI Adoption

A multiple regression analysis was performed to identify which factors most significantly influenced executive AI adoption. As shown in Table 4, trust (PEI), accuracy, and bias reduction emerge as significant predictors of AI adoption likelihood.

Table 4. Regression Model Output: AI Adoption Predictors

Variable	Coefficient (β)	Standard Error	t-Value	p-Value
Intercept (β_0)	−0.135	0.042	−3.21	$p < 0.01$
PEI (β_1)	0.72	0.14	5.14	$p < 0.001$
Accuracy (β_2)	0.51	0.09	4.78	$p < 0.001$
BiasReduction (β_3)	0.38	0.11	3.43	$p < 0.005$

Key Takeaways from Regression Analysis

- Model Fit ($R^2 = 0.86$): 86% of the variation in AI adoption likelihood is explained by trust calibration, accuracy improvements, and bias reduction.
- PEI (Trust Calibration) is the dominant predictor of AI adoption, with a coefficient of $\beta_1 = 0.72, p < 0.001$.
- AI adoption likelihood is significantly influenced by decision accuracy ($\beta_2 = 0.51$, $p < 0.001$) and cognitive bias reduction ($\beta_3 = 0.38, p < 0.005$).
- The statistical significance of all predictors confirms the robustness of the AI-Augmented Leadership framework in fostering executive confidence in AI-driven decision-making.

The analysis goes beyond descriptive comparisons by applying ANOVA and regression to quantify the causal drivers of AI adoption. In particular, trust calibration emerges as the dominant predictor ($\beta = 0.72$, $p < 0.001$), providing empirical evidence for the role of perceived explainability in executive AI uptake—a novel insight in leadership augmentation research.

4.4 Monte Carlo Convergence: AI Performance Over Time

AI adoption in executive decision-making is contingent on trust calibration, which evolves over multiple decision cycles. The study modeled trust using the Perceived

Explainability Index (PEI) and evaluated AI adoption likelihood over time. Table 5 tracks how improvements in AI explainability increase the Perceived Explainability Index (PEI) and corresponding AI adoption rates across iterations.

Table 5. AI Trust Calibration & Adoption Likelihood Over Iterations

Iteration	Trust Score (PEI)	AI Adoption Likelihood (%)
10	0.45	20%
50	0.72	65%
100	0.88	90%

The findings suggest that as AI explainability improves, trust in AI-driven recommendations increases proportionally. At early iterations, the PEI score remains low (0.45), corresponding with limited AI adoption (20%). However, as the AI system provides more interpretable and reliable decision insights, trust scores rise to 0.88, leading to a 90% AI adoption rate by the 100th iteration.

4.5 M&A Case Study: Practical Validation of NOVA + DAMA

To demonstrate the practical application of the NOVA + DAMA Framework is applied to optimize decision-making in a mergers & acquisitions (M&A) scenario. The goal is to determine whether an executive team should pursue a full acquisition, partial stake, joint venture, or alternative acquisition strategy, integrating AI-driven insights with human expertise while mitigating cognitive biases.

Step 1: Assigning Decision Weights ($_{1}$, $_{2}$, $_{3}$) Based on Mathematical Findings

The mathematical findings from Monte Carlo simulations (Table 4) and regression modeling (Table 5) provided the foundation for assigning decision weights in the DAMA framework. The decision weight assignments in the DAMA framework ($_{1}$, $_{2}$, $_{3}$) are computationally derived from prior simulation findings. Specifically, the regression analysis in Sect. 4.4 confirms that AI trust calibration (PEI) and cognitive bias mitigation significantly influence decision efficacy, while Monte Carlo simulations validate the incremental impact of AI augmentation. The decision weight allocations in Table 6 are computationally grounded in simulation and regression findings, reflecting realistic executive-AI interaction dynamics.

Table 6. AI-Augmented Decision Weight Assignments

Decision Component	Symbol	Assigned Weight ()	Rationale
AI-Generated Insights		0.45	AI provides strong financial modeling but lacks qualitative judgment.

(*continued*)

Table 6. *(continued)*

Decision Component	Symbol	Assigned Weight ()	Rationale
Human Heuristics		0.35	Executive intuition is crucial but prone to biases.
Bias Compensation Factor	()	0.20	Bias adjustments prevent overconfidence in valuation.

Thus, the refined DAMA decision function becomes:

$$D^* = 0.45A + 0.35H \; - 0.20(D) + (A, H) \cdot T(A) \tag{5}$$

where:

- $C(A, H)$ represents counterfactual strategy discovery, identifying optimal alternative paths.
- $T(A)$ represents trust calibration, ensuring AI explainability.

Step 2: Probabilistic Modeling of Counterfactual Strategy Discovery

Monte Carlo simulations were conducted to evaluate alternative strategies. The expected success probability and risk factor were derived using historical executive decision outcomes. Table 7 presents the expected success probabilities and associated risks for four strategic M&A options analyzed using Bayesian counterfactual modeling.

Table 7. Counterfactual Strategy Modeling in M&A Decisions

Scenario	Probability of Success (())	Risk Factor (())
Full Acquisition	55%	High (0.75)
Partial Stake (51%)	72%	Moderate (0.40)
Joint Venture	68%	Moderate (0.50)
Alternative Acquisition	50%	High (0.80)

A Bayesian decision analysis framework refines this evaluation by iterating over multiple simulated risk scenarios, reinforcing Scenario 2 (Partial Stake, 51%) as the most statistically optimal choice. Using Bayesian decision analysis, the model refines the optimal counterfactual path as

$$\text{Arg max}[(D) - R(D)] \tag{6}$$

which selects Scenario 2 (Partial Stake Acquisition, 51%) as the most strategically sound option.

Step 3: Trust Calibration via Perceived Explainability Index (PEI, ())

Executives evaluate AI recommendations based on their explainability and interpretability. The computation of the Perceived Explainability Index (PEI), as detailed in Table 8, incorporates AI transparency, interpretability, and justification strength.

Table 8. Trust Calibration Metrics in AI Decision-Augmentation

Explainability Metric	Weight ()	Score (0–1)
Transparency	0.40	0.85
Interpretability	0.35	0.75
Justification Strength	0.25	0.65

The final Perceived Explainability Index (PEI) is computed as:

$$T(A) = (0.40 \times 0.85) + (0.35 \times 0.75) + (0.25 \times 0.65) = 0.77 \tag{7}$$

Since $T(A) > 0.70$, the AI recommendation is accepted without override by human executives.

Final AI-Augmented Decision Calculation ($D*$)

Using the optimized DAMA model, the final decision is calculated as:

$$D^* = 0.45(0.72) + 0.35(0.85) - 0.20(0.40) + (0.72 \times 0.77) \tag{8}$$

$$D^* = 0.324 + 0.297 - 0.08 + 0.554 = 1.095 \tag{9}$$

Since $D* > 1.00$, the AI-Augmented Leadership system approves the Partial Stake (51%) Acquisition as the optimal strategy.

4.6 Strategic Insights

- Bias Mitigation: NOVA + DAMA achieves over 62% reduction in cognitive bias.
- Decision Accuracy: Outperforms human-only models by up to 26.1%.
- AI Trust Calibration: PEI is a decisive adoption lever ($\beta = 0.72$, $p < 0.001$).
- Scalability: The model generalizes across domains like M&A, crisis response, and market entry.

These findings establish AI-Augmented Leadership not as a futuristic concept, but as a pragmatic, mathematically validated paradigm ready for enterprise deployment. By harmonizing explainable AI with executive intuition, the NOVA + DAMA framework sets a new benchmark in leadership decision science.

5 Conclusions and Strategic Implications

This study establishes a rigorous foundation for AI-Augmented Leadership, integrating Multi-Agent AI Optimization through the NOVA Framework and DAMA Model. By combining cognitive modeling, Bayesian bias correction, trust calibration via the Perceived Explainability Index (PEI), and simulation-driven validation, the research demonstrates how AI can enhance—not replace—executive decision-making. The study's methodological innovation lies in combining mathematical modeling with cognitive alignment principles — a synthesis rarely seen in existing AI decision-augmentation research.

Simulation results confirm that AI-Augmented Leadership significantly improves decision quality. Trust in AI rises with explainability, bias is reduced by over 60%, and decision efficacy improves through dynamic adjustments of human heuristics. NOVA + DAMA consistently outperforms both human-only and traditional AI approaches.

Strategic Implications

- Corporate Governance: AI-augmented models offer strategic foresight in M&A, risk mitigation, and crisis response.
- AI Policy & Ethics: Organizations must integrate trust-based metrics like PEI into governance frameworks.
- Executive Development: Hybrid training is essential to foster human-AI collaboration, positioning AI as a cognitive partner.

Future research should validate these models using real-world executive data, refine dynamic trust metrics, and enhance reinforcement learning capabilities within NOVA. Cross-industry studies will help expand the framework's external validity.

AI-Augmented Leadership is no longer a futuristic ideal, it is a strategic imperative. As organizations confront increasing uncertainty, AI must be designed for explainability, adaptability, and synergy with human judgment. This paradigm marks a pivotal shift toward intelligent, trustworthy, and resilient executive decision-making.

References

Brynjolfsson, E., McAfee, A.: The business of artificial intelligence. Harv. Bus. Rev. (2017). https://hbr.org/2017/07/the-business-of-artificial-intelligence

Doshi-Velez, F., Kim, B.: Towards a rigorous science of interpretable machine learning. arXiv:1702.08608 (2017)

Gunning, D., Aha, D.W.: DARPA's explainable artificial intelligence (XAI) program. AI Mag. **40**(2), 44–58 (2019). https://doi.org/10.1609/aimag.v40i2.2850

Kahneman, D.: Thinking, Fast and Slow. Farrar, Straus and Giroux, New York (2011)

Lowe, R., Wu, Y., Tamar, A., Harb, J., Abbeel, P., Mordatch, I.: Multi-agent actor-critic for mixed cooperative-competitive environments. In: Advances in Neural Information Processing Systems (NeurIPS) (2017)

Metropolis, N., Ulam, S.: The Monte Carlo Method. Journal of the American Statistical Association, **44**, 335–341 (1949). https://doi.org/10.1080/01621459.1949.10483310

Puterman, M. L.: Markov Decision Processes: Discrete Stochastic Dynamic Programming. John Wiley & Sons Ltd., New York (1994). https://doi.org/10.1002/9780470316887

Rai, A.: Explainable AI: from black box to glass box. J. Acad. Mark. Sci. **48**, 137–141 (2020). https://doi.org/10.1007/s11747-019-00710-5

Rogers, E.M.: Diffusion of Innovations, 5th edn. Free Press, New York (2003)

Russell, S.J., Norvig, P.: Artificial Intelligence: A Modern Approach, 4th edn). Pearson (2021). https://doi.org/10.1109/MSP.2017.2765202

Silver, D., et al.: Mastering the game of go with deep neural networks and tree search. Nature. **529**(7587), 484–489 (2016)

Sutton, R.S., Barto, A.G.: Reinforcement Learning: An Introduction. MIT Press, Cambridge (2018)

Tversky, A., Kahneman, D.: Judgment under uncertainty: heuristics and biases. Science. **185**(4157), 1124–1131 (1974). https://doi.org/10.1126/science.185.4157.1124

Packages Design for Comorbid Psychiatric Disorders Based on Optimization of Disease Burden

Yaqi Zhang[1], Ying Chen[2], and Wenjie Yi[1](✉)

[1] College of Management, Shenzhen University, Shenzhen 518055, China
2300132019@email.szu.edu.cn, wenjie.yi@szu.edu.cn
[2] School of Artificial Intelligence, Shenzhen University, Shenzhen 518060, China

Abstract. In the context of achieving Universal Health Coverage, the allocation of healthcare resources faces complex decision-making challenges. A more scientifically sound and effective strategy is needed to balance the dynamic interplay between medical insurance cost control, clinical drug efficacy, and patient expenditure. Comorbid psychiatric disorders, as refractory diseases with high treatment costs, impose a heavy burden on both society and patients' families. This study proposes a Bi-Objective Optimization Model for the Burden of Depression and Anxiety (BDAA), aiming to optimize disease burden at both societal and household levels. To address the high-dimensional and non-convex model, a tailored solution framework named PSO for Burden of Depression and Anxiety (BDAAPSO) which is based on swarm intelligence algorithms is designed. The framework employs a tripartite population strategy for division of labor and collaboration, incorporating a particle adjustment mechanism to enhance both convergence speed and solution diversity in Pareto front exploration. The proposed burden optimization approach provides a scientifically grounded and practically viable decision-making tool for mental health resource allocation, contributing to the dynamic equilibrium between medical equity and fund sustainability under the Sustainable Development Goal 3.

Keywords: Service package · Depression · Constrained optimization · Burden of Disease

1 Introduction

Comorbidity is highly prevalent among individuals with mental disorders [1]. Empirical evidence indicates that approximately 50% of patients with mental disorders experience two or more psychiatric diagnoses within a 12-month period [2], with the comorbidity of depression and anxiety being the most common [3]. A large-scale study in the Netherlands reported that 67% of patients diagnosed with depression also met the criteria for anxiety disorders at onset, and up to 75% experienced both conditions at some point during their lifetime [4]. The comorbidity of depression and anxiety encounter multiple challenges in clinical practice, resulting in a generally low rate of guideline-concordant treatment among affected patients [5]. Patients with comorbid depression

X. Xie et al. (Eds.): ICSS 2025, LNOR, pp. 223–235, 2026.
https://doi.org/10.1007/978-3-032-13116-4_18

and anxiety often suffer from exacerbated symptoms and prolonged illness duration, leading to increased labour force loss and greater utilization of healthcare resources. Simultaneously, the high cost of long-term pharmacological treatment imposes substantial financial burdens on families. Therefore, optimizing the disease burden of patients with comorbid depression and anxiety has become a critical issue in public health and health insurance policy.

The current research on disease burden primarily focuses on burden assessment. Years of Life Lost (YLL) and Years Lived with Disability (YLD) are the main indicators used to measure disease burden. Existing studies based on the Global Burden of Disease project and regional research have quantified the disease burden associated with depression, anxiety, and their comorbidity. These studies have also revealed trends and disparities at the global, regional, and national levels [6, 7]. Disease burden optimization involves systematically quantifying and analysing the socioeconomic and health impacts of diseases. It employs the formulation of resource allocation strategies using mathematical models and optimization methods, with the aim of maximizing health benefits while minimizing costs. Existing research on disease burden optimization predominantly adopts single-objective optimization methods. For example, Strand et al. used cost-effectiveness analysis (CEA) to assess the economic feasibility of expanding basic neuropsychiatric services in Ethiopia [8]. They employed the cost per Disability-Adjusted Life Years as an indicator to develop a priority procurement list for antidepressants and expand mental health service coverage, thereby effectively reducing the disease burden.

In summary, existing research on disease burden has predominantly centered on the assessment domain. In contrast, the relatively scarce studies on disease burden optimization predominantly resort to Cost-Effectiveness Analysis (CEA) for decision-making. Regrettably, this approach is inadequate in harmonizing the efficiency of health insurance fund utilization with the financial burden borne by households. Thus, this study innovatively proposes a multi-objective optimization model that balances the dual dimensions of disease burden at both the societal and household levels. At the societal level, the model aims to maximize the clinical effectiveness of interventions, enhance reductions in YLLs and YLDS, and improve the efficiency of healthcare fund utilization. At the household level, the objective is to minimize out-of-pocket (OOP) expenses, alleviate financial hardship, and reduce the risk of catastrophic health expenditures. Multi-objective optimization problem is mostly an NP-hard problem, which highlights the complexity of finding an optimal solution. In response to the challenge, this study improves the standard Multi-Objective Particle Swarm Optimization (MOPSO) [9] algorithm by integrating problem-specific characteristics. These adjustments improve its adaptability to dynamic objectives across different scenarios and significantly boost its ability to identify feasible solutions. This optimization model provides a dynamic decision-making tool for the precise allocation of mental health resources, with significant practical and application value.

2 The Theoretical Model of BDAA

A Bi-Objective Optimization Model for the Burden of Depression and Anxiety (BDAA) is developed in this study. The model focuses on the design of service packages for comorbid psychiatric disorders. In the BDAA model, decision variable x_i is assigned a value of 1 if the given medication i is included in the optimized social service package for comorbid psychiatric disorders; otherwise, it is set to 0. The notation used in the mathematical formulation is summarized in Table 1.

Table 1. Sets and parameters

Symbols	Definitions
I	Set of medications in the service package
B	Set of unit costs for medications
R	Set of reimbursement ratios for medications
ED	Set of clinical efficacy values for antidepressants
EA	Set of clinical efficacy values for antianxiety medications
DD	Set of daily dosages for antidepressants
DA	Set of daily dosages for antianxiety medications
C	Set of feasible medication combinations for comorbid treatment
P	Set of average treatment durations per medication
$\overline{u}$	Upper limit of regional insurance expenditure for depression prevention
$\overline{h}$	Average household income in the region
s	Threshold for catastrophic health expenditure
De_g	Estimated number of depression patients in the region
An_g	Estimated number of anxiety patients in the region

2.1 Societal-Level Disease Burden Objective

The societal burden of disease is commonly measured using YLL and YLD. Antidepressant medications with better clinical efficacy can reduce patients' risk of suicide and improve their functional and work capacities, thereby decreasing both YLL and YLD. Therefore, this study defines the optimization objective for the societal dimension as the maximization of treatment efficacy. Lofgren et al. (2021) employed a simplified health-state model with assumed intervention effects to simulate the projected population health and fiscal risk protection (FRP) impacts of each intervention [10]. To enhance the model's real-world applicability, the therapeutic efficacy of pharmacological treatments for depression and anxiety was evaluated using standardized metrics: (1) odds ratios (ORs) for active medications versus placebo [11], and (2) Cohen's d values [12]. These parameters quantify the potential disease burden reduction achievable through antidepressant therapies at the societal level. Measuring the optimization effect of service packages on the social disease burden by using the intervention effects of drugs can, to some extent, avoid the ethically debatable issues that may arise when monetizing human health to measure the impact of diseases on different populations [13].

$$\max\ S = \sum_{i \in I} ed_i x_i + \sum_{i \in I} ea_i x_i \tag{1}$$

2.2 Household-Level Disease Burden Objective

Some psychiatric medications used in clinical treatment have been included in the national health insurance reimbursement list. The reimbursement rate of insured drugs r_i serves as a direct link between medical insurance cost control and household financial burden. It can reduce the unit OOP cost of medications $b_i(1 - r_i)$ to varying degrees and lower the risk of catastrophic health expenditures for patients' families. Therefore, this study uses the household OOP cost of medications to characterize the mental health burden at the household level.

$$\min\ F = \sum_{i \in I} x_i b_i (1 - r_i) \max(dd_i, da_i) p_i \tag{2}$$

In addition, this study reflects practical decision-making considerations such as government budget constraints and household financial risk within the model's constraints. The threshold for catastrophic health expenditure s is set at 40%, based on the definition provided by the World Health Organization. The average household income $\overline{h}$ is obtained from the official data released by the city's statistical department.

2.3 Practical Constraints in Optimizing the Burden of Comorbid Psychiatric Disorders

Equation (3) ensures that a decision is made for each candidate medication in the pool regarding whether it is included in the intervention package. For research informing health benefit package decisions in low- and middle-income countries (LMICs), careful consideration of health sector budget constraints is paramount [14]. Accordingly, Eqs. (4) and (5) represent the budget constraints of local governments for mental health services. Considering the complexity of actual medication use among patients with mental disorders, these constraints account for an extreme scenario in which all patients switch medications. To simplify the model, it is assumed that the treatment expenditures of patient families are direct drug costs. Equation (6) imposes a constraint to safeguard household financial stability, aiming to prevent catastrophic health expenditures resulting from OOP medication costs.

$$\frac{x_i}{\max(1, x_i)} = 1, \forall i \in I \tag{3}$$

$$\sum_{i \in I} x_i b_i r_i dd_i p_i De_g \le \overline{u}, \forall i \in I \tag{4}$$

$$\sum_{i \in I} c_i x_i b_i r_i \cdot \max(dd_i, da_i) p_i \cdot \max\left(De_g, An_g\right) \leq \overline{u}, \forall i \in I \tag{5}$$

$$\frac{\sum_{i \in I} c_i x_i b_i (1-r_i) \cdot \max(dd_i, da_i) \cdot p_i + \sum_{i \in I} (1-c_i) x_i b_i (1-r_i) dd_i p_i}{\overline{h}} \leq s, \forall i \in I \tag{6}$$

3 Data

To enhance practical relevance of the problem-solving process and evaluate the scalability and robustness of the proposed algorithm, this study utilizes empirical data from a first-tier city in China to construct instances of varying sizes, namely BDAA_Small, BDAA_Middle, and BDAA_Large. The fundamental parameter settings for each instance are summarized in Table 2. Among them, the prevalence rate of depression is taken from the average value of cities with comparable economic development levels in the 2023 Blue Book of China's Mental Health [15]. The prevalence rate of anxiety disorder is obtained from a national study in 2021 [16].

Table 2. Basic parameters of instances with different scales

Instance	Budget (10,000 CNY)	Population (10,000 persons)	Depression Incidence Rate	Anxiety Incidence Rate	Medication Duration (days)
BDAA_Small	4818	152.1	8%	2.25%	14
BDAA_Middle	26268	1779.01	10%	2.25%	14
BDAA_Large	26268	1779.01	12%	3.48%	28

Selective serotonin reuptake inhibitors (SSRIs) and serotonin and norepinephrine reuptake inhibitors (SNRIs) are considered first-line pharmacological treatments [17]. Medication changes are common in the treatment process for patients with mental disorders, especially those with comorbid conditions. Therefore, it is essential that the service package includes a wide variety of medications. The candidate set of medications for the service package comprises 14 clinically proven SSRIs or SNRIs with high efficacy and accessibility. The basic parameters of these medications are presented in Table 3.

Table 3. Basic parameters of medications in the candidate set

Drug	Medication Name	OOP rate	CNY	Dep-OR	Dep-mg	Anx-d	Anx-mg
1	Amitriptyline Hydrochloride Tablets	20%	19.84	2.13	50	0	0
2	Clomipramine Hydrochloride Tablets	20%	15.70	1.49	75	0	0

(continued)

Table 3. (*continued*)

Drug	Medication Name	OOP rate	CNY	Dep-OR	Dep-mg	Anx-d	Anx-mg
3	Escitalopram Oxalate Tablets	20%	6.10	1.68	20	0.45	10
4	Fluoxetine Hydrochloride Capsules	20%	9.19	1.52	20	0.56	40
5	Fluvoxamine Maleate Tablets	40%	53.00	1.69	100	0.56	100
6	Sertraline Hydrochloride Tablets	40%	8.30	1.67	50	0.54	50
7	Citalopram Hydrobromide Tablets	40%	6.00	1.52	20	0.23	20
8	Paroxetine Hydrochloride Tablets	40%	10.57	1.75	20	0.42	20
9	Venlafaxine Hydrochloride Capsules	20%	11.00	1.78	75	0.33	75
10	Duloxetine Hydrochloride Enteric-coated Capsules	40%	19.60	1.85	60	0.53	60
11	Agomelatine Tablets	40%	137.60	1.65	50	0	0
12	Bupropion Hydrochloride Sustained-release Tablets	100%	37.00	1.58	150	0	0
13	Desvenlafaxine Succinate Extended-release Tablets	100%	89.60	1.49	50	0	0
14	Vortioxetine Hydrobromide Tablets	100%	9.87	1.66	10	0	0

4 Methods

4.1 BDAAPSO Algorithm Design

In this study, the social service package design problem, BDAA, is defined. Optimized based on the burden of comorbid mental illnesses, BDAA presents multiple optimization objectives and is constrained by real-world problems. As a result, it is classified as an NP-hard problem, highlighting the complexity of finding an optimal solution. In response to the challenges of solving the BDAA problem, this study proposes the BDAAPSO (PSO for Burden of Depression and Anxiety) algorithm. By integrating the characteristics of the BDAA model, a particle perturbation strategy and a particle correction strategy are introduced. BDAAPSO can meet the requirements for solving nonlinear and multi-objective optimization problems in the context of designing social service packages for comorbid mental illnesses.

The velocity and position update methods of the traditional Multi-Objective Particle Swarm Optimization (MOPSO) algorithm [9] are shown in Eqs. (7) and (8).

When addressing the BDAA problem, the Multi-Objective Particle Swarm Optimization (MOPSO) algorithm confronts the challenge of sparse feasible solutions. Moreover, due to the single evolution characteristic, MOPSO is highly prone to converge to a local optimum. Therefore, a perturbation strategy combining targeted perturbation and global perturbation is adopted to improve the solution efficiency of the algorithm.

$$v_{p1}^{d}(t+1) = w_1 v_{p1}^{d} + c_1 r_1 \left(Pbest_{p1}^{d} - x_{p1}^{d}(t)\right) + c_2 r_2 \left(Gbest^{t} - x_{p1}^{d}(t)\right) \tag{7}$$

$$x_{p1}^{d}(t+1) = x_{p1}^{d}(t) + v_{p1}^{d}(t+1) \tag{8}$$

First, perform sparse targeted perturbation on the all - zero velocity vector, as shown in Eq. (9). If the condition $\left\| v_i^{pert} \right\|_2 = 0$ still holds after adding the targeted perturbation, apply global perturbation, as shown in Eq. (10).

$$v_i^{pert} = v_i^{old} + 0.1 \bullet \sum\nolimits_{j \in J} \epsilon_j e_j \tag{9}$$

$$v_i^{pert} = 0.1 \cdot U(-1, 1)^{D} \tag{10}$$

When balancing the disease burden of depression in society and families, the BDAA model needs to take into account rather complex practical constraints. Under the condition of the initial solution assigned randomly, the feasible solution space of this model shows sparsity. Therefore, a correction mechanism has been designed for particles that are prone to violating the constraint limits, so as to increase the probability of the occurrence of feasible solutions and the computational efficiency. The particle correction strategy will be initiated when the medical insurance reimbursement amount exceeds the pre-set upper limit, or when the out-of-pocket expenses of the family exceed the health disaster threshold. First, a priority index sequence is formed based on the cost-effectiveness attributes of the drugs. Starting from the drug with the highest cost, the corresponding position in the drug distribution plan is successively marked as invalid x = 0, and at the same time, the cumulative cost of the excluded drugs is recorded. By preferentially eliminating drugs with high prices, this strategy minimizes the number of drugs excluded while optimizing the efficiency of resource allocation and maintaining the treatment coverage.

4.2 Experimental Studies

To comprehensively evaluate the performance of the BDAAPSO algorithm on the BDAA problem, five baseline algorithms are selected with consideration of their applicability, novelty, and diversity. Specifically, tDEACPBI [18], CCMO [19], ANSGAIII [20], SNS-GAII [21], and MOPSO [9] are adopted for comparison. The parameter settings for these algorithms are configured according to the recommendations provided in the original references, as implemented in the PlatEMO platform [22]. To guarantee fairness, the population size for all swarm intelligence algorithms is uniformly set to 30, and the maximum number of iterations is fixed as 10,000 to ensure adequate convergence. Each algorithm is independently run 30 times on each test function, and the mean and standard error of the results are reported.

Table 4. Hypervolume (HV) values of all algorithms on different scale test cases

Problem	tDEACPBI	CCMO	ANSGAIII
BDAA_Small	7.4964e + 3 (5.00e − 1) −	7.4965e + 3 (6.76e − 1) −	7.4966e + 3 (5.37e − 1) −
BDAA_Middle	7.4965e + 3 (5.96e − 1) −	7.4963e + 3 (6.41e − 1) −	7.4964e + 3 (5.38e − 1) −
BDAA_Large	7.4766e + 3 (1.23e + 0) −	7.4760e + 3 (1.00e + 0) −	7.4767e + 3 (1.13e + 0) −
+/−/=	0/3/0	0/3/0	0/3/0
Problem	**SNSGAII**	**MOPSO**	**BDAAPSO**
BDAA_Small	7.4965e + 3 (5.60e − 1) −	7.4967e + 3 (4.88e − 1) −	**9.7491e + 3 (1.77e − 1)**
BDAA_Middle	7.4967e + 3 (6.40e − 1) −	7.4965e + 3 (7.17e − 1) −	**9.7491e + 3 (1.40e − 1)**
BDAA_Large	7.4765e + 3 (9.88e − 1) −	7.4765e + 3 (1.21e + 0) −	**9.7482e + 3 (3.21e − 1)**
+/−/=	0/3/0	0/3/0	

Table 4 presents the HV values of the BDAAPSO algorithm and the five baseline algorithms across different problem scales. The results indicate that BDAAPSO consistently outperforms all comparison algorithms across all three instances, demonstrating stable and robust search capabilities within the solution space of the BDAA model. Moreover, BDAAPSO exhibits superior convergence toward the approximated true Pareto front while effectively preserving solution diversity.

Figure 1 presents the maximum runtime per execution for each algorithm. For small- and medium-scale instances, although the runtime of BDAAPSO differs slightly from that of the baseline algorithms, it remains within an acceptable range. In large-scale instances, the maximum runtime of BDAAPSO exceeds that of the comparison algorithms by less than 10%. These results suggest that BDAAPSO achieves a favorable balance between computational efficiency and solution quality. Moreover, as the problem scale increases, the relative runtime overhead of BDAAPSO exhibits a clear convergence trend, further demonstrating its scalability and robustness in addressing complex scenarios.

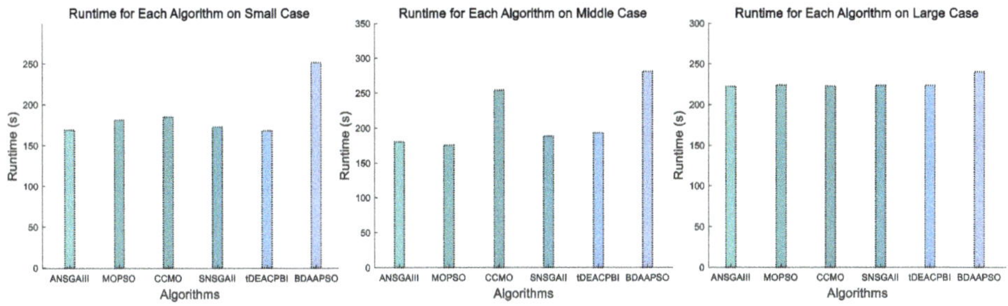

Fig. 1. Comparison of algorithm runtime across different test case scales

5 Results

The BDAA model demonstrates strong flexibility in addressing complex real-world scenarios, enabling the generation of diverse reference solutions for designing social service packages for comorbid psychiatric disorders under varying instance parameters. Table 5 summarizes the objective values of the nondominated solutions obtained for BDAA problems of different scales, along with the corresponding cost-effectiveness and the number of medication types associated with each solution.

Based on the nondominated solutions across the three problem scales, the recommendation frequencies of 14 antidepressant medications are summarized in Fig. 2. The frequencies vary considerably among the 14 medications, with Drugs 11, 12, and 13 being recommended significantly less often. From the perspective of disease burden, these three medications impose a relatively heavy financial load on patients' households due to their high prices and cost-sharing ratios. Specifically, Drug 11 (Agomelatine Tablets) has a cost-sharing ratio of 40%, while Drug 12 (Bupropion Hydrochloride Sustained-Release Tablets) and Drug 13 (Desvenlafaxine Succinate Sustained-Release Tablets) have cost-sharing ratios of 100%. Moreover, Agomelatine Tablets and Desvenlafaxine Succinate Sustained-Release Tablets are priced at 137.60 CNY and 89.60 CNY, respectively, making them the two most expensive medications among the 14 candidates. In terms of clinical efficacy, these medications show limited therapeutic effectiveness for depression and are not indicated for the treatment of patients with comorbid depression and anxiety.

By jointly considering medication costs and benefits, the BDAA model consistently minimizes the inclusion of these three drugs across different instances, demonstrating the soundness of its design. Operating within the feasible scope of medical insurance budgets, the model systematically evaluates medications based on cost-sharing ratios, prices, and clinical efficacy. As a result, it achieves a more effective balance between clinical value and economic accessibility. This capability further contributes to reducing the overall burden associated with comorbid psychiatric disorders.

Table 5. Summary of dominant solutions across different instances

Household Burden	Drug Efficacy	C/E	Type of Drugs
BDAA_Large			
52.51	−11.86	4.43	6
94.84	−20.11	4.72	10
134.41	−22.36	6.01	11
208.41	−23.94	8.71	12
387.61	−25.43	15.24	13
BDAA_Middle			
33.80	−16.37	2.06	8
57.27	−20.70	2.77	10
67.21	−22.36	3.01	11
104.21	−23.94	4.35	12
193.81	−25.43	7.62	13
BDAA_Small			
99.98	−21.80	4.59	11
101.57	−22.50	4.51	11

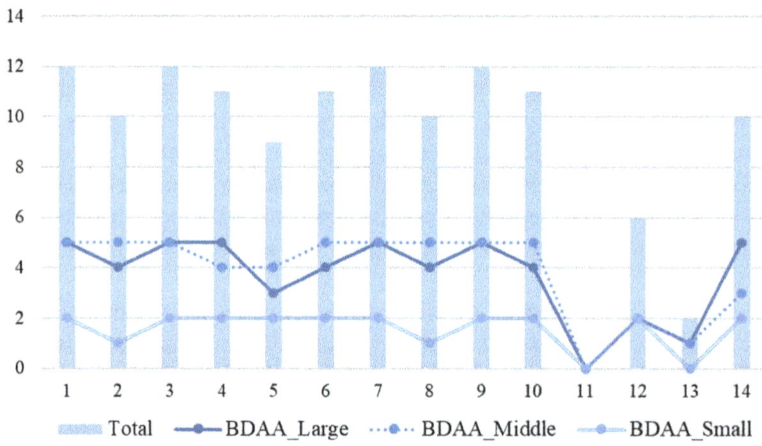

Fig. 2. Comparison of recommendation frequencies for 14 medications

6 Conclusion

This study adopts a balanced perspective rooted in public health complex systems theory, focusing on the optimization of disease burden. It innovatively constructs a dual-perspective society-household collaborative optimization framework, providing an operational methodology for the design of social service packages targeting comorbid psychiatric disorders. By developing the BDAA bi-objective optimization model, the study integrates social disease burden and household economic risk into a unified analytical system, achieving a dynamic balance between the efficiency of medical insurance fund utilization and patients' out-of-pocket expenses. The model offers a scientific basis for decision-making in managing the complex relationship between medical cost control, clinical efficacy, and patient expenditures.

However, this study has certain limitations. First, the current model is limited to the most common comorbidity of depression and anxiety, and does not fully capture the diversity of other complex comorbid psychiatric disorders. Second, the variables and quantitative indicators used to characterize disease burden from different perspectives are relatively simplified and do not sufficiently account for real-world factors such as adverse drug reactions and long-term treatment outcomes. To address these limitations, future research could advance in the following directions: expanding the set of candidate medications to encompass treatments for a broader spectrum of mental health conditions; and optimizing the model's variable design by incorporating dynamic parameters and machine learning algorithms to enhance its adaptability to complex healthcare environments. These efforts may contribute to the development of more refined and intelligent disease burden optimization models, thereby providing stronger scientific support for public health policy formulation and healthcare system improvement.

Acknowledgements. This work was supported by the National Natural Science Foundation of China (Project No.72401202), Shenzhen Higher Education Support Plan (Project No. 20231120174835002), and Postgraduate Independent Innovation Achievement Cultivation Project of Shenzhen University (No. 868-000002020220). The funders had no role in study design, data collection, data analysis, data interpretation, or writing of the report.

References

1. Kessler, R.C.: Lifetime and 12-month prevalence of DSM-III-R psychiatric disorders in the United States: results from the national comorbidity survey. Arch. Gen. Psychiatry. **51**, 8 (1994). https://doi.org/10.1001/archpsyc.1994.03950010008002
2. Cameron, O.G.: Understanding comorbid depression and anxiety. Psychiatr. Times. **24**(14), 51–51 (2007). https://link.gale.com/apps/doc/A172186316/AONE?u=anon~b09d9b74&sid=googleScholar&xid=d90aee0a
3. Felez-Nobrega, M., Haro, J.M., Koyanagi, A.: Multimorbidity, depression with anxiety symptoms, and decrements in health in 47 low- and middle-income countries. J. Affect. Disord. **317**, 176–184 (2022). https://doi.org/10.1016/j.jad.2022.08.110
4. Lamers, F., et al.: Comorbidity patterns of anxiety and depressive disorders in a large cohort study: The Netherlands study of depression and anxiety (NESDA). J. Clin. Psychiatry. **72**, 341–348 (2011). https://doi.org/10.4088/JCP.10m06176blu

5. Jha, M.K., et al.: Mental health-related disability days and costs among patients with treatment-resistant depression initiated on esketamine nasal spray and conventional therapies in the USA. Drugs Real World Outcomes. **12**, 35–43 (2025). https://doi.org/10.1007/s40801-024-00478-3
6. GBD 2019 Mental Disorders Collaborators: Global, regional, and national burden of 12 mental disorders in 204 countries and territories, 1990–2019: a systematic analysis for the Global Burden of Disease Study 2019. Lancet Psychiatry. **9**, 137–150 (2022). https://doi.org/10.1016/S2215-0366(21)00395-3
7. Tian, W.: et al., Burden of depressive and anxiety disorders in China and its provinces, 1990–2021: findings from the Global Burden of Disease Study 2021. Br. J. Psychiatry, 1–11 (2025). https://doi.org/10.1192/bjp.2024.267
8. Strand, K.B., Chisholm, D., Fekadu, A., Johansson, K.A.: Scaling-up essential neuropsychiatric services in Ethiopia: a cost-effectiveness analysis. Health Policy Plan. **31**, 504–513 (2016). https://doi.org/10.1093/heapol/czv093
9. Coello Coello, C.A., Lechuga, M.S.: MOPSO: a proposal for multiple objective particle swarm optimization. In: Proceedings of the 2002 Congress on Evolutionary Computation. CEC'02 (Cat. No.02TH8600), Honolulu, HI, USA, vol. 2, pp. 1051–1056. IEEE (2002). https://doi.org/10.1109/CEC.2002.1004388
10. Lofgren, K.T., Watkins, D.A., Memirie, S.T., Salomon, J.A., Verguet, S.: Balancing health and financial protection in health benefit package design. Health Econ. **30**(12), 3236–3247 (2021). https://doi.org/10.1002/hec.4434
11. Furukawa, T.A., et al.: Comparative efficacy and acceptability of first-generation and second-generation antidepressants in the acute treatment of major depression: protocol for a network meta-analysis. BMJ Open. **6**, e010919 (2016). https://doi.org/10.1136/bmjopen-2015-010919
12. Bandelow, B., Reitt, M., Röver, C., Michaelis, S., Görlich, Y., Wedekind, D.: Efficacy of treatments for anxiety disorders: a meta-analysis. Int. Clin. Psychopharmacol. **30**, 183–192 (2015). https://doi.org/10.1097/YIC.0000000000000078
13. Scovronick, N., Ferranna, M., Dennig, F., Budolfson, M.: Valuing health impacts in climate policy: ethical issues and economic challenges: study examines the ethical issues and economic challenges of valuing health impacts in climate policy. Health Aff. **39**(12), 2105–2112 (2020). https://doi.org/10.1377/hlthaff.2020.01117
14. Karsu, O., Morton, A.: Trading off health and financial protection benefits with multiobjective optimization. Health Econ. **30**, 55–69 (2021). https://doi.org/10.1002/hec.4176
15. Good Mood, Mental Health Branch of China Narcotics Association: The 2023 Blue Book of China's Mental Health (2023)
16. Tian, W., et al.: Burden of depressive and anxiety disorders in China and its provinces, 1990–2021: findings from the Global Burden of Disease Study 2021. Br. J. Psychiatry, 1–11 (2025). https://doi.org/10.1192/bjp.2024.267
17. Gosmann, N.P., et al.: Selective serotonin reuptake inhibitors, and serotonin and norepinephrine reuptake inhibitors for anxiety, obsessive-compulsive, and stress disorders: a 3-level network meta-analysis. PLoS Med. **18**, e1003664 (2021). https://doi.org/10.1371/journal.pmed.1003664
18. Ming, F., Gong, W., Wang, L., Gao, L.: A constraint-handling technique for decomposition-based constrained many-objective evolutionary algorithms. IEEE Trans Syst Man Cybern Syst. **53**, 7783–7793 (2023). https://doi.org/10.1109/TSMC.2023.3299570
19. Tian, Y., Zhang, T., Xiao, J., Zhang, X., Jin, Y.: A coevolutionary framework for constrained multiobjective optimization problems. IEEE Trans. Evol. Comput. **25**, 102–116 (2021). https://doi.org/10.1109/TEVC.2020.3004012
20. Jain, H., Deb, K.: An evolutionary many-objective optimization algorithm using reference-point based nondominated sorting approach, part II: handling constraints and extending to

an adaptive approach. IEEE Trans. Evol. Comput. **18**(4), 602–622 (2013). https://doi.org/10.1109/TEVC.2013.2281535
21. Kalyanmoy, D., Himanshu, J.: An evolutionary many-objective optimization algorithm using reference-point-based nondominated sorting approach, part I: solving problems with box constraints. IEEE Trans. Evol. Comput. **18**(4), 577–601 (2014). https://doi.org/10.1109/TEVC.2013.2281535
22. Tian, Y., Cheng, R., Zhang, X., Jin, Y.: PlatEMO: a MATLAB platform for evolutionary multi-objective optimization. IEEE Comput. Intell. Mag. **12**, 73–87 (2017). https://doi.org/10.1109/MCI.2017.2742868

Optimizing Bus Fleet Energy Composition: Balancing Operator Costs and Carbon Emissions in Transit Network

Xinyi Zhang[1], Xi Lin[1], Xumei Chen[1](✉), and Yutong Wang[2]

[1] Key Laboratory of Transport Industry of Big Data Application Technologies for Comprehensive Transport, Ministry of Transport, Beijing Jiaotong University, Beijing 100044, P.R. China
{24125710,24125702,xmchen}@bjtu.edu.cn
[2] Beijing Transport Development Centre, Beijing 101160, P.R. China
16120899@bjtu.edu.cn

Abstract. The transition to new energy in the public transportation sector has become a critical strategy for achieving sustainable urban development and reducing carbon emissions. This study analyses the impact of bus fleet energy composition, specifically the proportions of battery electric buses (BEBs), liquefied natural gas (LNG) buses, and diesel buses, on operator costs and carbon emissions in urban transit networks. A multi-objective optimization model is proposed to minimize passenger travel costs, operator costs, and carbon emissions, while incorporating constraints on the minimum proportions of buses with different energy types to reflect the trend of transitioning to new energy in the public transportation industry. And the NSGA-II algorithm is employed to solve the model. A case study of Beijing's central urban area is conducted to validate the model, which demonstrates that increasing the proportion of BEBs with conventional electricity from 75% to 94% reduces carbon emissions by 0.68%, while adopting green electricity for BEBs achieves a more substantial 56% reduction. Sensitivity analysis highlights the trade-offs between emission reductions and operator costs, revealing that carbon trading revenues alone are insufficient to offset the higher costs of BEBs adoption.

Keywords: transit network design · fleet composition · low-carbon benefits · operator costs

1 Introduction

In recent years, the energy transition in public transportation has emerged as a global trend. In China, cities such as Shenzhen and Guangzhou have already completed the full transition of their bus fleets to alternative fuel vehicles. As a fundamental requirement of modern public transportation systems, the adoption of alternative fuel buses, especially electric buses, significantly alleviates environmental impacts and plays a critical role in reducing carbon emissions and mitigating dependence on conventional energy sources

X. Xie et al. (Eds.): ICSS 2025, LNOR, pp. 236–249, 2026.
https://doi.org/10.1007/978-3-032-13116-4_19

[1]. However, the energy transition also presents challenges for transit network design studies that account for network structure. An important issue that remains to be further explored is how to configure transit networks rationally under the ongoing trend of energy transition with the aim of balancing operator costs and carbon emissions.

Early studies on transit network design that account for network structure have developed optimization models with the primary objectives of minimizing economic costs. Chien et al. [2] initiated systematic research on grid transit networks to determine optimal stop spacing, route spacing and headway by minimizing total costs for operators and passengers. A hybrid network structure integrating grid and radial features was proposed by Daganzo [3], aiming to minimize the combined operator and passenger costs by adjusting stop spacing, grid coverage, and headway. Chen et al. [4] developed a continuum approximation framework to optimize transit networks combining grid layout in the core and hub-and-spoke design in peripheral areas, with service boundaries, spacings, and headways jointly optimized to minimize total system costs. The increasing focus on low-carbon development in the public transportation sector has led to an increase in research that integrates carbon emission considerations. Based on Daganzo's framework, emission constraints were incorporated into the model by Griswold et al. [5]. The result demonstrated that minimizing total costs could simultaneously reduce vehicular emissions. Duran-Micco et al. [6] formulated a bi-objective TNDFSP model that explicitly considers both emissions and travel time, introducing a memetic algorithm and a heterogeneous fleet structure to explore trade-offs between environmental impact and service performance. Barraza and Estrada [7] proposed an analytical model for designing BEB networks and applied it in a case study to compare system costs and emissions across eight scenarios with different vehicle capacities, powertrains, and charging strategies. Li et al. [8] developed a continuous approximation model to optimize stop spacing, headway, and charging range for the electric bus network. The model was numerically validated by comparison with a diesel bus system. While previous studies have developed optimization models considering both operator costs and carbon emissions, the impact of bus fleet energy composition remains to be further explored. Furthermore, the potential benefits of carbon trading have rarely been incorporated into the analyses.

Therefore, this study focuses on the impact of fleet composition with different energy type on operator costs and carbon emissions in urban transit networks, aiming to providing support for public transportation planning and policy formulation. First, an analysis of the hierarchical network layout for gridiron road systems is conducted. Subsequently, a mathematical model for transit network optimization is formulated, in which route spacing, stop spacing, headway, and the proportions of buses with different energy types are considered as decision variables, aiming to minimize passenger travel costs, operator costs, and carbon emissions. In addition, an NSGA-II algorithm is designed to solve the optimization problem. Finally, a case study of the central area of Beijing within the Fourth Ring Road is conducted to validate the proposed model, with a sensitivity analysis performed from the perspectives of carbon emissions, operator costs, and bus fleet energy composition.

The remainder of the paper is organized as follows. The next section is methodology, including problem description, model formulation, and solving algorithm design. The third section presents a case study, followed by the conclusions in the final section.

2 Methodology

2.1 Problem Description

a. Hierarchical network layout for gridiron road systems

Hierarchical trunk-feeder configurations are widely regarded as effective for bus networks in the gridiron road system. In accordance with these configurations, the proposed model adopts a two-tier structure that integrates arterial bus routes and feeder bus routes to form a hybrid grid and hub-and-spoke network.

As illustrated in Fig. 1, the bus system is abstracted as a rectangular service area with dimensions $D_L \times D_W$ (km). The entire region is divided into a core region and a central region, where the core region has a length of d_L (km) and a width of d_W (km). Bus stops are modeled as uniformly distributed with constant spacing, where both horizontal and vertical intervals are set to S (km).

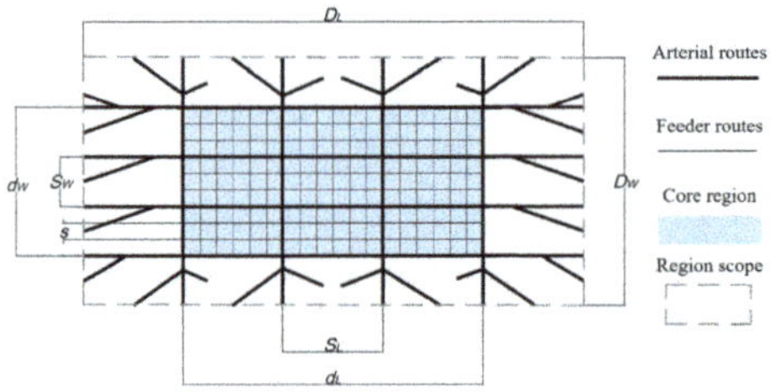

Fig. 1. Structure of bus network

In the core region, the arterial and feeder bus routes are arranged in a grid pattern. The spacing between arterial routes is S_L (km) in the north-south direction and S_W (km) in the east-west direction. In central region, the transit network is formed by branches extending from arterial routes, and maintains the same spacing intervals as in the core region.

b. Model assumption

We analyze the spatiotemporal passenger flow characteristics within the study area to provide empirical support for determining area attribute parameters in the model. Based on the empirical results, we formulate the following assumptions regarding bus line operations and passenger demand across the entire bus network [7, 11].

(1) All bus routes operate bidirectionally.
(2) $S_L = n_1 S$ and $S_W = n_2 S$ ($n_1, n_2 \in Z$).
(3) The headway during peak hours is higher than that during off-peak hours.

(4) The passenger travel demand during peak hours within the region is Λ (persons/h), while the passenger travel demand during off-peak hours is λ (persons/h). Under the supply-oriented model, passenger trips are assumed to follow a random distribution in time and an independent uniform distribution in space.
(5) Passengers consistently select the nearest accessible bus stops to both their origins and destinations. They randomly select walking directions to bus stops, with equal probability assigned to all possible directions. Also, they select bus routes minimizing transfer frequency, with all transfers occurring at the first available transfer station.
(6) The proportion of electric buses cannot be lower than the current level of 75%.

2.2 Model Formulation

The optimization of bus network design can be categorized into two primary types based on the guiding approach [9], including (1) demand-oriented and (2) supply-oriented optimization. Correspondingly, optimization models can be divided into discrete numerical models and continuous analytical models [12]. Analytical models assume that passenger travel demand is uniformly distributed in space and occurs randomly over time. This allows for data aggregation and the formulation of objective functions that minimize both operator and passenger costs, without requiring extensive OD data. These models are relatively simple to solve and exhibit strong applicability. Moreover, analytical models that incorporate the topological structure of bus networks offer good continuity and global optimization potential. Therefore, this study adopts an analytical modeling approach for bus network optimization. A representative example is the analytical model proposed by Daganzo [3], which aims to minimize the total system cost comprising both operator and passenger costs by optimizing decision variables such as bus stop spacing and headway.

However, public transit network optimization problem inherently involves competing objectives, such as minimizing passenger travel time costs, reducing operator costs, and lowering carbon emissions. Single-objective optimization may exhibit high sensitivity to predefined weights and fail to capture the inherent trade-offs. To address this trade-off issue effectively, it is essential to establish a multi-objective optimization model.

In view of the ongoing transition of the transit industry toward new energy sources, this study develops a multi-objective optimization model that includes passenger travel costs, operator costs, and carbon emissions. The proportions of buses with different energy types are considered as decision variables to quantitatively analyze the impact of bus fleet energy composition on the operator costs and carbon emissions.

a. **Input Parameters**

In the optimization model developed in this study, the input parameters include (1) regional attributes, (2) passenger travel characteristics, (3) bus operation parameters, and (4) constraint-related parameters. These parameters are presented in detail in Table 1.

b. **Optimization Objective**

The decision variables of the model include stop spacing (S), arterial route spacings (S_L and S_W), headway during peak hours and off-peak hours (H_1 and H_2 respectively), as

Table 1. Notation and Parameter Descriptions

Notation	Description	Unit
D_L, D_W	Length and width of the entire region	km
d_L, d_W	Length and width of the core region	km
P_1, P_2	Number of peak hours and off-peak hours	–
Λ_i	Passenger travel demand during peak hours within the zone i	persons/h
λ_i	Passenger travel demand during off-peak hours within the zone i	persons/h
μ_{pi}	Passenger time value coefficient, where $i = \{1, 2, 3, 4\}$ corresponds to riding, walking, waiting, and transfer, respectively	CNY/h
v_f	Passenger walking speed	km/h
v_a	Average operating speed of buses	km/h
c_{di}	Operating costs per vehicle-kilometer, where $i = \{1, 2, 3\}$ corresponds to BEBs, LNG buses, diesel buses	CNY/(veh·km)
c_{ti}	Operating costs per vehicle-hour where $i = \{1, 2, 3\}$ corresponds to BEBs, LNG buses, diesel buses	CNY/(veh·h)
γ	Carbon trading price	CNY/t
e_{vi}	Carbon emissions per unit time for a single bus where $i = \{1, 2, 3\}$ corresponds to BEBs, LNG buses, diesel buses	t/h
β_i	Proportion of different energy types of buses where $i = \{1, 2, 3\}$ corresponds to BEBs, LNG buses, diesel buses	–
H_{min}, H_{max}	Minimum and maximum headway	min
S_{min}, S_{max}	Minimum and maximum stop spacing	km

well as the proportion of BEBs, LNG buses, and diesel buses (β_1, β_2, and β_3 respectively). The optimization objective is to minimize the passenger travel cost, the carbon emissions, and the operator costs.

(1) The passenger travel cost Z_1 is calculated as the sum of the time spent in different travel stages multiplied by their corresponding time value coefficients, as shown in Eq. (1). The total travel time of passengers includes in-vehicle time T, walking time T_f, waiting time T_w, and transfer time T_t. The calculations are shown in Eqs. (2), (3), (4) and (5), respectively.

$$\min Z_1 = \mu_{p1}T + \mu_{p2}T_f + \mu_{p3}T_w + \mu_{p4}T_t \tag{1}$$

$$T = \left(\frac{\lambda_1}{\lambda_1 + \lambda_2}\left(\left(\frac{d_L}{2}\right)^2 + \left(\frac{d_W}{2}\right)^2\right)^{\frac{1}{2}} + \frac{\lambda_2}{\lambda_1 + \lambda_2}\frac{(D_L + D_W)}{4}\right)/v_a \tag{2}$$

$$T_f = 0.4S/v_f \tag{3}$$

$$T_w = 0.5\frac{H_1P_1 + H_2P_2}{P_1 + P_2} \tag{4}$$

$$T_t = 0.02\left(1 - e^{-0.5(n_1+n_2)}\right) \tag{5}$$

(2) The carbon emissions Z_2 is expressed as Eq. (6). R represents the total length of bus routes as shown in Eq. (7). N_{bus} represents the total number of buses in operation per unit time as shown in Eq. (8).

$$\min Z_2 = \sum\nolimits_{i=1}^{3} \beta_i e_{vi} N_{bus} \tag{6}$$

$$R = 2(D_L + D_W)\left(\frac{1}{n_1S} + \frac{1}{n_2S}\right) \tag{7}$$

$$N_{bus} = \frac{R}{P_1 + P_2}\left(\frac{\Lambda_1}{\lambda_1}\frac{P_1}{H_1v_a} + \frac{P_2}{H_2v_a}\right) \tag{8}$$

(3) The operator costs Z_3 is formulated in Eq. (9). c_{di} and c_{ti} can be derived from Eqs. (10) and (11), respectively.

$$\min Z_3 = \sum_{i=1}^{3} \beta_i (c_{di}R + c_{ti}N_{bus}) \tag{9}$$

$$c_{di} = \frac{c_{Ni}e_{Ni} + c_{ei}}{l_i} + c_{Si}t_{Si} \tag{10}$$

$$c_{ti} = \frac{c_{Zi}}{t_{Zi}} + \frac{c_{Wi}}{t_{Wi}} \tag{11}$$

where c_{Zi} denotes the vehicle procurement cost (CNY/veh), t_{Zi} represents the operational lifespan of the bus (h), c_{Wi} is the average staff wage (CNY), t_{Wi} refers to the working hours of staff (h), c_{Ni} denotes the energy price (CNY/L, CNY/kg, or CNY/kWh), e_{Ni} represents the energy consumption per unit distance (L/km, kg/km, or kWh/km), c_{ei} is the additional/replacement battery cost for electric buses (CNY), l_i refers to the total service mileage of the bus (km), c_{Si} is the cost per maintenance session for a bus, measured in yuan per occurrence (CNY/time), and t_{Si} represents the bus maintenance cycle, measured in occurrences per kilometer (times/km).

c. **Constraints**

The proposed model is subject to the following constraints. Eq. (12) defines the constraint for bus stop spacing. Eqs. (13) and (14) specifies the headway constraint during peak hours and off-peak hours. Equation (15) represents the integer constraints for n_1 and n_2. Equation (16) defines the constraint that the total proportion of the three types of buses is 1. Equation (17) specifies the upper and lower bounds for the proportions of each bus type.

$$S_{min} \leq S \leq S_{max} \tag{12}$$

$$H_{min} \leq H_1 \leq H_{max} \tag{13}$$

$$2H_{min} \leq \mathrm{H}_2 \leq 2H_{max} \tag{14}$$

$$n_1, n_2 \in \mathbb{Z} \tag{15}$$

$$\beta_1 + \beta_2 + \beta_3 = 1 \tag{16}$$

$$\beta_1 \in [0.75,1], \beta_2 \in [0,0.20], \beta_3 \in [0,0.05], \tag{17}$$

2.3 Solving Algorithm Design

The Non-dominated Sorting Genetic Algorithm II (NSGA-II) is particularly effective for solving multi-objective optimization problems due to its ability to maintain a diverse set of Pareto-optimal solutions while ensuring convergence near the true Pareto front [11]. The Pareto-optimality principle naturally accommodates the inherent trade-offs in transit planning, such as operator costs versus carbon emissions. This algorithm preserves elitism by merging of parent and offspring populations to prevent the loss of high-quality solutions. It also incorporates fast non-dominated sorting, enabling efficient Pareto ranking and reducing computational complexity compared to traditional methods.

Furthermore, the use of the crowding distance mechanism promotes diversity in solutions, preventing premature convergence and ensuring a well-distributed Pareto front, which is critical for identifying a range of viable solutions that balance costs and emission reductions. The decision variables in the proposed model include both discrete and continuous variables. NSGA-II's flexibility in handling mixed-variable optimization ensures efficient exploration of the solution space. Therefore, we employ the NSGA-II algorithm to solve the proposed multi-objective optimization model in this study. The implementation details of NSGA-II are presented below.

(1) Initialize Population. An initial population P of size N is generated, and the fitness values of individuals in P is then evaluated. The three objective functions in the proposed model serve as fitness functions in the algorithm.
(2) Non-Dominated Sorting. Non-dominated sorting is performed on P to divide it into Pareto fronts
(3) Crowding Distance Assignment. Crowding distance values are assigned to individuals within each Pareto front.
(4) Evolutionary Loop. In each generation, genetic operators such as selection (e.g., tournament selection), crossover, and mutation are applied to generate an offspring population Q. The parent population P and offspring population Q are then combined

into R (R = P∩Q). Non-dominated sorting is performed on R to divide it into Pareto fronts. Individuals for the next generation P are selected from R by adding individuals from Pareto fronts in order, while using crowding distance to maintain diversity if a front exceeds the remaining slots. Finally, the fitness of the new population P is evaluated.

(5) Termination. The algorithm stops when the termination criteria are met (e.g., maximum generations or convergence). The algorithm outputs all non-dominated solutions in the final population as the approximate Pareto frontier.

3 Case Study

3.1 Parameters Setting

To validate the proposed optimization model, we conduct a case study of Beijing's central urban area within the Fourth Ring Road (Fig. 2). The area within the Second Ring Road is defined as Zone 1 (i.e. core region), while the area between the Second and Fourth Ring Roads is designated as Zone 2 (i.e. central zone). Both zones are simplified as rectangular regions for analysis.

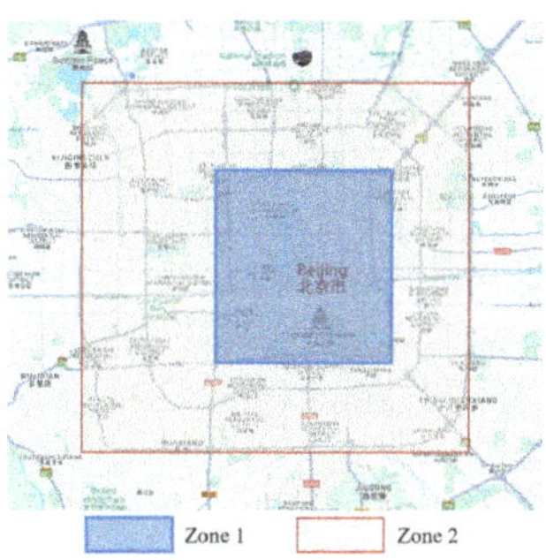

Fig. 2. Area of case study

The actual OD distribution patterns in the study region are statistically analyzed through the matching of IC card data with corresponding transit station locations. We process three foundational datasets including IC card data, station data, and route data. The IC card data reflect passenger travel behaviors including card ID, bus route, vehicle number, boarding and alighting stations, as well as corresponding timestamps. The data processing consists of two main steps. First, quality control procedures are applied to clean the raw IC card data. The processed IC card records are spatially integrated with route and station data to determine passenger trip origin-destination distributions.

Spatially, passenger flows exhibit a centralized pattern, showing a significant difference from a uniform distribution. Temporally, passenger demand is predominantly concentrated during the morning (7:00–9:00) and evening (17:00–19:00) peak hours. Based on the analysis of one-week IC card data, the values of area attribute parameter for the model are determined as presented in Table 2. The values of passenger travel

Table 2. Values of area attribute parameters

Parameters	Value	Unit
D_L, D_W	16, 16	km
d_L, d_W	8, 8	km
P_1, P_2	4, 10	h
λ_1, λ_2	30000, 90000	persons/h
Λ_1, Λ_2	40000, 120000	persons/h

Table 3. Values of passenger travel attribute parameters

Parameters	Value	Unit
μ_{p1}	13.5	CNY/h
μ_{p2}	42.19	CNY/h
μ_{p3}	33.75	CNY/h
μ_{p4}	40.5	CNY/h
v_f	4.5	km/h

attribute parameters presented in Table 3 are determined through existing research on passenger's travel behavior and time value in Beijing.

The values of transit operation parameters (Table 4) are derived from comprehensive data sources including (1) urban road speed limit regulations, (2) the Urban Road Engineering Design Code in China, (3) subsidy policy documents, (4) vehicle scrappage regulations, (5) energy prices for various types of buses, (6) relevant maintenance and cost benchmark, and (7) previous studies on carbon emissions of different types of buses [10]. In addition, we assume that BEBs charge during nighttime hours which leads to an increase in labor costs. The charging electricity price is calculated as the average of peak and off-peak rates. The values of constraint parameters H_{min}, H_{max}, S_{min}, and S_{max}, are 3, 7.2, 0.3, and 0.8 respectively.

Table 4. Values of transit operation parameters

Parameters	Value	Unit
v_a	25	km/h
c_{d1}, c_{d2}, c_{d3}	1.59, 1.46, 2.94	CNY/(km·h)
c_{t1}, c_{t2}, c_{t3}	109.49, 72.79, 72.04	CNY/(veh·km)
e_{v1}, e_{v2}, e_{v3}	0.0138, 0.0140, 0.0196	t/h
γ	80	CNY/t

3.2 Results

a. **Optimization results**

The proposed multi-objective optimization model is solved using the NSGA-II. The NSGA-II algorithm is implemented using MATLAB's built-in gamultiobj function. The parameters are configured as follows. Population size = 100, maximum generations = 200, and Pareto fraction = 0.35. Other parameters remain at their default values.

To demonstrate the importance of considering the proportions of different energy types of buses in the bus network optimization model, this study establishes a baseline model that does not account for the proportions of different energy types of buses. The Pareto optimal solution sets obtained from solving the two models are then compared, as shown in Fig. 3. The solutions of the proposed model achieve lower operator costs for similar passenger travel costs, while also reducing carbon emissions, as indicated by the smaller bubble sizes. It can also be observed that after incorporating the proportions of different energy types of buses, the extreme values of the objective function for operator costs in the Pareto optimal solutions become larger. The underlying reason is that the proposed model incorporates a lower-bound constraint on the proportion of BEBs, which is set at the current level.

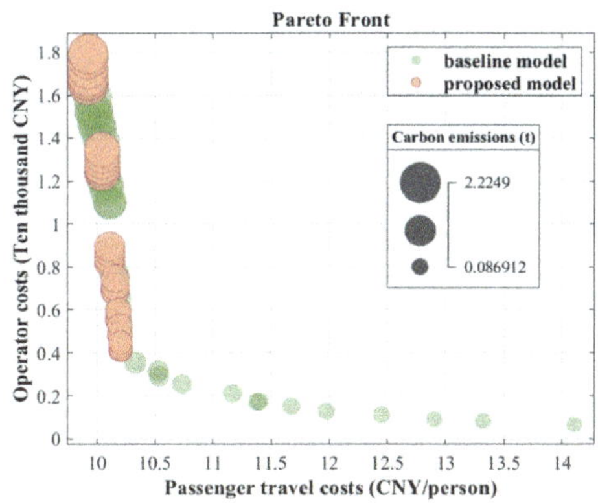

Fig. 3. Comparison of Pareto Fronts for Multi-Objective Optimization Models

The values of decision variable corresponding to one Pareto-optimal solution are (1) $S = 0.3$ km, (2) $n_1 = 1$ and $n_2 = 2$, (3) $H_1 = 3$ min and $H_2 = 6$ min, and (4) $\beta_1 = 0.94$, $\beta_2 = 0.05$, and $\beta_3 = 0.01$. The decision variables, excluding the proportions of different energy types of bus, are set to the values of the decision variables in the aforementioned Pareto optimal solutions. Then the comparison of objective function values under different proportions is shown in Table 5.

The passenger travel costs remain constant across all scenarios since the proportions of bus energy types do not directly affect this objective.

The carbon emissions during the operation of electric buses powered by green electricity are zero. Based on previous research [10], it can be calculated that the carbon emissions per hour for a single electric bus powered by green electricity are 0.0055 t.

When BEBs are powered by green electricity, carbon emissions are significantly reduced by 56%. In comparison, the carbon reduction benefit from increasing the proportion of conventionally-powered electric buses is relatively limited, showing only a 0.68% decrease. Compared to merely increasing the proportion of BEBs, switching to green electricity yields significantly more substantial carbon reduction benefits. The adoption of green electricity for BEBs provides more immediate climate benefits (a reduction of 44% carbon emissions) with the advantage of being implementable without changing current proportion of different types of bus.

The operator costs increase by 7.8% in the optimized results compared to the current level, which reflects the higher costs associated with increasing the proportion of BEBs.

Table 5. Comparison of objective function values under different proportions of bus energy types

		Passenger Travel Costs (CNY/person)	Carbon Emissions (t/h)	Operator Costs (CNY/h)
BEBs powered by conventional electricity	Current Level	10.0381	1.6472	$1.2212*10^4$
	Optimized result	10.0381	1.6360	$1.3173*10^4$
BEBs powered by green electricity	Current Level	10.0381	0.9215	$1.2212*10^4$
	Optimized result	10.0381	0.7234	$1.3173*10^4$

b. **Sensitivity analysis**

To further explore the impact of the proportion of BEBs on different three objectives, a sensitivity analysis is conducted. As the proportion of BEBs gradually increases, we assume that diesel buses are replaced with BEBs first, followed by LNG buses. As shown in Fig. 4, during the process of increasing the proportion of electric buses from 75% (current level) to 100%, the carbon reduction benefits from replacing diesel buses with electric buses are significantly greater than replacing LNG buses.

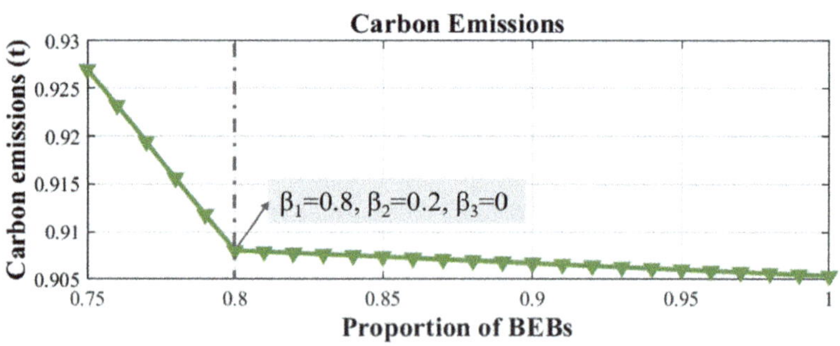

Fig. 4. Variation of carbon emission with different proportion of BEBs

In recent years, carbon trading practices in urban public transportation have been increasing. The potential revenue from emission reductions due to increased electric bus adoption are analyzed (Fig. 5). The carbon emission reduction from increasing the proportion of BEBs is calculated using the current emission level as the benchmark. This reduction is then multiplied by the carbon trading price γ to determine the carbon trading revenue. The result indicates that the increase in carbon trading revenue is far from sufficient to offset the operator cost increase due to the transition to electric buses.

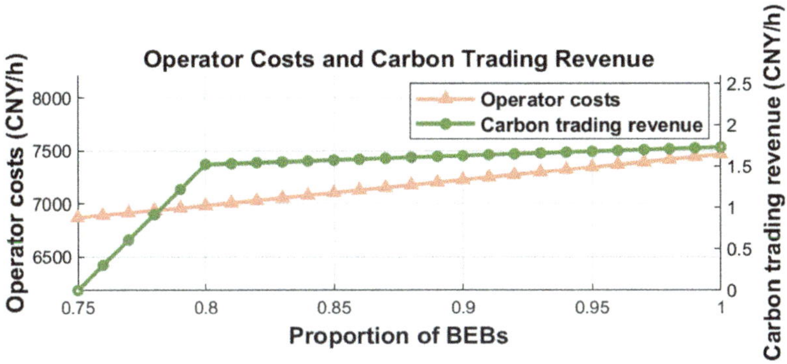

Fig. 5. Variation of operator costs and carbon trading revenue with different proportion of BEBs

Additionally, an analysis is conducted on the carbon emission reduction of BEBs with and without the use of green electricity. The results indicate that without green electricity, the carbon reduction effect of BEBs is relatively limited. In contrast, when powered by green electricity, BEBs achieve substantially greater reductions in carbon emissions, with values dropping from 0.52 t/h to 0.36 t/h (Fig. 6).

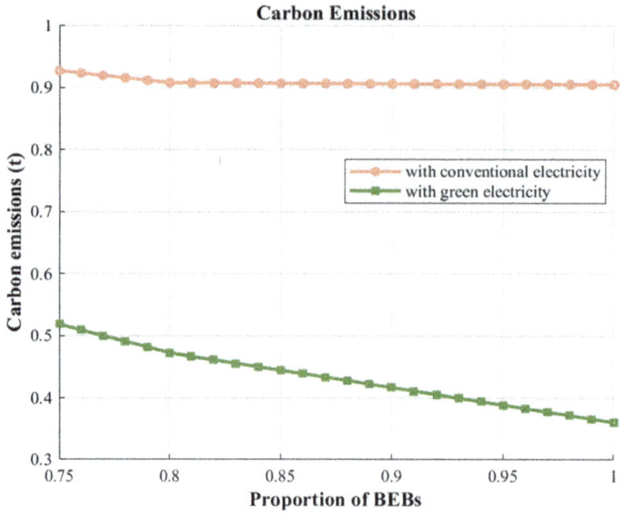

Fig. 6. Variation of carbon emissions with different proportion of BEBs and different types of energy

4 Conclusions

The transition to new energy in the public transportation sector has become a critical strategy for promoting sustainable urban development and reducing carbon emissions. In this context, this study investigates how different energy compositions of bus fleets affect operator costs and carbon emissions in urban transit networks. A multi-objective optimization model is developed to minimize passenger travel costs, operator costs, and carbon emissions, while considering constraints on the minimum proportion of each bus type to reflect current energy transition policies. The model is solved using the NSGA-II algorithm and validated through a case study of Beijing's central urban area. The analysis leads to several important findings.

(1) Increasing the proportion of BEBs to 94% while maintaining 5% LNG buses and 1% diesel buses reduces carbon emissions by 0.68%, but it will lead to an 8.3% increase in operator costs. These results suggest that carbon emission reduction through fleet electrification alone may be economically challenging.
(2) The study reveals that powering electric buses with green electricity can achieve a 44% emission reduction without altering the current proportions of different energy types buses. This insight proves that sustainable power sources play a fundamental role in reducing carbon emissions in the public transportation industry.

Based on the conclusions, it is suggested that the transit industry (a) prioritize replacing diesel buses due to their higher emission intensity, (b) promote the adoption of electric buses alongside the development of a green electricity supply system, and (c) increase operational subsidies to mitigate the additional costs of promoting electric buses.

Future research could further incorporate a wider range of sustainable energy sources, such as hydrogen energy, and conduct case studies on cities with a lower level of bus electrification.

Acknowledgements. The authors acknowledge that this paper is prepared based on the National Natural Science Foundation of China (NSFC) under grant #72271020. Thanks are also given to the anonymous reviewers for their helpful comments on this paper.

References

1. Pternea, M., Kepaptsoglou, K., Karlaftis, M.G.: Sustainable urban transit network design. Transp. Res. Part A Policy Pract. **77**, 276–291 (2015)
2. Chien, S., Schonfeld, P.: Optimization of grid transit system in heterogeneous urban environment. J. Transp. Eng. **123**(1), 28–35 (1997)
3. Daganzo, C.F.: Structure of competitive transit networks. Transp. Res. B Methodol. **44**(4), 434–446 (2010)
4. Chen, J., Liu, Z., Wang, S., Chen, X.: Continuum approximation modeling of transit network design considering local route service and short-turn strategy. Transp. Res. Part E Logist. Transp. Rev. **119**, 165–188 (2018)
5. Griswold, J.B., Sztainer, T., Lee, J., Madanat, S., Horvath, A.: Optimizing urban bus transit network design can lead to greenhouse gas emissions reduction. Front. Built Environ. **3** (2017). https://doi.org/10.3389/fbuil.2017.00005
6. Duran-Micco, J., Vermeir, E., Vansteenwegen, P.: Considering emissions in the transit network design and frequency setting problem with a heterogeneous fleet. Eur. J. Oper. Res. **282**, 580–592 (2020)
7. Barraza, O., Estrada, M.: Battery electric bus network: efficient design and cost comparison of different powertrains. Sustainability. **13**, 4745 (2021)
8. Li, X., Li, Y., Liu, W., Yuan, Y.: Optimal design of pure battery electric bus system on the grid network. Transportmetrica A Transp. Sci. **20** (2024)
9. Mees, P.: A Very Public Solution. Melbourne University Publishing, Melbourne (2015)
10. Badia, H., Estrada, M., Robusté, F.: Competitive transit network design in cities with radial street patterns. Transp. Res. B Methodol. **59**, 161–181 (2014)
11. Deb, K.: Multi-objective Optimization Using Evolutionary Algorithms. Wiley, Chichester (2001)
12. Zhang, Y. Research on the carbon emission reduction potential of urban new energy public transport system. PhD thesis. Jilin University (2023)

The Algorithmic Reconfiguration of Qualitative Inquiry: Navigating AI-Driven Efficiency and Interpretive Richness

Hootan Kamran[1(✉)], Atanaz Dorrani[2], and Houman Kamran[3]

[1] Northeastern University, Boston, USA
hkamranhabibkhani@northeastern.edu
[2] City of Mississauga, Mississauga, Canada
atanaz.dorrani@mississauga.ca
[3] University of the Pacific, Stockton, USA
hkamranhabibkhani@pacific.edu

Abstract. Artificial Intelligence (AI) profoundly reshapes research methodologies, offering opportunities and challenges. This paper critically examines the **algorithmic reconfiguration of qualitative inquiry**, focusing on the tension between **AI-driven efficiency** and **interpretive richness**. It explores how AI capabilities in large-scale data processing, pattern recognition, multimodal analysis, and cross-lingual understanding alter qualitative research. While AI enhances speed, cost reduction, and overcomes language/survey barriers, these efficiencies risk superficiality, algorithmic bias, and ethical dilemmas. The paper discusses strategies for navigating this tension, evolving researcher skills, and the imperative for **human-centricity** to ensure AI genuinely augments insights, particularly within Service Science and digital transformation.

Keywords: Artificial Intelligence · Qualitative Research · Quantitative Research · Research Methodology · Algorithmic Reconfiguration · Interpretive Richness · AI Ethics · Service Science · Digital Transformation · Semantic Embeddings · AI-Driven Efficiency · Human-AI Collaboration · Open-Ended Surveys

1 Introduction: AI's Seismic Shift in Research Methodologies

AI is a central force in academic research [1,2]. Its strengths in handling large datasets, pattern identification, automation, and cross-language understanding via semantic embeddings [3] prompt method revision [4]. These shifts align with INFORMS ICSS 2025's "AI, Society, and Digital Transformation" theme, positioning Service Science to both shape and gain from AI research.

X. Xie et al. (Eds.): ICSS 2025, LNOR, pp. 250–261, 2026.
https://doi.org/10.1007/978-3-032-13116-4_20

AI in research offers faster, scalable data analysis, yet challenges traditional methods [5,6]. AI tools redefine data collection and interpretation, raising questions about knowledge creation and validation [7].

Historically, natural sciences favored quantitative methods. In social sciences/humanities, the **qualitative-quantitative debate** became central, as numbers alone couldn't capture human experience or context [20]. Qualitative research explored depth/meaning; quantitative sought breadth/generalizability [8–10]. AI enhances both by analyzing large texts or detecting numerical patterns [11–13], prompting methodological reassessment [14,15].

This paper investigates qualitative inquiry's algorithmic reconfiguration: the tension between AI efficiency (overcoming survey/language barriers) and interpretive richness. Uncritical AI use may compromise nuanced understanding. This is vital for Service Science, ensuring AI augments research integrity.

2 Evolving Research Definitions in the Age of AI

2.1 Traditional Distinctions

Qualitative research deeply understands human experiences, contexts, and meanings, emphasizing interpretation and purposive samples, rooted in interpretivism/constructivism [16–18]. Data demands human effort. **Quantitative research** focuses on numerical data, statistical analysis, generalizable findings, and objectivity, using large samples for population inferences [19,20].

2.2 AI Blurring Boundaries

AI integration changes research methods, blurring qualitative-quantitative boundaries, fostering a flexible methodological environment. AI enhances human thinking, supporting large-scale qualitative analysis and adding depth to quantitative.

Qualitative at Scale: AI, via **Natural Language Processing (NLP)**, ML, speech recognition, and computer vision, enables analysis of vast qualitative data (e.g., interviews, open-ended surveys, social media) at unprecedented speed/scale [21,22]. This includes automated transcription [34]; NLP for thematic [25,26] and sentiment analysis [27]; and computer vision for visual data [28,29]. These expand qualitative project scope, enabling thematic analysis of open-ended survey questions at scale, allowing richer participant expressions and reducing bias [30].

Quantitative with Depth: AI enhances quantitative methods by uncovering subtle patterns, segmenting datasets, and detecting anomalies for deeper qualitative-style investigation. This allows more nuanced interpretations, approaching rich contextual insight. AI redefines research expertise, extending beyond technical execution to critical judgment, ethics, and interdisciplinary integration [31,32]. AI's ability to convert words into **semantic embeddings**

within a unified mathematical space enables advanced **cross-lingual analysis** and reveals deeper semantic relationships, fostering global research.

AI thus broadens quantitative analysis, supporting new exploration/interpretation, blurring methodological boundaries, and shifting data/analysis understanding.

2.3 Shifting Nature of "Data" and ""nalysis"

AI changes how 'data' is understood, enabling large-scale analysis of unstructured and multimodal sources (visual, multilingual, open-ended survey responses). Driven by **Generative AI**, this expands research topics, introduces synthetic datasets, and tailors AI training corpora. While making complex analysis accessible, it demands new methodological approaches for data quality, AI-generated interpretation validation, and ethical management of bias/data access.

AI also reshapes 'analysis' by automating routine tasks and enhancing human cognitive abilities, balancing machine efficiency and human insight. "AI co-scientist" systems generate hypotheses and design experiments, shifting the researcher's role to critical curator, validator, and interpreter [33]. This requires new skills like prompt engineering and rigorous validation, alongside commitment to transparency, accountability, and research integrity.

3 Algorithmic Reconfiguration of Qualitative Inquiry: Efficiency Gains and Interpretive Tensions

AI offers significant efficiency in qualitative research, but these benefits pose concerns for interpretive richness, contextual depth, and methodological rigor.

3.1 AI-Driven Efficiency: Expanding Horizons

AI improves qualitative research operational efficiency through:

- **Accelerated Data Processing:** Reduces time for transcription [34], initial coding [35], literature searches, enabling larger studies [37].
- **Cost Efficiencies:** Automation can reduce costs vs. manual/traditional software [36].
- **Enhanced Accessibility:** User-friendly AI platforms simplify technical processes for basic thematic/sentiment analysis [38].
- **Managing Large-Scale Data:** AI allows analysis of qualitative datasets at scales impractical manually [39].
- **Transforming Survey Methodology:** AI tools enable quantitative analysis of open-ended survey questions, collecting richer participant expressions without sacrificing sample size or predefined response bias [40–42].

3.2 The Tension: Efficiency vs. Interpretive Richness and Methodological Rigor

AI efficiencies challenge qualitative research values:

- **Epistemological and Methodological Challenges:** AI efficiency risks superficial interpretations, missing human contextual nuances [14]. Algorithms detect patterns but struggle with underlying reasons and causal explanations [44,45]. Skepticism exists about AI's reflexivity or grasp of rich context [43,46,47]. AI can reproduce embedded biases, lacking ethical judgment or self-correction [48,49]. Thus, AI supports but doesn't replace human interpretive, contextual, and reflexive strengths.

  Over-reliance on efficiency can lead to "qualitative-washing," where AI use falsely implies rigor despite lacking depth [50]. Algorithmic patterns might be mistaken for meaningful insights, undermining authentic qualitative research. Computational analysis offers coverage but often lacks interpretive depth and nuanced understandings [43,51,52]. This risks reintroducing positivist assumptions, distorting qualitative insight [53,54]. Algorithmic speed may diminish traditional practices essential for understanding the "why" and "how" [55,56].

  Biased training data can lead to amplified biases in AI qualitative analysis, misrepresenting marginalized perspectives [57]. Qualitative research requires theoretical knowledge, methodological rigor, ethical sensitivity, and reflexivity—capabilities AI lacks [58]. Human oversight and critical evaluation are essential.
- **Ethical Imperatives in AI-Assisted Qualitative Research:** Integrating AI requires thoughtful attention to ethical complexities:
  - **Data Privacy, Security, Ownership:** Using third-party AI platforms for sensitive data raises concerns [59].
  - **Informed Consent:** Robust consent is essential for participants to understand AI processing of their data [60].
  - **Transparency and Accountability:** AI tool use must be disclosed. AI's contribution needs accurate recognition, avoiding false attribution of insight/authorship [61].

AI's potential in qualitative research depends on balancing computational power for efficiency/scale with deep, nuanced, and ethically responsible interpretation. Critical awareness of these trade-offs is crucial.

Table 1 summarizes this balance:

Table 1. AI in Qualitative Research: Efficiency Gains vs. Interpretive Challenges

Aspect	AI Efficiency/Capability	Challenges to Interpretive Richness & Rigor	Key Evidence
Data Collection (Trans.)	Faster, cheaper audio/video trans	Accuracy issues, non-verbal cues, privacy	[34]
Survey Design	Large-scale open-ended questions	AI misinterpretation, superficial themes, bias	[40,42]
Data Analysis (Coding)	Automated coding, pattern spotting	Superficiality, nuance loss; AI struggles with meaning, reflexivity, bias	[14,35,43]
Accessibility	Lower cost tools, user-friendly	Uncritical use, "qual-washing"	[37,38,50]
Speed & Scale	Large qual data analysis, rapid processing	"Data deluge," depth vs. scale, decontextualized patterns	[39,43]
Language Barriers	Multilingual data analysis via translation	Translation nuances, cultural context loss, biased models	[3]
Interpretive Depth	Highlights areas for deeper human investigation	AI lacks empathy, reflexivity, deep contextual understanding	[14,44]
Ethical Integrity	Assists in identifying sensitive data	Privacy/security risks, consent challenges, bias	[59,60]
Bias Mitigation	Potentially identifies certain biases	AI inherits/amplifies biases, skewed findings	[48,57]

4 AI's Algorithmic Reconfiguration: New Frontiers and Enduring Challenges

4.1 AI-Driven Hybridity: Synergies and Interpretive Depth

AI bridges qualitative and quantitative research by processing diverse data and detecting complex patterns, supporting innovative **mixed-methods designs**. This catalyzes deeper, integrated understandings.

"Hybrid Analysis Integration" connects statistical data with nuanced human experiences. AI processes open-ended survey responses, linking themes to quantitative variables. It supports cross-lingual analysis, relating qualitative feedback to quantitative metrics via semantic embeddings [62]. Efficiency must not compromise interpretive depth [43].

AI also fosters "AI-Driven Mixed-Method Approaches," like virtual focus groups, relying on linguistic nuances as mathematical vectors. Transparency in algorithmic processes is vital to evaluate AI's influence on interpretive richness [45].

Ultimately, AI supports "synergetic" research by enabling qualitative work at scale and integrating quantitative elements. Yet, pursuing synergy might lead to reductionism, flattening qualitative data for algorithmic models or weakly contextualizing quantitative findings [49]. Human expertise, critical reflection, and methodological rigor remain essential [56].

4.2 Augmenting Research Capabilities: Balancing Power with Prudence

AI enhances research precision and large-dataset processing, requiring interpretive caution. While it cleans noisy data and improves quality [63], its algorithmic standards may miss human-recognized subtleties [64]. This demands balancing AI efficiency with human insight.

AI constructs dynamic user personas and supports strategic thinking via scenario simulation and multilingual text analysis [65]. However, AI-generated outputs need scrutiny for underlying assumptions or biases, ensuring authentic human complexity and sound reasoning [66].

In quantitative research, AI enhances advanced analytics and pattern recognition [67]. The challenge is moving beyond surface correlations to meaningful, context-aware explanations, confirming patterns reflect valid causal relationships [44].

Responsible AI use depends on critical, reflexive human involvement. Researchers must recognize AI's limitations, scrutinize its outputs, and embed its use within well-founded methodological frameworks that preserve interpretive depth and guard against algorithmic bias [52,68].

4.3 Emerging Paradigms and Epistemological Reconfigurations

AI integration drives new research paradigms and reassesses epistemological foundations. **"Computational Social Science"** [69,70] raises concerns that computational efficiency might sacrifice interpretive richness [43]. "AI-Augmented Interpretivism" is debated: AI manages complex data, but maintaining interpretive depth, reflexivity, and context sensitivity is challenging, as algorithmic reasoning conflicts with nuanced human interpretation.

These developments challenge epistemological assumptions. Traditional qualitative validity/reliability [71] require rethinking, needing new frameworks to ensure AI findings reflect real-world phenomena, not distortions. This calls for AI transparency and active bias management [72]. AI's role in analysis raises questions about knowledge construction—whether AI is a neutral tool or an active influence risking human agency if outputs are uncritically accepted [48,73].

AI's data-driven nature may promote objectivity, conflicting with researcher subjectivity and reflexivity [53,74]. This "paradigm paradox" suggests AI could reinforce positivism yet catalyze new hybrid paradigms [75]. The core challenge is harnessing AI's efficiency without sacrificing the interpretive richness giving qualitative inquiry unique power.

Cultivating "**epistemic humility**" is essential [76,77]. Researchers must critically engage with AI outputs, understanding limitations and biases. This reflexive stance ensures efficiency doesn't lead to loss of genuine, rich understanding, preserving interpretive research integrity in an AI-mediated world.

Table 2 highlights AI's methodological evolution:

Table 2. AI-Driven Methodological Evolution: Efficiency vs. Interpretive Richness

Traditional Aspect	AI-Enabled Efficiency/Capability	Impact on Interpretive Richness/New Paradigm Feature	Key Enablers/Tools
Qual. Data Analysis	Auto. thematic/sentiment analysis, large dataset coding	AI-augmented interpretivism (risk of superficiality); Qual research at scale (risk of depth loss)	NLP, ML, GenAI, QDAS
Quant. Data Analysis	Advanced pattern recognition, predictive analytics, anomaly detection	Enhanced predictive modeling; Deeper quant insights if contextualized (else decontextualized)	ML (deep learning), NLP, Big Data
Mixed-Methods Design	Seamless integration of diverse data (incl. cross-lingual)	Hybrid Analysis Integration; AI-driven adaptive mixed methods (richer insights if depth maintained)	AI for data fusion, NLP (semantic embeddings)
Data Scale Lim.	Large-N qual studies feasible; Deeper seg. in large-N quant studies	Broader but shallower understanding if scale trumps depth	Cloud computing, AI-powered QDAS
Interpretive Focus	AI pattern-finding augments human interpretation	Shift to human validation of AI outputs; Risk of AI shaping interpretation	Interactive AI tools, GenAI
New Data Sources	Analysis of massive, unstruct., real-time, multilingual data	Comp. Social Science; Multimodal/multilingual analysis (richer data but demands sophisticated frameworks)	Web scraping with AI, Cross-lingual embeddings

5 Navigating the Reconfigured Landscape: Implications and Imperatives

AI's algorithmic reconfiguration of qualitative inquiry presents a pivotal tension: AI-driven efficiency vs. interpretive richness. Navigating this demands understanding implications and committing to key imperatives.

Implications for Research Practice and Service Science. AI-infused methodologies necessitate continuous, critical evaluation of AI tool use, ensuring they augment, not overshadow, human interpretive skills. Researchers must manage outputs, validate AI-assisted findings, and maintain methodological transparency. For Service Science, AI's capacity to analyze vast customer data (open-ended surveys, multilingual feedback, multimodal interactions) offers unprecedented opportunities to deepen understanding of customer experiences, identify service failures, and catalyze innovation. The central imperative: AI efficiency must not lead to superficial understanding of complex service ecosystems. Human experience richness in service must be preserved, demanding AI insights are contextualized, critically appraised, and validated through deep human understanding. Research informing societal understanding must be equitable, unbiased, and reflect diverse human experiences.

Imperatives for Responsible AI Integration. To harness AI's potential while safeguarding research integrity and interpretive richness, core imperatives emerge:

- **Human-in-the-loop:** AI must assist, not replace, human researchers. Critical judgment, ethical oversight, and final interpretation remain human responsibilities.
- **Transparency:** Clear disclosure of AI tool function, limitations, and output generation/validation processes is non-negotiable for trustworthiness.
- **Proactive bias detection and mitigation:** Vigilance is essential in selecting tools, interrogating outputs, and ensuring AI applications avoid unfair, skewed, or decontextualized interpretations, especially for marginalized groups or complex constructs like meaning vectors and cross-lingual analyses.

The Call for Responsible Innovation and Evolving Standards. AI's rapid evolution necessitates a proactive, collaborative academic stance. Forums like INFORMS ICSS 2025 are vital for fostering critical discourse and guiding responsible AI innovation. There's a pressing need for new or adapted methodological/ethical standards for AI-infused research, including "meta-research" into AI's impacts on methodologies, data quality, ethics, and epistemology. Such efforts ensure scientific integrity and humanistic values remain central to this digital transformation. The goal: AI's algorithmic reconfiguration of qualitative inquiry should lead to genuinely enhanced, richer understanding, not just accelerated data processing or superficial engagement.

6 Conclusion: Human-Centricity and Interpretive Richness in AI-Infused Qualitative Inquiry

AI's algorithmic reconfiguration of qualitative inquiry demands sophisticated, critical human engagement. The path forward hinges on a synergistic partnership: harnessing AI efficiencies to deepen, not diminish, interpretive richness.

Successfully navigating this terrain requires expanding researcher competencies: **AI literacy** and **critical algorithmic appraisal**. More profoundly, it amplifies uniquely human skills: deep critical thinking, nuanced contextual interpretation, ethical reasoning, and robust reflexivity. These guide AI application, interrogate outputs, and ensure data narratives remain authentically human and methodologically sound.

The enduring imperative is unwavering **human-centricity**. AI must remain a tool for theoretically grounded, ethically robust, meaningful research questions aimed at rich, contextualized insights. This requires continuous reflexive practice, scrutinizing researcher positionality and how human-AI interaction shapes research and knowledge. The ultimate objective is valid, ethically sound, and profoundly insightful understanding, not just efficient data processing.

AI presents a dual aspect: a powerful enabler offering unprecedented efficiency and scale, yet simultaneously challenging interpretive depth and methodological rigor. The future of qualitative research lies in a collaborative paradigm where AI's computational power supports nuanced, critical, and ethical human engagement. Challenges AI introduces often compel re-examination of long-standing qualitative inquiry practices. This journey, steered by human intellect and commitment to preserving interpretive richness, promises more profound understandings of complex social realities and service systems. Future scholarly work should pioneer frameworks and best practices for ethical, effective, and epistemologically sound human-AI collaboration, ensuring AI truly augments qualitative inquiry's depth and reach.

References

1. Wang, H., Fu, T., Du, Y., et al.: Scientific discovery in the age of artificial intelligence. Nature **620**(7972), 47–60 (2023)
2. Lu, C.: Rethinking artificial intelligence from the perspective of interdisciplinary knowledge production. AI Soc. **39**(3), 3059–3060 (2024)
3. García-Ferrero, I., Agerri, R., & Rigau, G.: A common semantic space for monolingual and cross-lingual meta-embeddings. arXiv preprint arXiv:2001.06381 (2020)
4. Rolnik, Z.: The Impact of Artificial Intelligence on academic research. Univ. Libr. Innov. Res. Stud. **1**(1), 9–11 (2024)
5. Pal, S.: A paradigm shift in research: exploring the intersection of artificial intelligence and research methodology. Int. J. Innov. Res. Eng. Multidisc. Phys. Sci. **11**(3), 1–7 (2023)
6. Pal, S.: Artificial intelligence in the research panorama: a dissection of global propensities and quintessential practices. Int. J. Res. Publ. Rev. **4**(5), 3439–3442 (2023)
7. Kulkarni, M., Mantere, S., Vaara, E., et al.: The future of research in an artificial intelligence-driven world. J. Manag. Inq. **33**(3), 207–229 (2024)
8. B?ban, A.: Reconceptualization of the division between quantitative and qualitative research methods. Cognitie Creier Comportament/Cogn. Brain Behav. **12**(4) (2008)
9. Mackenzie, N., Knipe, S.: Research dilemmas: Paradigms, methods and methodology. Issues Educ. Res. **16**(2), 193–205 (2006)
10. Brannen, J.: Mixing methods: the entry of qualitative and quantitative approaches into the research process. Int. J. Soc. Res. Methodol. **8**(3), 173–184 (2005)
11. Gamboa, A.J.P., Díaz-Guerra, D.D.: Artificial intelligence for the development of qualitative studies. LatIA **1**, 4 (2023)
12. Jalali, M.S., Akhavan, A.: Integrating AI language models in qualitative research: replicating interview data analysis with ChatGPT. Syst. Dyn. Rev. **40**(3), e1772 (2024)
13. Feuston, J.L., Brubaker, J.R.: Putting tools in their place: the role of time and perspective in human-AI collaboration for qualitative analysis. Proc. ACM Human-Comput. Interact. **8**(CSCW1), 1–28 (2024)
14. Williams, R.T.: Paradigm shifts: exploring AI's influence on qualitative inquiry and analysis. Front. Res. Metr. Anal. **9**, 1331589 (2024)
15. Hudoud, A.: Integrating artificial intelligence into research methodology: examining potential bias and mitigation strategies. Arab J. Qual. Assur. High. Educ. **18**(64) (2025)
16. Denzin, N.K., Lincoln, Y.S. (eds.): The SAGE handbook of qualitative research, 4th edn. SAGE Publications (2011)

17. Creswell, J.W., Poth, C.N.: Qualitative Inquiry and Research Design: Choosing Among Five Approaches, 4th edn. SAGE Publications (2016)
18. Lincoln, Y.S., Lynham, S.A., Guba, E.G., et al.: Paradigmatic controversies, contradictions, and emerging confluences, revisited. In: Denzin, N.K., Lincoln, Y.S. (eds.) The Sage Handbook of Qualitative Research, 4th edn., pp. 97–128. Sage Publications (2011)
19. Creswell, J.W., Creswell, J.D.: Research Design: Qualitative, Quantitative, and Mixed Methods Approaches, 5th edn. Sage Publications (2017)
20. Bryman, A.: Social Research Methods, 5th edn. Oxford University Press, Cambridge (2016)
21. Phillips, D.C., Burbules, N.C.: Postpositivism and educational research. Rowman & Littlefield (2000)
22. Hitch, D.: Artificial intelligence augmented qualitative analysis: the way of the future? Qual. Health Res. **34**(7), 595–606 (2024)
23. Kim, J.Y., Liu, C., Calvo, R.A., et al.: A comparison of online automatic speech recognition systems and the nonverbal responses to unintelligible speech. arXiv preprint arXiv:1904.12403 (2019)
24. Eftekhari, H.: Transcribing in the digital age: Qualitative research practice utilizing intelligent speech recognition technology. Eur. J. Cardiovasc. Nurs. **23**(5), 553–560 (2024)
25. Towler, L., Bondaronek, P., Papakonstantinou, T., et al.: Applying machine-learning to rapidly analyze large qualitative text datasets to inform the COVID-19 pandemic response: comparing human and machine-assisted topic analysis techniques. Front. Public Health **11**, 1268223 (2023)
26. Bennis, I., Mouwafaq, S.: Advancing AI-driven thematic analysis in qualitative research: a comparative study of nine generative models on Cutaneous Leishmaniasis data. BMC Med. Inf. Decis. Mak. **25**(1), 124 (2025)
27. Eberhardt, S.T., Schaffrath, J., Moggia, D., et al.: Decoding emotions: exploring the validity of sentiment analysis in psychotherapy. Psychother. Res. **35**(2), 174–189 (2025)
28. Bernasco, W., Hoeben, E.M., Koelma, D., et al.: Promise into practice: application of computer vision in empirical research on social distancing. Sociol. Methods Res. **52**(3), 1239–1287 (2023)
29. Bucy, E.P.: Politics through machine eyes: what computer vision allows us to see. J. Visual Polit. Commun. **10**(1), 59–68 (2023)
30. Manstein, S. M., Shiah, E., Laikhter, E.: Surveys and questionnaires: design, measures, and classic examples. In Translational Surgery, pp. 273–280. Elsevier (2023)
31. Lin, X.: Cognitio emergens: agency, dimensions, and dynamics in human-AI knowledge co-creation. arXiv preprint arXiv:2505.03105 (2025)
32. Zary, N., Lemkhente, Z.: Balancing efficiency and depth: a systematic review of Artificial Intelligence's impact on research competencies across the research lifecycle. arXiv preprint arXiv:2502.12345 (2025)
33. Gottweis, J., Weng, W.-H., Daryin, A., Tu, T., Palepu, A., Sirkovic, P., et al.: Towards an AI co-scientist. arXiv preprint arXiv:2502.18864 (2025)
34. Mojadeddi, Z., Rosenberg, J.: Automated transcription of interviews in qualitative research using artificial intelligence: a simple guide. J. Surg. Res. Pract. **5**, 1–6 (2024)
35. Rietz, T., Maedche, A.: Cody: an AI-based system to semi-automate coding for qualitative research. In: Proceedings of the 2021 CHI Conference on Human Factors in Computing Systems, pp. 1–14. Association for Computing Machinery (2021)

36. Saxena, K., Diamond, R., Conant, R.F., Mitchell, T.H., Gallopyn, G., Yakimow, K.E.: Provider adoption of speech recognition and its impact on satisfaction, documentation quality, efficiency, and cost in an inpatient EHR. AMIA Summ. Translat. Sci. Proc. **2018**, 186 (2018)
37. Dobbs, D., Yauk, J., Meng, H., Jalil, T.: The use of artificial intelligence transcription technology in caregiver qualitative data collection. Innov. Aging **8**(Suppl. 1), 459 (2024)
38. Overney, C., Saldías, B., Dimitrakopoulou, D., Roy, D.: Sensemate: an accessible and beginner-friendly human-AI platform for qualitative data analysis. In: Proceedings of the 29th International Conference on Intelligent User Interfaces, pp. 922–939. Association for Computing Machinery (2024)
39. Martin, S., Beecham, E., Kursumovic, E., Armstrong, R.A., Cook, T.M., Déom, N., et al.: Comparing human vs. machine-assisted analysis to develop a new approach for Big Qualitative Data Analysis. medRxiv preprint (2024)
40. Chyung, S.Y., Kennedy, M., Campbell, I.: Evidence-based survey design: the use of ascending or descending order of Likert-type response options. Perform. Improv. **57**(9), 9–16 (2018)
41. Bee, D.T., Murdoch-Eaton, D.: Questionnaire design: the good, the bad and the pitfalls. Arch. Dis. Childhood-Educ. Pract. **101**(4), 210–212 (2016)
42. O'Cathain, A., Thomas, K.J.: 'Any other comments?' Open questions on questionnaires – a bane or a bonus to research? BMC Med. Res. Methodol. **4**, 1–7 (2004)
43. Kitchin, R.: Big data, new epistemologies and paradigm shifts. Big Data Soc. **1**(1), Article 2053951714528481 (2014)
44. Pearl, J.: The seven tools of causal inference, with reflections on machine learning. Commun. ACM **62**(3), 54–60 (2019)
45. Pasquale, F.: The Black Box Society: The Secret Algorithms that Control Money and Information. Harvard University Press, Boston (2015)
46. Pillow, W.: Confession, catharsis, or cure? Rethinking the uses of reflexivity as methodological power in qualitative research. Int. J. Qual. Stud. Educ. **16**(2), 175–196 (2003)
47. Agar, M.H., et al.: The Professional Stranger: An Informal Introduction to Ethnography, vol. 276. Academic Press, Cambridge (1980)
48. Noble, S.U.: Algorithms of Oppression: How Search Engines Reinforce Racism. New York University Press, New York (2018)
49. O'Neil, C.: Weapons of Math Destruction: How Big Data Increases Inequality and Threatens Democracy. Crown (2017)
50. Nolan, R.: 'Can't help myself': on generative AI, the performance of qualitative research and slow scholarship. Qual. Res., Article 14687941241308696 (2025)
51. Lury, C., Wakeford, N.: Inventive Methods. Routledge, Abingdon (2012)
52. Boyd, D., Crawford, K.: Critical questions for big data: provocations for a cultural, technological, and scholarly phenomenon. Inf. Commun. Soc. **15**(5), 662–679 (2012)
53. Barfield, W.: Cyber-Humans: Our Future with Machines, vol. 291. Springer, Heidelberg (2015)
54. Guba, E.G., Lincoln, Y.S.: Competing paradigms in qualitative research. In: Denzin, N.K., Lincoln, Y.S. (eds.) Handbook of Qualitative Research, pp. 105–117. Sage Publications (1994)
55. Maxwell, J.A.: Qualitative Research Design: An Interactive Approach. Sage Publications, Thousand Oaks (2013)
56. Saldaña, J.: The Coding Manual for Qualitative Researchers, 4th edn. Sage Publications, Thousand Oaks (2021)

57. Narayanan Venkit, P., Gautam, S., Panchanadikar, R., Huang, T.-H., Wilson, S.: Unmasking nationality bias: a study of human perception of nationalities in AI-generated articles. In: Proceedings of the 2023 AAAI/ACM Conference on AI, Ethics, and Society, pp. 554–565. Association for Computing Machinery (2023)
58. Christou, P.A.: How to use artificial intelligence (AI) as a resource, methodological and analysis tool in qualitative research? Qual. Rep. **28**(7), 1998–2015 (2023)
59. Mozersky, J., Parsons, M., Walsh, H., Baldwin, K., McIntosh, T., DuBois, J.M.: Research participant views regarding qualitative data sharing. Ethics Human Res. **42**(2), 13–27 (2020)
60. Cohen, I.G.: Informed consent and medical artificial intelligence: what to tell the patient? Georgetown Law J. **108**, 1425 (2019)
61. Hosseini, M., Resnik, D.B., Holmes, K.: The ethics of disclosing the use of artificial intelligence tools in writing scholarly manuscripts. Res. Ethics **19**(4), 449–465 (2023)
62. Pires, T., Schlinger, E., Garrette, D.: How multilingual is multilingual BERT? arXiv preprint arXiv:1906.01502 (2019)
63. Monarch, R.M.: Human-in-the-Loop Machine Learning: Active learning and annotation for human-centered AI. Simon and Schuster (2021)
64. Kitchin, R.: Thinking critically about and researching algorithms. In: The Social Power of Algorithms, pp. 14–29. Routledge, Abingdon (2019)
65. Kaplan, A., Haenlein, M.: Siri, Siri, in my hand: who's the fairest in the land? On the interpretations, illustrations, and implications of artificial intelligence. Bus. Horiz. **62**(1), 15–25 (2019)
66. Makridakis, S.: The forthcoming Artificial Intelligence (AI) revolution: its impact on society and firms. Futures **90**, 46–60 (2017)
67. Jordan, M.I., Mitchell, T.M.: Machine learning: trends, perspectives, and prospects. Science **349**(6245), 255–260 (2015)
68. Rudin, C.: Stop explaining black box machine learning models for high stakes decisions and use interpretable models instead. Nat. Mach. Intell. **1**(5), 206–215 (2019)
69. Lazer, D., et al.: Computational social science. Science **323**(5915), 721–723 (2009)
70. Salganik, M. J.: Bit by Bit: Social Research in the Digital Age. Princeton University Press, Princeton (2019)
71. Lincoln, Y.S.: Naturalistic Inquiry, vol. 75. Sage, Thousands Oaks (1985)
72. Kaplan, A., Haenlein, M.: Rulers of the world, unite! The challenges and opportunities of artificial intelligence. Bus. Horiz. **63**(1), 37–50 (2020)
73. Hayles, N.K.: How we Became Posthuman: Virtual Bodies in Cybernetics, Literature, and Informatics. IOP Publishing (2000)
74. Hammersley, M.: What is Qualitative Research?. Bloomsbury Academic (2012)
75. Hesse-Biber, S.N., Johnson, R.B.: The Oxford Handbook of Multimethod and Mixed Methods Research Inquiry. Oxford University Press, Cambridge (2015)
76. Cetina, K. K.: Epistemic Cultures: How the Sciences Make Knowledge. Harvard University Press, Boston (1999)
77. Tufekci, Z.: Big questions for social media big data: representativeness, validity and other methodological pitfalls. In: Proceedings of the International AAAI Conference on Web and Social Media, vol. 8, no. 1, pp. 505–514 (2014)

Persuading AI Agents in a Queueing Game of Socially Scarce Resources Acquisition

Hongyi Liu(✉) and Qiaochu He

College of Business, Southern University of Science and Technology, Shenzhen, Guangdong 518000, China
12433211@mail.sustech.edu.cn, heqc@sustech.edu.cn

Abstract. Queueing models serve as effective proxies for the allocation of socially scarce resources, where customers sequentially receive limited services. To address this, service providers aiming to maximize social welfare can use quality-related signals to deter low-priority customers from queueing, thereby reserving access for high-priority counterparts. We conduct a behavioral economics experiment between the service provider (the environment) and customers (AI agents, or "suspects"). Our proposed framework integrates human-algorithm interaction in a transparent "white-box" setting. To explain the experimental results, we construct a theoretical model of information design to regulate scarce services allocation. Our research emphasizes the interaction between AI decision-making and its environment, highlighting how AI can overcome behavioral inefficiencies (e.g., selfish queueing) to foster cooperation and improve welfare.

Keywords: socially scarce resources · queueing games · Q-learning · information design

1 Introduction

The allocation of socially scarce resources remains a critical challenge in resource distribution systems. During the early phase of the COVID-19 pandemic, a surge in demand for face masks—amid near-fixed production capacity—exemplified severe supply-demand imbalances. According to WHO data, global mask demand increased 100-fold in 2020, accompanied by price spikes exceeding 2,000%. Similar patterns are observed in humanitarian operations, organ transplant waitlists, and more. Queueing models serve as effective proxies for such allocation systems, where customers sequentially receive limited services. Customers can be categorized into high- and low-priority segments based on the utility they derive from these scarce resources. High-priority customers (e.g., healthcare workers, the elderly, and essential workers) need immediate access, while low-priority customers (e.g., healthy individuals with minimal exposure) may stockpile for precautionary reasons. The key inefficiency in stratified queueing systems arises from negative externalities: when arrivals are stochastic, early low-priority customers may deplete resources, leaving high-priority customers underserved—ultimately diminishing social welfare.

X. Xie et al. (Eds.): ICSS 2025, LNOR, pp. 262–275, 2026.
https://doi.org/10.1007/978-3-032-13116-4_21

To address this, service providers aiming to maximize social welfare can use quality-related signals to deter low-priority customers from queuing, thereby reserving access for high-priority counterparts. We formalize this as a two-stage queueing game: in the first stage, the provider sends a quality signal, shaping customer beliefs about service quality; in the second, customers decide whether to join the queue in a discrete-time sequential game. We conduct a behavioral economics experiment between the service provider (the environment) and customers (AI agents, or "suspects"). Our proposed framework integrates human-algorithm interaction in a transparent "white-box" setting, training AI agents with behavioral considerations and managerial implications in mind. It suffices to model only low-priority customers, who determine threshold positions for opting out, while high-priority customers are assumed to always accept service. Through reinforcement learning, we simulate this environment using Q-learning agents, with parametric controls to ensure consistent behavior among high-priority customers. The transparency of the training and decision process defines the "white-box" nature of the experiment. This framework enables systematic analysis of low-priority agents' responses to quality signals and the impact on social welfare.

To explain these results, we construct a theoretical model of information design to regulate scarce resources allocation. As highlighted by Calvano et al. [1], AI-driven optimization through reinforcement learning in complex scenarios may inadvertently trigger systemic inefficiencies. Our research emphasizes the interaction between AI decision-making and its environment, highlighting how AI can overcome behavioral inefficiencies (e.g., selfish queuing) to foster cooperation and improve welfare. Similar to the equilibrium-driven mechanism in transportation systems proposed by Ning and Du [2], our framework enhances social welfare without suppressing rational user incentives. In extended settings, we explore stratified information interventions—sending accurate signals to high-priority customers and strategically noisy ones to low-priority customers—which further enhance social outcomes. Methodologically, we contribute an integrated, white-box analytical framework for studying human-algorithm interactions in scarcity-driven systems. Our findings offer practical insights for policymakers and service providers managing the allocation of limited resources.

The remainder of this paper is organized as follows. Section 2 reviews the relevant literature. In Sect. 3, we introduce the experimental setup of the reinforcement learning framework and present the results. Section 4 develops a theoretical model to explain the experimental results. Section 5 discusses extensions. Concluding remarks are provided in Sect. 6. All supporting materials and proofs are included in the appendix.

2 Literature Review

The allocation of socially scarce resources, such as medical supplies during crises, poses significant operational challenges due to supply-demand mismatches and behavioral inefficiencies. Traditional queuing models, rooted in operations research, focus on balancing capacity and stochastic demand [3]. However, these models often overlook the role of information asymmetry and user heterogeneity in shaping queuing behavior. For instance, Larson [3] emphasized the psychological dimensions of queuing, noting that perceived fairness and value inferences influence participation, yet his framework lacks mechanisms to address strategic resource depletion by low-priority customers—a critical issue observed during the COVID-19 mask shortage [4].

Advances in reinforcement learning (RL) offer tools to model and optimize dynamic interactions in queuing systems. Unlike classical optimization, RL agents learn adaptive policies through trial-and-error, mimicking human learning under uncertainty [5]. For example, Dulac-Arnold et al. [6] applied RL to vaccine distribution, where AI agents prioritized high-risk groups while dynamically adjusting signals to deter low-priority demand. Yet, most RL frameworks remain opaque, hindering trust in critical domains like healthcare.

The emerging field of explainable AI (XAI) addresses this limitation. Arrieta et al. [7] defined XAI as systems providing human-interpretable explanations for decisions, ensuring accountability in public resource allocation. In a queuing context, this translates to "white-box" RL agents that explicitly reveal how quality signals are generated and how user thresholds are calculated. For example, Grand-Clément et al. [8] developed interpretable bandit algorithms for rationing medical supplies, where decision rules (e.g., "prioritize users with comorbidities") are transparently communicated to stakeholders. Such approaches bridge the gap between algorithmic efficiency and ethical governance, a priority in post-pandemic policy design [9]. While XAI advances transparency in algorithmic decisions (e.g., triage rules), its application to modeling customer queuing behavior remains underexplored [7].

There are critical gaps persist in existing research:

1. Queueing models with AI agents. Current XAI frameworks focus on decision transparency but neglect dynamic modeling of customer queuing behavior (e.g., threshold-based opt-out) [7]. Developing human-centered AI models for queueing systems is challenging as existing models often fail to account for the roles of information asymmetry and user heterogeneity in shaping queuing behaviors [3, 4].
2. Opacity in AI Systems: RL solutions prioritize performance over transparency, limiting adoption in socially sensitive contexts [7].

Our work addresses these gaps by uniting interpretable RL with behaviorally grounded mechanism discovery, creating a transparent, adaptive system for scarce resource allocation.

3 AI Experiment

In our basic experiment, we conduct a standard queueing game experiment between the service provider (the environment) and customers (AI agents, or "suspects"). Through a Q-learning approach, we train AI agents during the repeated interactions in which the service provider broadcasts quality signals and customers queue to access socially scarce services. The model assumes that all low-priority customer agents share a common belief. The key parameters in the Q-learning model are outlined as follows:

3.1 Environment

In each round, the provider allocates a finite service capacity (denoted as k) to the system. A substantial influx of customers (λ agents, where $\lambda \gg k$) enters the queue following a stochastic arrival sequence, subsequently making sequential decisions under supply scarcity constraints. The condition $\lambda \gg k$ formally captures the systemic undersupply that characterizes socially scarce service markets.

The model assumes a heterogeneous customer base comprising two distinct types: high-priority (h) and low-priority (l) agents. A randomly selected agent is classified as high-priority with probability m and low-priority with probability $1 - m$. Concurrently, service provision operates under a two-tiered quality framework: high-quality (H) and low-quality (L) services. At the initiation of each round, the service quality level is stochastically determined via a binary process, remaining fixed throughout the round. Specifically, the system delivers high-quality service with probability p and low-quality service with probability $1 - p$. The valuation hierarchy adheres to the following structure:

1. Quality Dominance: For any given customer type $\theta \in \{h, l\}$, high-quality service yields strictly greater utility than low-quality service:

$$V_{Hh} > V_{Lh} \tag{1}$$

$$V_{Hl} > V_{Ll} \tag{2}$$

2. Type Dominance: For any given service quality $q \in \{H, L\}$, high-priority customer derives higher utility than low-priority customer:

$$V_{Hh} > V_{Hl} \tag{3}$$

$$V_{Lh} > V_{Ll} \tag{4}$$

Customers evaluate their expected payoffs and decide sequentially whether to accept the service. Upon completion of this decision-making process, individuals who opt out of service depart from the queue, and those who accept are served under a First-In, First-Out (FIFO) policy. The payoff structure is as follows:

1. Accepting customers with available capacity:
   A customer of type $\theta \in \{h, l\}$ accepting service of quality $q \in \{H, L\}$ receives a net payoff:

$$\pi(q, \theta) = V_{q\theta} - w \cdot c \tag{5}$$

where $V_{q\theta}$ denotes the type- and quality-specific service utility, w represents cumulative waiting time, initialized at $w = 1$ and incremented by 1 per served agent; c is the marginal cost of waiting time.

2. Accepting customers with exhausted capacity:
   A customer accepting service after capacity depletion receives $\pi = 0$.
3. Outside option:
   A customer choosing not to accept service receives a reservation utility normalized to $\pi_{\text{out}} = 0$.

Customers are modeled as risk-averse decision-makers, so they accept the service if and only if $E[\pi(q, \theta)] > 0$. For model simplicity, we assume high-priority customers will always choose to accept service:

$$V_{Lh} > c \cdot k \tag{6}$$

Given the fixed decision-making behavior of high-priority customers, the analytical focus is narrowed to the strategic interplay between the service provider and low-priority customers. The service provider aims to maximize the average social welfare, defined as the average sum of all customers' utilities in each round. Low-priority customers' objective is to maximize their individual utility. The service provider plans to persuade low-priority customers to abstain from service through signals related to service quality to achieve higher social welfare.

The model is structured as a two-stage signaling game with imperfect information. In Stage 1, the service provider establishes a public signaling mechanism to strategically influence low-priority customers' decisions. At the initiation of each round, service quality $q \in \{H, L\}$ is stochastically determined, with its realization remaining private to the service provider. Based on the quality state q, the provider broadcasts a uniform signal $\sigma \in \{\sigma_H, \sigma_L\}$ to all queued customers: when $q = L$, the provider deterministically transmits $\sigma = \sigma_L$; when $q = H$, the provider emits $\sigma = \sigma_H$ with probability x and $\sigma = \sigma_L$ with probability $1 - x$. In Stage 2, low-priority customers sequentially decide whether to accept or decline service through a protocol that integrates their prior belief p, the observed signal σ, and their positional index i in the queue. Critically, this decision-making process occurs under imperfect information—customers lack knowledge of others' choices and must act solely on individualized observational data.

This study employs a reinforcement learning framework to simulate the strategic dynamics between a service provider and low-priority customers in a queueing system. The service provider exogenously sets the signaling parameter x. Low-priority customer behavior is modeled through a homogeneous population of Q-learning agents, designed to approximate sequential decision-making under imperfect information. A critical simplification in this framework assumes that all low-priority customers adhere to a shared public belief, enabling the use of a single Q-table to represent their collective policy. This assumption effectively abstracts away individual belief heterogeneity, focusing instead on population-level behavioral convergence.

AI experiment will conduct multiple rounds, with the timeline for each round as follows (Table 1):

Table 1. Timeline of a single round

Stage	Event
0	The service quality of this round is randomly generated
1	Customers arrive sequentially in random order, with customer types are randomly generated
2	The service provider emits a signal based on the selected information rule and the service quality of this round.

(continued)

Table 1. (*continued*)

Stage	Event
3	Customers receive signals and make decisions sequentially
4	After all customers make decisions, accepting customers are served sequentially
5	Compute each customer's payoff and social welfare

3.2 Q-learning

The decision-making behavior of low-priority customers is modeled through a Q-learning algorithm, with all agents sharing a centralized Q-table to reflect homogeneous strategic learning under public belief assumptions. The state space is defined as $s \in \{\text{signal type, position}\}$, where signal type corresponds to the quality-related signal σ broadcast by the service provider, and position denotes the agent's ordinal rank in the queue. The action space comprises a binary choice: $a \in \{\text{take service, not take service}\}$. To balance exploration and exploitation, an ε-greedy policy is implemented, where ε represents the probability of selecting the greedy action (exploitation). During initial training phases, ε is set to a low value to prioritize extensive exploration of state-action pairs. As training progresses, ε is systematically annealed and fixed at 1 in later stages, enforcing deterministic decision-making and ensuring behavioral stability for robust statistical analysis.

The service provider's strategic signaling mechanism is parameterized by the variable x, which governs the precision of quality-related information disclosure. To analyze the welfare implications of varying signal precision, x is systematically varied across 21 discrete configurations $x \in \{0,0.05, 0.10, \ldots,0.95,1\}$. For each parameter configuration x, the Q-learning agents undergo extensive training until Q-table convergence is achieved.

3.3 Experiment Outcomes

3.3.1 Low Value Customer's Decisions at Fixed x

Table 2. Fully-converged Q-table of low-priority agents

State	Take service	Not take service
1, high type signal	11.0000	0.0
2, high type signal	9.0043	0.0
3, high type signal	7.0087	0.0
4, high type signal	5.0140	0.0
5, high type signal	3.0171	0.0
6, high type signal	0.0337	0.0
7, high type signal	0.0002	0.0

(*continued*)

Table 2. (*continued*)

State	Take service	Not take service
8, high type signal	0.0000	0.0
9, high type signal	0.0000	0.0
10, high type signal	0.0000	0.0
1, low type signal	3.6768	0.0
2, low type signal	1.6773	0.0
3, low type signal	−0.2935	0.0
4, low type signal	−1.3932	0.0
5, low type signal	−2.5590	0.0
6, low type signal	−2.8841	0.0
7, low type signal	−2.3760	0.0
8, low type signal	−1.3873	0.0
9, low type signal	−0.5717	0.0
10, low type signal	0.0000	0.0

Note: $\lambda = 10$, $k = 5$, $c = 2$, $p = 0.5$, $m = 0.5$, $V_{Hh} = 30$, $V_{Lh} = 15$, $V_{Hl} = 13$, $V_{Ll} = 2$

The reinforcement learning framework adopts a simplified structure analogous to the multi-armed bandit problem, focusing on state-dependent action-value estimation rather than complex sequential decision trees. In this setup, the fully converged Q-table encapsulates the learned policy, where each entry $Q\ (s, a)$ represents the empirically-derived expected long-term return for taking action a in state s (Table 2).

3.3.2 Impact of x-Value on Social Welfare

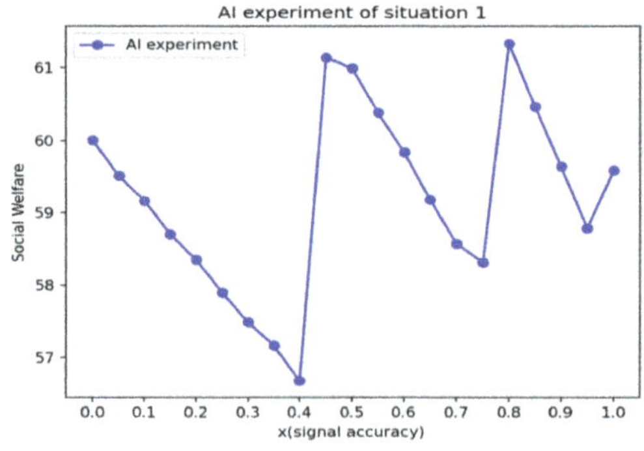

Fig. 1. Impact of x-value on social welfare
Note: $\lambda = 10$, $k = 5$, $c = 2$, $p = 0.5$, $m = 0.5$, $V_{Hh} = 30$, $V_{Lh} = 15$, $V_{Hl} = 13$, $V_{Ll} = 2$

As x increases from 0 to 1, the expected social welfare exhibits a piecewise linear trend characterized by altering phases of gradual declines and vertical surges (Fig. 1).

4 Theoretical Model

To elucidate the empirically observed decision-making patterns of low-priority customers and the associated social welfare dynamics, we construct a theoretical model that closely mirrors the experimental framework. The model retains all core specifications of the experimental setup—including signaling mechanisms, utility hierarchies, and behavioral protocols—with a single divergence: whereas the experiment operationalizes a finite queue length, the theoretical abstraction assumes an infinite queue length.

4.1 Information Design

The service provider operates in a capacity-constrained environment characterized by acute supply scarcity ($\lambda \gg k$), where only customers positioned at the forefront of the queue can access services. To maximize social welfare—equivalent to prioritizing service allocation to high-priority customers—the provider employs strategic information design to persuade low-priority customers occupying advantageous queue positions to voluntarily exit. Low-priority customers, acting as rational utility maximizers, base their decisions on the expected payoff:

$$E[\pi\ (q, l)] = E[V_{ql} - w \cdot c] \tag{7}$$

By selectively obfuscating or attenuating perceived service quality through signaling, the provider strategically reduces low-priority customers' expected utility, inducing them to decline service when $E[\pi\ (q, l)] \leq 0$. This targeted information design mitigates queue-position externalities, reallocating scarce capacity to posterior high-priority customers and thereby enhancing aggregate welfare.

The service provider emits two distinct quality signals, $\sigma \in \{\sigma_H, \sigma_L\}$, to modulate customers' expectations of service value. Signal σ_H is designed to amplify perceived service quality by indicating a higher likelihood of high-quality service, thereby increasing the expected valuation. Conversely, signal σ_L attenuates perceived quality by suggesting a greater probability of low-quality service, reducing the expected valuation (Fig. 2).

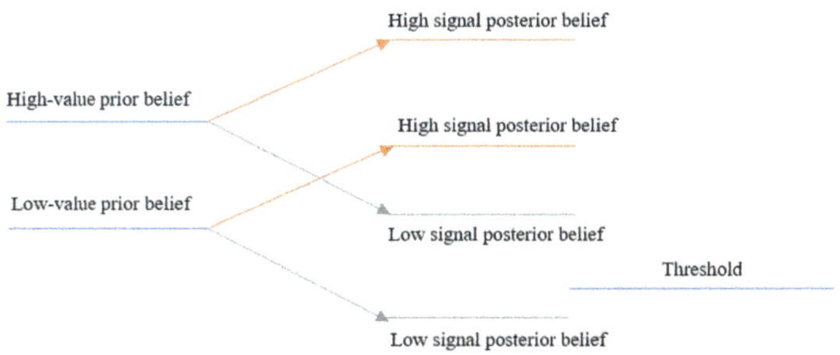

Fig. 2. The mechanism of information design

The service provider's signaling strategy is as follows: When the actual service quality is high ($q = H$), the provider emits signal σ_H with probability x and signal σ_L with probability $1 - x$. When service quality is low ($q = L$), the provider emits signal σ_H with probability y and signal σ_L with probability $1 - y$. Intuitively, for the maximization of social welfare, service providers have no incentive to send signal σ_H when the service quality is low ($q = L$); this can also be mathematically proven. Thus, welfare-optimality necessitates $y^* = 0$, reducing the signaling rule to the following: When the actual service quality is high ($q = H$), the provider emits signal σ_H with probability x and signal σ_L with probability $1 - x$. When service quality is low ($q = L$), the provider always emits signal σ_L. The provider's objective reduces to identifying the optimal signal precision $x^* \in [0,1]$ that maximizes aggregate welfare $W(x)$.

4.2 Persuasion Strategy

Persuasion strategy means all potential low-priority customers choose not to take service starting from a certain position (denoted as a). The service provider optimizes x to maximize social welfare through the persuasion strategy. The range for x to induce exit of customers with expected waiting time w is:

$$x \geq \frac{p * V_{Hl} + (1 - p) * V_{Ll} - c * w}{p * (V_{Hl} - c * w)} \tag{8}$$

Because of the waiting time monotonicity, if such x can successfully persuade a low-priority customer with waiting time w to exit, it must persuade all subsequent and potential low-priority customers (with waiting time $\geq w$) to exit. Thus, each value of x corresponds to a persuasion strategy. The model accommodates a maximum of $k + 1$ distinct persuasion strategies, ranging from initiating persuasion at the first position to the $(k + 1)$-th position, with the exact number contingent upon the specific parameter assignments. Under the parameter assumptions of the model, determining the optimal x is equivalent to identifying the optimal a. Therefore, we can take the parameter a (representing the starting position of the persuasion strategy) as the independent variable instead of x to maximize social welfare, which simplifies the computation.

When determining the optimal value of x (or a), there is a trade-off between signal strength and frequency: increasing x enhances the persuasive power of low signals, encouraging exits from earlier queue positions to boost social welfare, but simultaneously reduces the likelihood of low signals occurring, creating a counteracting effect that constrains welfare improvement.

4.3 Expected Waiting Time

The public belief assumption ensures the existence of a critical threshold position (denoted as o) in the queue, such that all potential low-priority customers will choose to accept service before o and decline service afterward. Let i denote a customer's queue position. The expected waiting time $w(i)$ is defined as:

$$w(i) = \begin{cases} i & , if\ i < o \\ (o - 1) + (i - o) * m + 1 & , if\ i \geq o \end{cases} \tag{9}$$

In fact, the threshold o is just a of the corresponding persuasion strategy.

4.4 Expected Social Welfare

Due to the stochastic nature of the environment, even under identical information rules, social welfare may exhibit significant variations across different rounds. Consequently, the service provider aims to maximize the expected social welfare. The discrepancy in expected social welfare between the theoretical model (assuming an infinite queue length) and AI experiments (with finite queue length) arises from whether the service capacity is fully utilized. When $\lambda \gg k$, the gap between them approaches 0 and can be negligible. Therefore, for simplicity, the theoretical analysis adopts the infinite queue length assumption. The expected social welfare $W(x)$ is:

$$\begin{aligned} W(x) = (1-p)\,[(a-1)mV_{Lh} + (a-1)\,(1-m)V_{Ll} + (k-a+1)V_{Lh}] \\ + p(1-x)\,[(a-1)mV_{Hh} + (a-1)\,(1-m)V_{Hl} + (k-a+1)V_{Hh}] \\ + px\,[kmV_{Hh} + k(1-m)V_{Hl}] - \tfrac{k(1+k)}{2}\mathrm{c} \end{aligned} \quad (10)$$

subject to the original constraints from Eqs. (1), (2), (3), (4) and (6), augmented by:

$$0 \le p, x, m \le 1 \quad (11)$$

$$0 < c, k \quad (12)$$

$$1 \le a \le k + 1 \quad (13)$$

$$\mathrm{a} = \mathrm{Min}\left[\left\lceil \frac{p * v_{Hl} * (1-x) + (1-p) * v_{Ll}}{c * [p * (1-x) + (1-p)]} \right\rceil, k+1\right] \quad (14)$$

4.5 Equilibrium Concept

The equilibrium behavior of customers under the information rule x is structured as follows:

1. High-priority Customers:
   Always accept service regardless of queue position i.
2. Low-priority Customers:
   When receiving a high signal, they accept service if their position $i \le k$, defaulting to rejection for $i > k$ (payoff-irrelevant, yielding zero utility). When receiving a low signal, they accept service only if $i < a$, strictly rejecting for $i \ge a$.

The equilibrium is thus fully defined by the pair $\{x^*, a^*\}$, where x^* governs the optimal information design, and a^* captures the endogenous behavioral threshold for low-signal responses.

5 Analysis

5.1 The Optimal Information Design

The service provider seeks an optimal $x \in [0,1]$ to maximize social welfare. The model's high-dimensional parameter space generates complex analytical solution characteristics. The study employs numerical analysis and related methodologies (e.g., heatmap visualization) to generate representative analytical results, which elucidate the core insights of the research.

Proposition 1
The optimal information design is related to the proportion of high-quality services, and the difference in perceived value of high- and low-priority customers for distinct service qualities, but in a great many cases, it is not related to the proportion of high-priority customers.

5.2 The Explanation on the Impact of Information Precision(x) on Social Welfare

Social welfare follows a piecewise linear trajectory as x varies over the interval [0,1], with each linear segment corresponding to intervals where the persuasion strategy remains fixed. Within each interval, expected welfare monotonically decreases as x rises due to diminishing low-signal transmission probabilities, while the expected number of persuaded low-signal individuals stays constant. At interval boundaries, discrete upward jumps in welfare occur when the marginal gain from increased persuasion success outweighs the reduced likelihood of low-signal scenarios.

6 Extension

In the main model, since high-priority customers always derive sufficient utility from service ($V_{Lh} > c{\cdot}k$), signals only affect low-priority customers' decisions. In the extension, the service utility for high-priority customers will be reduced, they may also be persuaded to abstain from accepting service. Thereby, the service provider's signalling strategy will change. Our extended analysis will build upon this foundation to enable more comprehensive investigations.

6.1 The Optimal Information Design under Extended Conditions

Proposition 2
When high-priority customers may also be persuaded to abstain from accepting service, the service provider should truthfully disclose quality information to maximize social welfare.

The Nash equilibrium of the expanded conditions under the information rule x is structured as follows:

1. High-priority Customers:
   When receiving a high signal, they accept service if their position $i \le k$, defaulting to rejection for $i > k$ (payoff-irrelevant, yielding zero utility). When receiving a low signal, they accept service only if $i < a_1$, strictly rejecting for $i \ge a_1$.
2. Low-priority Customers:
   When receiving a high signal, they accept service if their position $i \le k$, defaulting to rejection for $i > k$ (payoff-irrelevant, yielding zero utility). When receiving a low signal, they accept service only if $i < a_2$, strictly rejecting for $i \ge a_2$.

The equilibrium is thus fully defined by the pair $\{x^*, a_1{}^*, a_2{}^*\}$, where x^* governs the signal design, and $a_1{}^*$, $a_2{}^*$ capture the endogenous behavioral threshold for low-signal responses of high- and low-priority customers. $a_1{}^* \ge a_2{}^*$, as high-priority customers' elevated utility parameters render them less responsive to persuasion aimed at abandoning service acquisition.

The calculation of expected $a_1{}^*$, $a_2{}^*$ and social welfare is not elaborated here for brevity. Derivative analysis implies that when high-priority customers are also susceptible to being persuaded to abandon service acquisition, truthful quality disclosure constitutes the optimal information design strategy (Fig. 3).

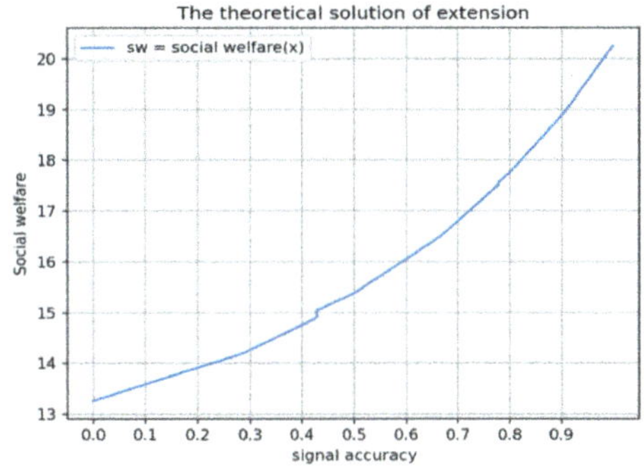

Fig. 3. Impact of x-value on social welfare under extended conditions
Note: $\lambda = 10$, $k = 5$, $c = 2$, $p = 0.5$, $m = 0.5$, $V_{Hh} = 15$, $V_{Lh} = 3$, $V_{Hl} = 13$, $V_{Ll} = 2$

The monotonic increase in social welfare can be theoretically explained through two mechanisms. First, the downward adjustment of high-priority customers' value parameters reduces the disparity in social welfare outcomes between high- and low-priority segments when service is acquired. Second, enhanced social welfare fundamentally depends on high-priority customers' ability to make utility-maximizing decisions. As x increases, decision accuracy for high-priority customers is improved systematically.

6.2 Heterogeneous Signal

When the service utility of high-priority customers falls below a critical threshold, homogeneous information design in the principal-agent framework fails to achieve efficacy. Under the assumption that the service provider can discern customer types, it strategically employs heterogeneous signaling tailored to distinct segments, formalized through an optimal information rule denoted as $\{x_1{}^*, x_2{}^*\}$. This rule ensures precision-adjusted signals are allocated to high- and low-priority customers, respectively, thereby restoring allocation efficiency even under diminished high-priority utility conditions.

Proposition 3 *When the service provider sends heterogeneous signals to high- and low-priority customers, social welfare maximization requires sending fully accurate signals to high-priority customers while potentially distorting signals for low-priority customers* $(0 \le x_2^* < x_1^* = 1)$.

The expected social welfare function under heterogeneous signals is derived in an approach analogous to the baseline experiment and obtained the optimal solution $(0 \le x_2^* < x_1^* = 1)$. By comparing the expected social welfare calculation functions of both approaches, it can be observed that, compared with homogeneous signaling, heterogeneous signaling significantly improves social welfare.

7 Conclusions

This study investigates the role of AI in achieving implicit cooperation and social welfare within resource-constrained service systems. Specifically, we focus on the interaction between decision-making agents and service environments, examining how artificial intelligence can be leveraged to counteract behavioral inefficiencies, such as selfish queuing. We develop a two-stage sequential game model wherein service providers use quality-based signaling mechanisms to encourage self-selection, nudging low-priority customers to abstain from service consumption to reduce the negative externalities for their higher-value counterparts.

Employing a behavioral economics experiment framework, we model service providers as environmental agents and customers as AI agent "suspects". Reinforcement learning simulations with Q-learning agents allow us to quantify welfare outcomes under varying information designs. Our results highlight that optimal signaling configurations depend on key structural parameters: the availability of high-quality service and the perceived heterogeneity in service value. Notably, in many scenarios, the optimal configuration is robust to changes in the proportion of high-priority customers.

The theoretical model further shows that when high-priority customers demonstrate flexibility in their value perception, truthful signaling becomes a dominant strategy. We extend the analysis to settings where user types are observable, finding that heterogeneous signaling—accurate for high-priority customers, noisy for low-priority customers—significantly outperforms uniform strategies in terms of social welfare. These findings have direct implications for the design of service systems, suggesting that transparent and adaptive information policies can effectively balance equity and efficiency in human-algorithm interactions.

References

1. Calvano, E., Calzolari, G., Denicolò, V., Pastorello, S.: Artificial intelligence, algorithmic pricing, and collusion. Am. Econ. Rev. **110**(10), 3267–3297 (2020)

2. Ning, Y.Q., Du, L.L.: Robust and resilient equilibrium routing mechanism for traffic congestion mitigation built upon correlated equilibrium and distributed optimization. Transp. Res. B Methodol. **168**, 170–205 (2023)
3. Larson, R.C.: Perspectives on queues: social justice and the psychology of queueing. Oper. Res. **35**(6), 895–905 (1987)
4. Lu, Y., Musalem, A., Olivares, M., Schilkrut, A.: Measuring the effect of queues on customer purchases. Manag. Sci. **59**(8), 1743–1763 (2013)
5. Sutton, R.S., Barto, A.G.: Reinforcement Learning: An Introduction. 2nd edn. Adaptive Computation and Machine Learning series, MIT Press (2018)
6. Dulac-Arnold, G., et al.: Challenges of real-world reinforcement learning: definitions, benchmarks and analysis. Mach. Learn. **110**(9), 2419–2468 (2021)
7. Arrieta, A.B., et al.: Explainable artificial intelligence (XAI): concepts, taxonomies, opportunities, and challenges toward responsible AI. Inf. Fusion. **58**, 82–115 (2020)
8. Grand-Clément, J., et al.: Interpretable machine learning for resource allocation with application to ventilator triage. arXiv preprint arXiv:2110.10994(2021)
9. Tsai, T.C., et al.: Algorithmic fairness in pandemic forecasting: lessons from COVID-19. npj Digit. Med. **5**, 59 (2022)

Comorbidity Patterns Before and After Juvenile Idiopathic Arthritis Diagnosis

Kazi Arman Ahmed[1], Hamid Hadidi[2], Mohammad Fili[1](✉), Dursun Delen[2], Guiping Hu[1], and Lizhi Wang[1]

[1] School of Industrial Engineering and Management, Oklahoma State University, Stillwater, OK 74078, USA
{kazi.ahmed,mohammad.fili,guiping.hu,lizhi.wang}@okstate.edu
[2] Department of Management Science & Information Systems, Oklahoma State University, Stillwater, OK 74078, USA
{hamidreza.hadidi,dursun.delen}@okstate.edu

Abstract. Juvenile idiopathic arthritis (JIA) is a chronic inflammatory disease that begins before the age of 16, characterized by persistent joint pain, swelling, and stiffness. In this study, we utilized the Oracle Real-World Data (formerly Cerner Real-World Data), to examine the demographic characteristics of JIA patients and to identify patterns of comorbidities both before and after diagnosis. Our analysis revealed that individuals identified as Black or African American are diagnosed with JIA at significantly older ages compared to other racial groups, highlighting a critical disparity given the importance of early detection for better clinical outcomes. Pre-diagnostic comorbidity analysis indicated that the presence of musculoskeletal pain, particularly when occurring alongside joint disorders, abdominal pain, or respiratory symptoms, could serve as early clinical indicators for JIA. Post-diagnosis, the comorbidity burden shifted from isolated musculoskeletal complaints to broader multi-system involvement, with joint, skin, and respiratory conditions becoming predominant.

Keywords: Comorbidities · Electronic health record (EHR) · Juvenile idiopathic arthritis (JIA) · Oracle Real-World Data

1 Introduction

Juvenile idiopathic arthritis (JIA), formerly known as Juvenile Rheumatoid Arthritis (JRA), is a chronic, autoimmune inflammatory disorder that begins before the age of 16, lasts more than six weeks, and has no identifiable cause to date [1]. As the most common pediatric rheumatic disease, this condition affects approximately 220,000 children and adolescents under the age of 18 in the United States, corresponding to a prevalence of 305 per 100,000 [2]. Although it is primarily diagnosed during childhood, a significant proportion of patients continue to experience active disease into adulthood, with associated long-term disability and reduced quality of life [3, 4].

In recent decades, therapeutic advances, particularly the widespread use of biologic and non-biologic disease-modifying antirheumatic drugs (DMARDs), have improved

X. Xie et al. (Eds.): ICSS 2025, LNOR, pp. 276–288, 2026.
https://doi.org/10.1007/978-3-032-13116-4_22

disease control for many patients [5]. However, disease activity alone does not fully capture the burden experienced by children with JIA. Several studies have documented that even patients with minimal clinical symptoms may report suboptimal health-related quality of life, including physical, emotional, and psychosocial challenges [6, 7].

JIA is increasingly recognized as a systemic condition associated with a wide range of comorbidities, including disorders such as type 1 diabetes mellitus, autoimmune thyroiditis, celiac disease, and inflammatory bowel disease, uveitis, allergic rhinitis, and atopic dermatitis [8–11]. In a large US-based cohort study, Lovell et al. [9] reported that over half of the examined autoimmune and associated conditions had significantly higher prevalence in JIA patients than in the general pediatric population, with several conditions showing a more than 20-fold increase. In addition to physical comorbidities, mental health concerns are common in this population. Adolescents and young adults with JIA have been found to have a higher prevalence of depression, anxiety, and emotional or behavioral disorders compared to their healthy peers [12, 13].

One important area that has not been studied enough is the timing of when other comorbidities appear in relation to the JIA diagnosis [14]. Some of these conditions may develop before JIA is officially diagnosed. These early health problems might be connected to changes in the immune system that happen before arthritis begins and could serve as warning signs that a child is at higher risk of developing JIA. On the other hand, some comorbidities appear after the JIA diagnosis. These may be the result of long-term inflammation in the body, side effects from medications used to control arthritis, or other changes in the immune system over time. Being able to tell the difference between conditions that occur before JIA and those that develop afterward can help researchers and doctors better understand how the disease develops and progresses. It can also help identify patients who might need closer monitoring or additional support. Another research gap is that despite growing recognition of the importance of comorbidities in JIA, comprehensive evaluations of their prevalence in large, real-world clinical populations remain limited. Moreover, many existing studies are based on small cohorts, single-center data, or focus on a narrow subset of conditions. The use of electronic health record (EHR) data for large-scale epidemiological analysis offers an opportunity to identify patterns of co-occurring diseases in JIA with greater granularity and representativeness.

The objective of this study is to conduct a comprehensive analysis of JIA patients using a large EHR dataset. Specifically, this study aims to (1) characterize the demographic distribution of children and adolescents diagnosed with JIA, (2) identify the most common comorbidities recorded prior to JIA diagnosis, (3) determine the most prevalent comorbid conditions emerging after diagnosis, and (4) provide insights that can inform early screening, multidisciplinary management, and long-term care planning. Understanding the intersection between JIA, comorbidity burden, and demographic context is critical for improving clinical outcomes and quality of life in this vulnerable pediatric population.

The rest of the paper is organized as following: Sect. 2 describes the dataset, data preprocessing steps, and statistical analysis methods. Sect. 3 presents the results of our analysis. Finally, Sect. 4 summarizes the key findings and discusses directions for future research.

2 Data and Methods

This section outlines the methodological steps taken to analyze the comorbidity patterns and demographic disparities associated with JIA. Our approach is structured into three main components: data extraction and cleaning, statistical evaluation on the age and race groups, and comorbidity identification.

2.1 Data

The present study utilized the Oracle Real-World Data (formerly Cerner Real-World Data), a large, de-identified electronic health record system containing detailed clinical data from millions of patients across diverse healthcare settings. Oracle Real-World Data organizes clinical data into structured tables, such as disease conditions, demographics, and encounters. The conditions table records diagnoses, including 10th revision of International Classification of Diseases (ICD-10) codes and onset dates, enabling identification of patients with specific conditions. The demographics table contains patient attributes such as year of birth, gender, race, and ethnicity, providing essential context for analyzing health disparities and population-level trends. The encounters dataset documents healthcare interactions, including visit types, dates, providers, and associated facilities, offering a longitudinal view of patient care trajectories.

To prepare the data for the analysis, we followed a series of preprocessing steps including extracting, cleaning, and merging to identify patients with JIA and their associated clinical records. The preprocessing pipeline is outlined in Fig. 1. First, we used the ICD-10 codes corresponding to JIA/JRA [15] to identify patients diagnosed with the condition at any point in their lifetime. Although the preferred term is Juvenile Idiopathic Arthritis (JIA), the ICD-10 classification system still uses the term JRA; therefore, we included both terms in our search criteria. This initial step resulted in 35,129 unique patients. The processing steps for each dataset are as follows:

Demographics data: We filtered patients with a current age between 0 and 110 years and reformatted race and ethnicity fields, resulting in the removal of 4 patients ($n = 35{,}125$).
Condition data: For condition data, disease conditions were extracted using unique patient IDs, focusing specifically on Juvenile Idiopathic Arthritis (JIA) conditions, yielding 208,590 records (i.e., observations). Encounter IDs from this dataset were used to link to encounter records.
Encounter data: For encounter data, records were extracted and merged with the condition table through encounter IDs, then further merged with the demographics table using patient IDs. Service dates were used to determine the earliest encounter date for each patient, marking the first JIA diagnosis. The patient's date of birth and the first JIA diagnosis date were used to calculate the age at diagnosis.

Later, in the merged dataset, further data cleaning involved removing patients whose age at first JIA diagnosis was above 16 years [1]. The final dataset comprised 9,257 unique patients with 99,395 medical records. We utilized the Clinical Classifications Software Refined (CCSR) [16] tool to aggregate individual ICD-10 diagnosis codes into broader, clinically meaningful condition categories, facilitating more effective analysis of comorbidity patterns associated with JIA.

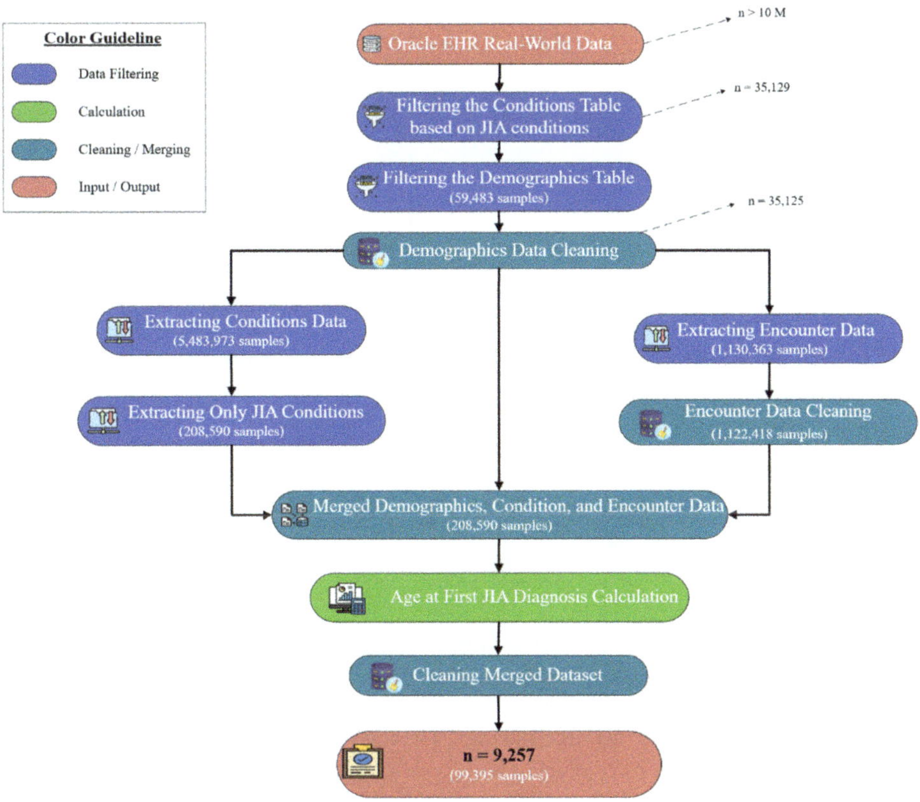

Fig. 1. Flowchart of the data processing pipeline. *n* is the number of unique patients at each stage while samples represent observations. In the presence of multiple records per patient, the number of samples would be higher than the unique patients.

To further explore the temporal dynamics of comorbidities associated with JIA, we analyzed the occurrence of diseases before and after JIA diagnosis. This process is detailed in Fig. 2. We began by extracting condition data (1,092,455 records) and encounter data (685,348 records) for the JIA-diagnosed patients. The condition data included all recorded diagnoses for each patient, identified by ICD-10 codes, while the encounter data provided timestamps of medical visits. These datasets were merged to create a unified dataset, linking patient conditions with their corresponding encounter dates. To ensure data quality and relevance, we cleaned the merged dataset by retaining only unique diseases per patient. From the disease name, we retained the first occurrence of each unique disease for each patient, avoiding redundancy from repeated diagnoses. This cleaning process resulted in a refined dataset of 236,519 samples. Next, we defined pre- and post-diagnosis cohorts by leveraging the Age at First JIA diagnosis. To analyze comorbidities within a specific temporal window, we utilized the Age at First JIA and the first encounter date for each disease to identify conditions occurring within 5 years before and 5 years after the first JIA diagnosis. This 5-year window was chosen to capture both immediate and longer-term comorbidities associated with JIA.

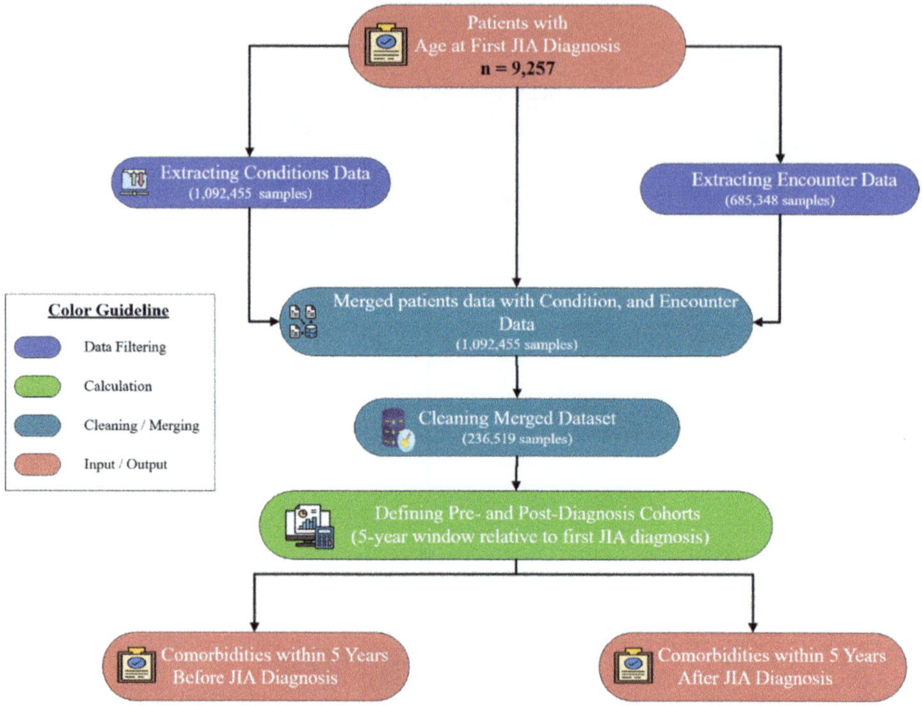

Fig. 2. Flowchart illustrating the methodology for analyzing pre- and post-JIA diagnosis comorbidities.

The analysis of patient demographics revealed that the age at first JIA diagnosis predominantly clusters within the later stages of childhood (Fig. 3A). Stratifying the cohort into developmental age groups demonstrated that the majority of diagnoses occurred during the "Grade School" (5–12 years) and "Teen" (12–16 years) periods, with minimal diagnoses in infancy ("Babies") and early childhood ("Toddlers" and "Preschoolers"). The mean age at diagnosis was 10.0 years, while the median was 11.0 years, both aligning with the "Grade School" category. This age distribution suggests that clinical onset of JIA often becomes apparent during middle to late childhood, rather than early life stages. The findings highlight a critical window for early surveillance and intervention during the transition into school years, emphasizing the importance of age-targeted screening efforts to improve early detection and outcomes in JIA management.

The geographic distribution of JIA patients across the United States is illustrated in the choropleth map in Fig. 3B. States such as Texas, California, New York, and Pennsylvania exhibited the highest counts of JIA patients, with Texas and California each reporting over 4,500 cases. In contrast, states in the Midwest and Mountain regions, such as North Dakota, South Dakota, and Montana, had the lowest counts, with fewer than 1,000 cases each. This pattern indicates a higher prevalence of JIA in more populous states.

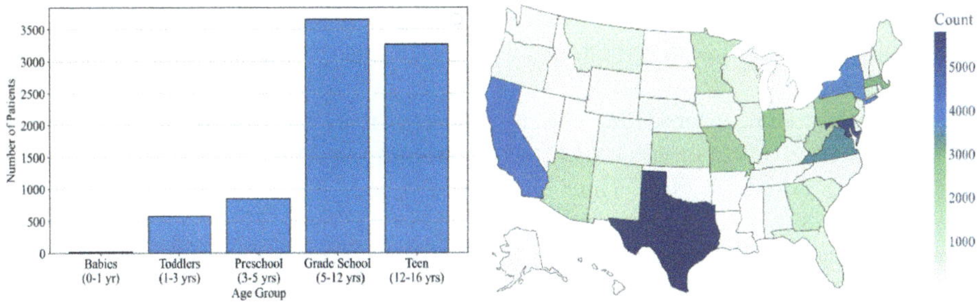

(A) (B)

Fig. 3. Overview of age at first diagnosis and geographic distribution of JIA patients. (**A**) Distribution of first JIA diagnosis across developmental age groups; (**B**) Choropleth map of the United States displaying the frequency of JIA patients by state. Patients may have multiple locations listed in their records.

2.2 Statistical Analysis

We used statistical analysis to compare the median age at first JIA diagnosis between different race groups. Since the distribution of age at diagnosis was left-skewed, we employed the non-parametric Kruskal-Wallis test to assess whether the distribution of age at first JIA diagnosis differed significantly across groups. Following a Kruskal-Wallis test, post-hoc pairwise comparisons were conducted using Dunn's test with Holm correction to identify specific racial groups exhibiting significant differences [17]. We used a significance level of 0.05 for statistical analysis.

2.3 Analysis of Pre- and Post-diagnosis Comorbidities

To analyze pre- and post-diagnosis comorbidities in JIA patients, we constructed a co-occurrence matrix within a five-year window before and after the initial diagnosis, separately to display the frequency with which different pairs of conditions occurred in the patient cohort. We then plotted the frequency of comorbidity combinations for top conditions using UpSet Plots.

3 Results

This section presents the findings derived from the statistical and visual analyses of our cohort. First, we explore demographic trends such as age at diagnosis and geographic distribution of JIA patients. Then, we examine significant differences in diagnosis age across racial groups. Lastly, we analyze the patterns of disease co-occurrence observed in the five years before and after the JIA diagnosis.

3.1 Demographic Analysis

The Kruskal–Wallis test was conducted to evaluate whether the age at first JIA diagnosis differed across racial groups. The test yielded a statistically significant result ($H = 25.13$,

$p = 0.0003$), indicating that at least one racial group differs significantly from others in terms of diagnosis age. The distribution of diagnosis age across racial groups is visualized in Fig. 4.

Among the racial groups, the "Other" category ($n = 900$) had the lowest median age at diagnosis (10 years), while the "Asian" group ($n = 251$) had the highest median age (12 years). Although the "White" ($n = 6{,}339$) and "Black or African American" ($n = 1{,}034$) groups shared a similar median age of 11 years, Dunn's test with Holm adjustment revealed a significant difference in the distribution between these groups ($p < 0.0001$). This difference was driven not by the central tendency but by the spread, with "Black or African American" patients being more frequently diagnosed at older ages, as indicated by their higher 25th percentile. Similarly, a significant difference was observed between "Black or African American" and "Other" groups ($p = 0.003$), again reflecting distributional disparities. Additionally, the "Native Hawaiian or Other Pacific Islander (NHPI)" group exhibited the highest 75th percentile age, highlighting a broader spread of diagnosis age in that population.

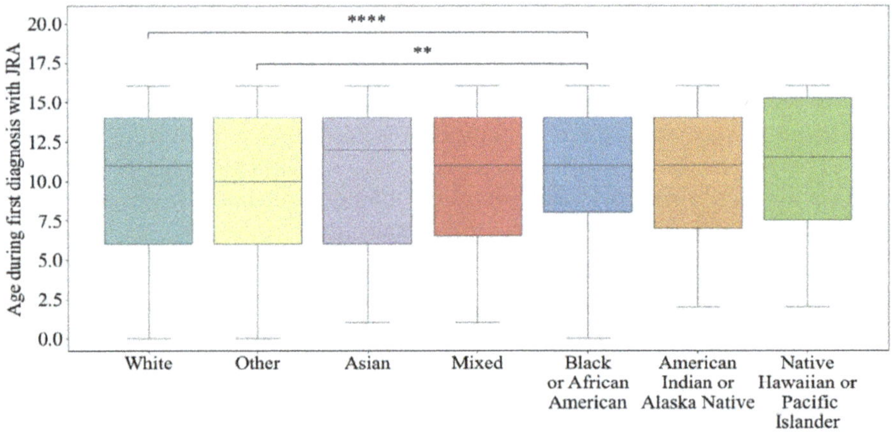

Fig. 4. Boxplot of age at first JIA diagnosis across racial groups, with pairwise comparisons using Holm-adjusted Dunn's test (****: p-value <0.0001; ***: p-value <0.001; **: p-value <0.01; *: p-value <0.05)

3.2 Pre-diagnosis Comorbidities

Analysis of disease co-occurrence patterns in the five years preceding JIA diagnosis revealed notable relationships among conditions frequently reported prior to JIA. The co-occurrence matrix for the top 16 comorbidities, visualized as a heatmap in Fig. 5, shows the frequency of pairwise comorbidity occurrences among JIA patients. The highest co-occurrence was between "Joint Disorder" and "Musculoskeletal Pain" ($n = 992$ or 10.72%), indicating that these conditions frequently co-occurred in the pre-JIA period. Other significant co-occurrences also observed between "Musculoskeletal Pain" and conditions like "Abnormal Findings" ($n = 790$ or 8.53%), "Upper Respiratory Infections" ($n = 503$ or 5.43%), and "Abdominal Pain" ($n = 500$ or 5.40%). The lowest co-occurrence was between "Otitis Media" and "Tendon & Synovial Disorders" (60 patients or 0.65%), suggesting less frequent overlap between these conditions.

To further explore comorbidity patterns, an Upset plot focusing on the top 5 pre-JIA comorbidities (i.e., Skin Inflammation, Abdominal Pain, Respiratory Symptoms, Joint Disorders, Musculoskeletal Pain) was generated (see Fig. 6). The plot illustrates the intersection sizes of these comorbidities across different combinations. The largest intersection was for patients with only "Musculoskeletal Pain" (n = 869 or 9.39%). The next most common was dual comorbidities: notably "Musculoskeletal Pain" with "Joint Disorders" (n = 595 or 6.43%), followed by other single conditions like "Joint Disorder" (n = 487 or 5.26%), and "Respiratory Symptoms" (n = 270 or 2.92%). The largest triple-comorbidity group was "Joint Disorders," "Musculoskeletal Pain," and "Abdominal Pain" (n = 101 or 1.09%). Notably, some other triple comorbidities, such as "Joint Disorders," "Musculoskeletal Pain," and "Respiratory Symptoms" (n = 91 or 0.98%), had higher occurrence frequencies than certain dual comorbidities, such as "Skin Inflammation" with "Respiratory Symptoms" (n = 60 or 0.65%).

These findings may suggest that musculoskeletal pain, especially when combined with joint disorders, abdominal pain, or respiratory symptoms, could raise clinical suspicion for JIA, prompting earlier investigation.

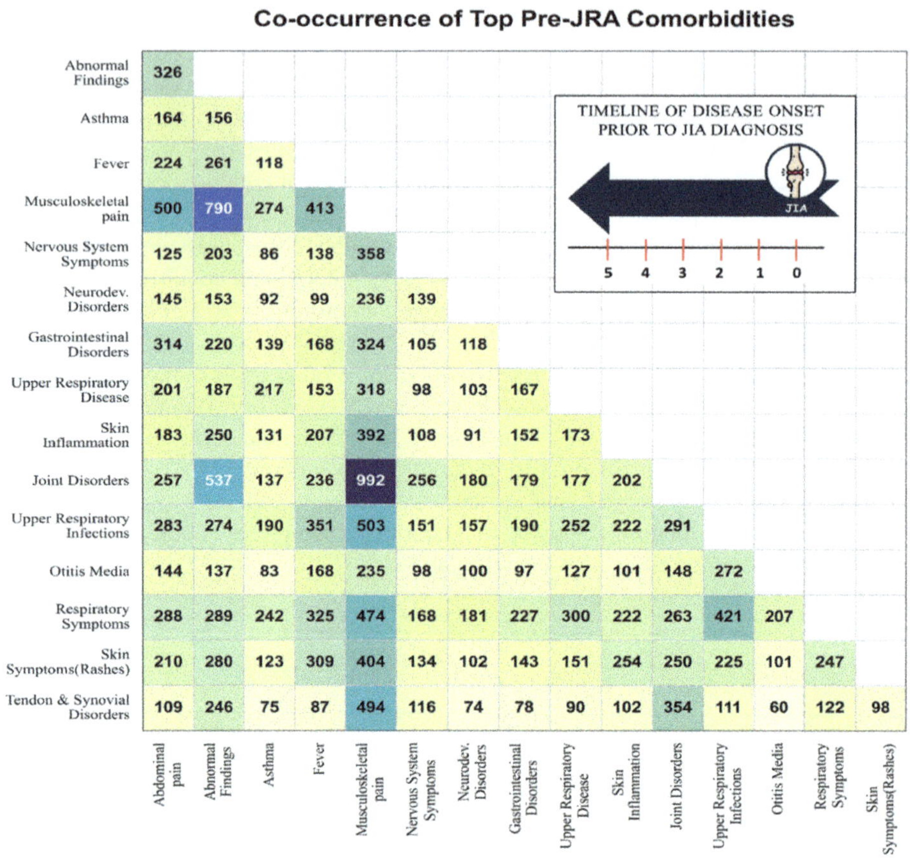

Fig. 5. Co-occurrence Matrix of Comorbidities within 5 years before the diagnosis.

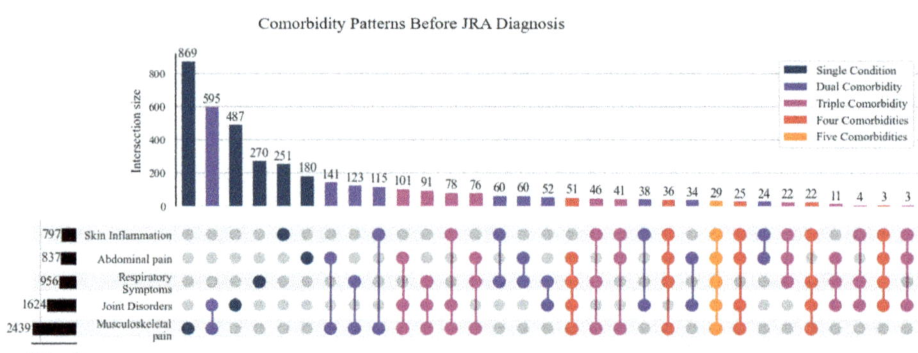

Fig. 6. UpSet plots displaying comorbidity combinations observed in the five years before JIA diagnosis.

3.3 Post-diagnosis Comorbidities

Analysis of disease co-occurrence patterns in the five years following JIA diagnosis, identified significant relationships among conditions frequently reported after JIA diagnosis (see Fig. 7). Several patterns diverge sharply from the pre-diagnosis period. "Musculoskeletal Pain" remains the hub of post-JIA morbidity, now pairing most often with "Abnormal pain" ($n = 1526$ or 16.5%), "Joint Disorder" ($n = 1382$ or 14.93%), "Abdominal Pain" ($n = 900$ or 9.72%), "Skin Inflammation" ($n = 788$ or 8.51%) and "Respiratory Symptoms" ($n = 718$ or 7.75%). New skeletal complications such as Osteoarthritis emerges, which co-occurs with "Musculoskeletal Pain" ($n = 481$ or 5.2%), and with "Joint Disorders" ($n = 376$ or 4.06%), reflecting early degenerative change once JIA is established. Skin involvement also intensified in this phase. For instance, "Skin Inflammation" co-occurred with "Joint Disorders" in 222 patients and with "Respiratory Symptoms" in 389. Several new disease categories that were not prominent in the pre-diagnosis phase appear post-diagnosis in significant numbers, including "Acquired Deformities", "Immune Disorders", "Spondylopathies/Spondyloarthropathy", and "Uveitis & Ocular Inflammation".

The UpSet plot (Fig. 8) reinforces this transition from isolated complaints to complex multimorbidity. The single most common post-diagnosis condition is "Musculoskeletal Pain" ($n = 1220$ or 13.18%), yet the second most frequent pattern already involves a dual combination, "Musculoskeletal Pain" and "Joint Disorders", ($n = 706$ or 7.63%). Triple intersections such as musculoskeletal pain, joint disorders and skin inflammation involve 148 children; most of the quadruple and even five-way combinations occur in up to 70 patients, more than twice the frequency observed before diagnosis.

After a JIA diagnosis, comorbidity burden shifts from isolated musculoskeletal complaints to multi-system clusters dominated by joint, skin and respiratory involvement. Persistent musculoskeletal pain coupled with rising skin or respiratory symptoms should trigger aggressive disease-control strategies and early referral to dermatology or pulmonology. Recognizing these post-diagnosis clusters enables clinicians to anticipate complications, individualize follow-up schedules, and initiate preventive therapy before irreversible damage occurs.

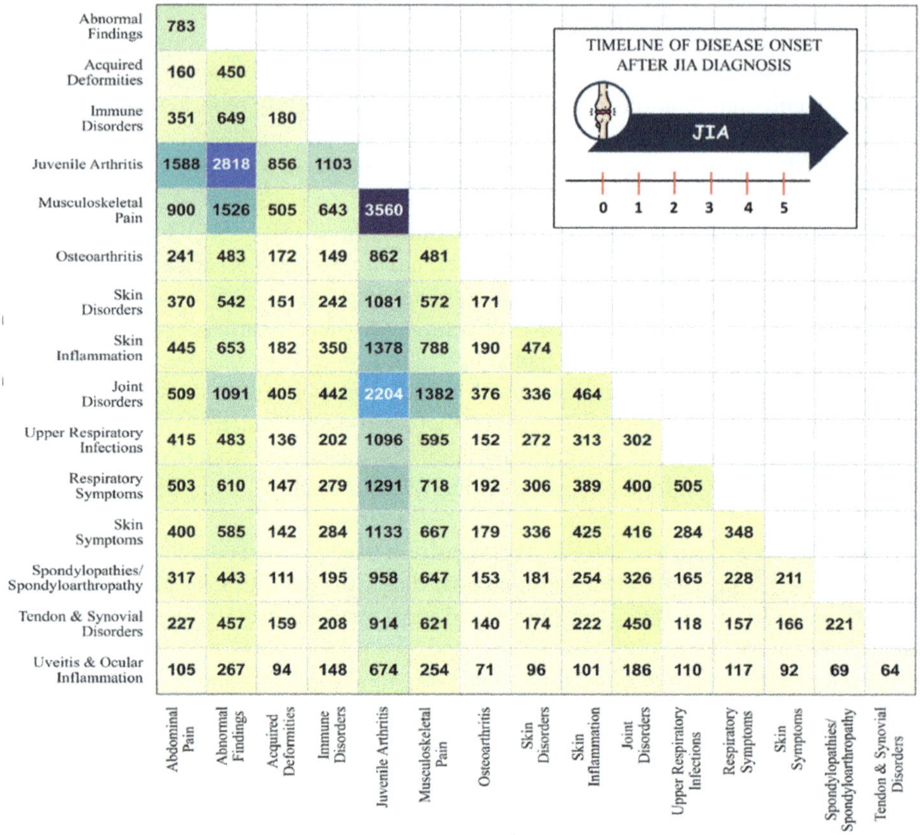

Fig. 7. Co-occurrence Matrix of Comorbidities within 5 years after JIA diagnosis.

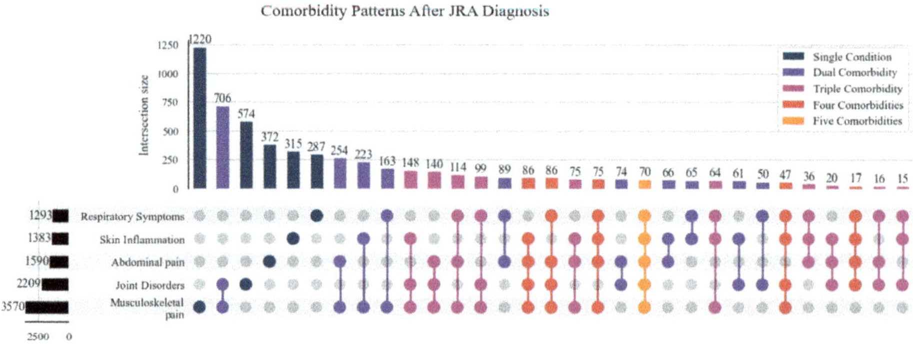

Fig. 8. UpSet plots displaying comorbidity combinations observed in the five years after JIA diagnosis.

4 Conclusion

In the present study, we utilized Oracle Real-World Data to analyze comorbidity patterns among patients diagnosed with Juvenile Idiopathic Arthritis across a decade-long window spanning five years before and after diagnosis. While analyzing patient demographics, we observed significant variation in the age distribution at first JIA diagnosis across racial groups. Notably, Black or African American patients were more frequently diagnosed at older ages, as reflected in higher quartile values. These findings suggest a need for race-aware screening strategies to ensure timely diagnosis across all populations. Our comorbidity analysis revealed distinct patterns preceding and following JIA diagnosis. In the pre-diagnosis period, musculoskeletal pain emerged as a central feature, frequently co-occurring with joint disorders, abdominal pain, and respiratory symptoms. These clusters, particularly musculoskeletal pain alongside joint, abdominal, or respiratory conditions, may represent early systemic manifestations of JIA, underscoring their potential utility as clinical indicators for earlier diagnostic evaluation. During post-diagnosis, the comorbidity landscape transitioned from localized musculoskeletal complaints to multi-system involvement, dominated by persistent joint disorders, escalating skin inflammation, and respiratory symptoms, alongside new-onset complications such as acquired deformities, immune disorders, and uveitis. This shift reflects both the systemic progression of JIA and possible sequelae of chronic inflammation or treatment effects.

Despite the valuable insights generated, this study has several limitations. First, this study did not perform patient de-duplication across hospitals and healthcare systems, which may introduce bias if individuals received care at multiple institutions and should be considered when interpreting the results. Second, the study focused only on a ten-year observation window, capturing five years after and before diagnosis, which may overlook long-term complications. Third, some potentially important factors such as disease severity, medication, environmental factors, and genetic information were not included in the analysis. Future research should build upon these results by incorporating such covariates to better understand their influence on disease progression. Integrating a broader range of clinical and biological data could enable the development of more robust predictive models that quantify the risk of specific comorbidities and guide personalized management strategies for patients with Juvenile Idiopathic Arthritis.

Acknowledgments. The authors gratefully acknowledge Oracle for providing access to Real World Data, a large EHR data warehouse, in support of this research. Any opinions, findings, and conclusions or recommendations expressed in this material are those of the authors and do not necessarily reflect the views of Oracle Cerner.

References

1. Ravelli, A., Martini, A.: Juvenile idiopathic arthritis. Lancet. **369**, 767–778 (2007)
2. Lites, T.D., Foster, A.L., Boring, M.A., Fallon, E.A., Odom, E.L., Seth, P.: Arthritis among children and adolescents aged <18 years—United States, 2017–2021. MMWR Morb. Mortal Wkly. Rep. **72**, 788–792 (2023). https://doi.org/10.15585/mmwr.mm7229a3

3. Minden, K., et al.: Long-term outcome in patients with juvenile idiopathic arthritis. Arthritis Rheum. **46**, 2392–2401 (2002). https://doi.org/10.1002/art.10444
4. Zak, M., Pedersen, F.K.: Juvenile chronic arthritis into adulthood: a long-term follow-up study. Rheumatology. **39**, 198–204 (2000). https://doi.org/10.1093/rheumatology/39.2.198
5. Ringold, S., et al.: 2019 American College of Rheumatology/Arthritis Foundation guideline for the treatment of juvenile idiopathic arthritis: therapeutic approaches for non-systemic polyarthritis, Sacroiliitis, and Enthesitis. Arthritis Care Res. **71**, 717–734 (2019). https://doi.org/10.1002/acr.23870
6. Seid, M., Opipari, L., Huang, B., Brunner, H.I., Lovell, D.J.: Disease control and health-related quality of life in juvenile idiopathic arthritis. Arthritis Care Res. **61**, 393–399 (2009). https://doi.org/10.1002/art.24477
7. Gutierrez-Suarez, R., et al.: Health-related quality of life of patients with juvenile idiopathic arthritis coming from 3 different geographic areas. The PRINTO multinational quality of life cohort study. Rheumatology. **46**, 314–320 (2006). https://doi.org/10.1093/rheumatology/kel218
8. Horneff, G., et al.: Incidence, prevalence, and comorbidities of juvenile idiopathic arthritis in Germany: a retrospective observational cohort health claims database study. Pediatr. Rheumatol. **20**, 100 (2022). https://doi.org/10.1186/s12969-022-00755-x
9. Lovell, D.J., Huang, B., Chen, C., Angeles-Han, S.T., Simon, T.A., Brunner, H.I.: Prevalence of autoimmune diseases and other associated conditions in children and young adults with juvenile idiopathic arthritis. RMD Open. **7**, e001435 (2021). https://doi.org/10.1136/rmdopen-2020-001435
10. Schenck, S., et al.: Comorbidity of type 1 diabetes mellitus in patients with juvenile idiopathic arthritis. J. Pediatr. **192**, 196–203 (2018). https://doi.org/10.1016/j.jpeds.2017.07.050
11. Simon, T.A., Harikrishnan, G.P., Kawabata, H., Singhal, S., Brunner, H.I., Lovell, D.J.: Prevalence of co-existing autoimmune disease in juvenile idiopathic arthritis: a cross-sectional study. Pediatr. Rheumatol. **18**, 43 (2020). https://doi.org/10.1186/s12969-020-00426-9
12. Milatz, F., Albrecht, K., Minden, K., Marschall, U., Klotsche, J., Callhoff, J.: Mental comorbidities in adolescents and young adults with juvenile idiopathic arthritis: an analysis of German nationwide health insurance data. Pediatr. Rheumatol. **22**, 10 (2024). https://doi.org/10.1186/s12969-023-00948-y
13. Borchers, A.T., Selmi, C., Cheema, G., Keen, C.L., Shoenfeld, Y., Gershwin, M.E.: Juvenile idiopathic arthritis. Autoimmun. Rev. **5**, 279–298 (2006). https://doi.org/10.1016/j.autrev.2005.09.011
14. Kearsley-Fleet, L., et al.: Burden of comorbid conditions in children and young people with juvenile idiopathic arthritis: a collaborative analysis of 3 JIA registries. Rheumatology. **61**, 2524–2534 (2022). https://doi.org/10.1093/rheumatology/keab641
15. 2025 ICD-10-CM Codes M08*: Juvenile arthritis, https://www.icd10data.com/ICD10CM/Codes/M00-M99/M05-M14/M08-. Last accessed 23 Apr 2025
16. Clinical Classifications Software Refined (CCSR): https://hcup-us.ahrq.gov/toolssoftware/ccsr/ccs_refined.jsp. Last accessed 27 Apr 2025
17. Riffenburgh, R.H.: Chapter summaries. In: Statistics in Medicine, pp. 533–580. Elsevier (2006). https://doi.org/10.1016/B978-012088770-5/50067-8

Validation of Survey Measures for Behavioral Customer Engagement with Textual Content from Customer Reviews

Toshikuni Sato(✉) and Takumi Kato

School of Commerce, Meiji University, Chiyoda, Tokyo 101-8301, Japan
{tsato,takumi_kato}@meiji.ac.jp

Abstract. The availability of numerous customer engagement scales, coupled with frequent modification or integration to suit particular research objectives, has resulted in a relative paucity of studies focusing on their rigorous validation. This study contributes to the understanding of this issue from a novel perspective: nomological validation through analysis of integrated heterogeneous data. This methodology involves analyzing the association between the textual content of customer reviews and scaled response data on customer engagement to examine whether the estimated relationships align with prior theoretical or empirical expectations. For the empirical analysis, these data were collected through a survey of hotel users. The review text was transformed into 1- and 2-gram terms, and their relationships with latent factors indicative of customer engagement were examined. As a result, theoretically and empirically sound terms related to customer engagement were identified, but several terms yielded inconsistent findings. This study thus focused on incongruent findings on satisfaction and dissatisfaction, which are typically considered closely related to behavioral customer engagement, and discusses the challenges of the present survey measures. This holds potential for broader implications and contributes to the validation and use of other customer engagement scales.

Keywords: customer engagement · customer experience · text analysis · construct validation

1 Introduction

Measuring customer or consumer engagement is of importance to both academic researchers and industry practitioners, as evidenced by the extensive efforts dedicated to developing questionnaire-based scales [1–3]. Researchers can select appropriate approaches based on their aims by considering the subtle differences in the conceptualization of customer engagement [4]. However, the existing confusion surrounding the definition of customer engagement and the adaptation of these scales creates a less-than-desirable situation [3], even though acknowledging the inherent challenges in developing and selecting a completely faultless measure. Existing, well-established scales are often used under the assumption that they accurately measure the intended constructs, which

X. Xie et al. (Eds.): ICSS 2025, LNOR, pp. 289–303, 2026.
https://doi.org/10.1007/978-3-032-13116-4_23

results in insufficient testing of the validity of each measurement method. For instance, the issues concerning the dimensionality of the latent factors underlying a scale and the ability to identify highly (or lowly) engaged customers have been revealed based on comparative scale analyses employing confirmatory factor analysis (CFA) and item response modeling across several consumer engagement scales [5]. Moreover, beyond evaluating a scale's isolated performance and validity, verifying the core theoretical underpinnings and key characteristics (hallmarks) of customer engagement within a nomological network is also crucial [3].

To address these challenges, this study discusses the validity of a customer engagement scale from a novel perspective through simultaneous analysis of heterogeneous data. Specifically, along with the collection of scaled response data, customer reviews were also gathered from the same respondents via a survey to investigate the alignment between review content and the underlying customer engagement tendencies. As explained in the next section, customer experience is generally recognized as a significant antecedent of customer engagement, while customer reviews hold the potential to yield rich insights into the specifics of these experiences. Acknowledging the significance of customer engagement manifested through concrete actions (e.g., repurchase behavior and word-of-mouth communication) that can generate substantial value for firms, this study adopts a well-established scale [6] that focuses on the behavioral dimension of customer engagement. Nevertheless, this methodological framework has the potential for broader applicability to various customer engagement scales, as well as to other pertinent measurement scales.

2 Background

2.1 Customer Experience and Engagement

The influence of customer experience on customer engagement is both intuitive and common, as supported by substantial evidence. Extremely positive or negative experiences with the focal company or competitors can potentially influence customer engagement behaviors, such as initiating brand communities, exploring alternatives, and positive or negative word-of-mouth [7]. Theoretical relationships and conceptual frameworks are discussed by several researchers [e.g., 8–10], suggesting that customer experience has a direct, indirect, or cyclical relationship with customer engagement. Empirical evidence indicates that attitudinal engagement can be both an antecedent and a consequence of customer experience [e.g., 11–13], whereas behavioral customer engagement is often demonstrated as an eventual outcome of customer experience [e.g., 14–16].

2.2 Customer Experience and Review Text

Given that even review readers (i.e., consumers) can discern others' service experiences and evaluations from reviews [17], reviews and word-of-mouth serve as significant informational resources for observing and understanding customer experiences to enhance product and service design strategies [18]. In the text analysis of customer reviews, clustering or dimensionality-reduction techniques are commonly employed to extract specific clusters, topics, or factors. For instance, seven factors (i.e., dining, value, family friendliness, amenities, employee attitude, tangible service, and cleanliness) were identified in bed-and-breakfast reviews (Ctrip) [19] and five topics (amenities, location, transactions, value, and experience) were extracted from hotel reviews (TripAdvisor) [20]. While similar topics or factors may be found using different methodologies, more unique topics, such as general recommendation, late/evening check-in, and sleep condition/bed can also emerge [21], or even a large number of unlabeled clusters can be categorized as positive or negative contexts [22]. These results indicate that the content of customer reviews provides a wealth of information for understanding customer experience. Hence, customer reviews can be adapted to measure customer experience.

3 Methodology

3.1 Proposed Approach

In the empirical analysis, a neural network model with penalty methods [e.g., 23–26] was employed for the joint analysis of survey responses (i.e., construct measurements) and textual information (i.e., customer reviews). The penalties of standardization for latent units and a block diagonal structure for the weight matrix introduce a CFA specification into the neural network, which enabled us to estimate the relationship between high-dimensional single terms and latent constructs estimated by the measurement scale [24, 26]. Therefore, this method can also be considered as a large-scale nomological network and validation effort. The applicability of this model for the validation of measurement scales has been discussed in the literature [26], as it helps examine anomalies or biases in survey responses. Additionally, this study employed 2-gram (bigram) text transformation to improve the interpretability of the results. An n-gram is a sequence of n terms, where "n" indicates the number of terms. For example, the 2-gram, "I think it was a great hotel," is divided into "I think," "think it," "it was," "was a," "a great," and "great hotel." A set of combined 1-gram (unigram) and 2-gram terms forming a document-term matrix was used to compare the model performance with the original document-term (i.e., 1-gram) matrix.

The empirical analysis included the following steps: (1) item scores for construct measurements paired with hotel reviews were collected; (2) a standard CFA model was estimated using the item scores, as well as a penalized neural network model using both the item scores and text information; (3) the performance of the penalized neural network model was evaluated using the root mean squared error (RMSE) and mean absolute error (MAE); and (4) the estimated factor scores were compared between the CFA and penalized neural network models using the correlation coefficient.

3.2 Data Collection and Preprocessing

Data were collected through a survey conducted from November 10 to November 17, 2022, with the support of a Japanese marketing research company. A sample of 3,000 respondents who had stayed at a hotel within the last month was collected and balanced for gender and age groups (20 s, 30 s, 40 s, 50 s, and 60 s) with equal representation in each category. Construct measurements were based previous studies conducted in contexts similar to ours (e.g., hotel and tourism industries). Customer engagement comprises four latent constructs (purchase, referral, word-of-mouth, feedback) with 16 items [6, 14], as shown in Table 1. For predictive validation, customer loyalty [27] was measured using four items: (1) I would say positive things about this hotel brand/hotel to other people, (2) I would recommend this hotel brand/hotel to someone who seeks my advice, (3) I would encourage friends and relatives to do business with this hotel brand/hotel, and (4) I would do more business with this hotel brand/hotel in the next few years. A five-point Likert scale (from "1 = strongly disagree" to "5 = strongly agree") was used to measure all item scores.

Hotel reviews were also collected through open-ended questions together with item scores. These text reviews were transformed into a document–term matrix (i.e., a bag of words) based on 1- and 2-grams. Common stop words were first removed, and then nouns and adjectives were extracted. The 1-gram comprised nouns and adjectives, while the 2-gram consisted of noun-adjective pairings. Terms that appeared only once were considered rare and were excluded. The 1- and 2-grams consisted of 1,062 and 162 terms, respectively.

Responses with potentially improper answers were deleted under the following conditions: (1) having at least one improper answer to the reverse-coded questions, (2) providing the same score for all questions, (3) having no answer or a refusal answer to the open-ended questions, and (4) having a short text length (individual total number of terms less than half of the average: 3.717). Finally, the sample comprised 1,289 customers (59.58% female, 40.42% male, 16.52% in their 20 s, 20.33% in their 30 s, 19.71% in their 40 s, 24.59% in their 50 s, and 18.85% in their 60 s).

3.3 Settings for Model Learning

Two types of models, 1- and 2-gram models, are defined as a simple two-layer neural network (input, one latent unit, and output layers). According to [24, 26], for customers $i = 1, 2, \ldots, n$, suppose $\mathbf{y}_i = [y_{1,i}, y_{2,i}, \ldots, y_{p,i}]^T$ is a vector of P-dimensional objective variables (measurement item scores), and $\mathbf{x}_i = [x_{1,i}, x_{2,i}, \ldots, x_{q,i}]^T$ is a vector of Q-dimensional predictor variables (bag of words). Assuming a simple two-layer (one-hidden layer) neural network, suppose $\mathbf{z}_i = [z_{1,i}, z_{2,i}, \ldots, z_{k,i}]^T$ is a vector of K-dimensional latent variables (latent factor scores), $\boldsymbol{h}_i = h(\mathbf{z}_i) = [h(z_{1,i}), h(z_{2,i}), \ldots, h(z_{k,i})]^T$ is a vector of K-dimensional hidden units, and h is an activation function, then the standard multi-output neural network model is given as follows:

$$\mathbf{y}_i = \underset{(P \times K)}{\mathbf{W}^{(2)}} \boldsymbol{h}_i, \tag{1}$$

$$\boldsymbol{h}_i = h(\mathbf{z}_i), \tag{2}$$

$$\mathbf{z}_i = \underset{(K \times Q)}{\mathbf{W}^{(1)}} \mathbf{x}_i, \tag{3}$$

where $\mathbf{W}^{(1)}$ and $\mathbf{W}^{(2)}$ are $K \times Q$ and $P \times K$ weight parameter matrices, respectively. The 1- and 2-gram models employ bag of 1-gram (1,062 terms) and 1- and 2-grams

Table 1. Construct measurements

Purchase (PUR)		**Word-of-mouth (WOM)**	
PUR1	I will continue buying the services of this hotel brand/hotel in the near future.	WOM1	I do not actively discuss this hotel brand/hotel on any media. (R)
PUR2	My purchases with this hotel brand/hotel make me content.	WOM2	I love talking about my experience with this hotel brand/hotel.
PUR3	I do not get my money's worth when I use this hotel brand/hotel. (R)	WOM3	I discuss the benefits that I get from this hotel brand/hotel with others.
PUR4	Using the services of this hotel brand/hotel makes me happy.	WOM4	I am a part of this hotel brand/hotel and mention it in my conversations.
Referral (REF)		**Feedback (FB)**	
REF1	I will promote this hotel brand/hotel if there are the monetary referral benefits provided by the hotel brand/hotel.	FB1	I would like to provide feedback about my experiences with this hotel brand/hotel to the firm.
REF2	In addition to the value derived from the service, the monetary referral incentives will also encourage me to refer this hotel brand/hotel to my friends and relatives.	FB2	I would like to provide suggestions for improving the performance of this hotel brand/hotel.
REF3	I will enjoy referring this hotel brand/hotel to my friends and relatives if there are monetary referral incentives.	FB3	I would like to provide suggestions/feedback about the new products/services of this hotel brand/hotel.
REF4	Given that I use this hotel brand/hotel, I will refer my friends and relatives to this hotel brand/hotel if there are monetary referral incentives.	FB4	I would like to provide feedback/suggestions for developing new products/services for this hotel brand/hotel.

(1,224 terms) as inputs, respectively. The other common settings are as follows: (1) the latent units (i.e., latent factors) are defined as four-dimensional latent behavioral customer engagements with 16 outputs (i.e., measurement items), as shown in Table 1; (2) the latent units use the hyperbolic tangent (tanh) function as an activation function; and (3) the Adam optimizer is used for model learning and the learning rate is set to 0.1.

Furthermore, three penalty functions were applied as follows:

$$f(\mathbf{Z}) = \lambda_1 \left\| I_K \odot \left\{ \tfrac{1}{n^*} \mathbf{Z}\mathbf{Z}^T \right\} - I_K \right\|_2^2 + \lambda_2 \left\| \overline{\mathbf{Z}} \right\|_2^2, \quad (4)$$

$$f\left(\mathbf{W}^{(2)}\right) = \lambda_3 \left\| \mathbf{W}^{(2)} - B \odot \mathbf{W}^{(2)} \right\|_2^2, \quad (5)$$

$$f\left(\mathbf{W}^{(1)}\right) = \lambda_4 \left\| \mathbf{W}^{(1)} \right\|_2^2, \quad (6)$$

where, I_K is the $K \times K$ identity matrix; $\mathbf{Z} = [\mathbf{z_1}, \mathbf{z_2}, \ldots, \mathbf{z_{n^*}}]$ is the $K \times n^*$ matrix of latent factors (constructs); n^* is a batch-size or total sample size; $\frac{1}{n^*}\mathbf{Z}\mathbf{Z}^T$ is the $K \times K$ covariance matrix of $\mathbf{Z}$; $\overline{\mathbf{Z}}$ is the $K \times 1$ mean vector of $\mathbf{Z}$; B is a $P \times K$ block-diagonal matrix whose element $b_{p,k}$ takes 1 if the item q is a corresponding measure of latent construct k and 0 for otherwise; and $\lambda_1, \lambda_2, \ldots, \lambda_4$ are tuning parameters. These penalty functions provide standardization of the latent variables (Eq. 4), a block-diagonal structure for the weight matrix between the latent units and outputs (i.e., the item-weight matrix) (Eq. 5), and an L2 (Ridge) norm for the weight matrix between the input and latent units (i.e., the term-weight matrix) (Eq. 6), respectively. The prior parameters were given as $\lambda_1 = 100$, $\lambda_2 = 100$, $\lambda_3 = 500$, and $\lambda_4 = 0.0001$.

For learning the models, the samples were randomly divided into three parts: (1) training sample ($N_{train} = 1{,}031$), (2) validation sample ($N_{val} = 129$), and (3) test sample ($N_{test} = 129$). Then the test sample was fixed, but each model learning process was repeated 200 times with different initial values and reshuffled the combinations of the training and validation samples. This is because the parameter estimates of the neural network models potentially depend on the initial values and splitting the dataset (i.e., training, validation, or test samples). Therefore, the means and standard deviations (SDs) of the 200 estimates were used to evaluate robustness. The standard measurement model specified in the CFA model was estimated to obtain the factor scores of the latent constructs corresponding to the training, validation, and test samples. These factor scores were compared with those estimated using a penalized neural network model.

4 Empirical Analysis and Discussion

To leverage textual data for the scale validation of behavioral customer engagement, the initial steps involved evaluating the performance of the models, followed by assessments of the validity of text-based constructs. Text-based constructs are represented by latent factors (hidden units) within a penalized neural network model linking text terms to item scores (i.e., 1- and 2-gram models), and text-based factor scores are estimated using the trained model and text term inputs. If previous theoretical and empirical assumptions regarding the relationships between customer experience and engagement are supported, consistent relationships should be found, such that a positive experience drives positive engagement behaviors, as evidenced by the links between text terms (i.e., customer experience) and the latent factors of customer engagement.

4.1 Model Validation

To examine the fit and performance of the 1- and 2-gram models, Table 2 reports the mean RMSE and MAE for each sample. As discussed in Sect. 3.3, the test sample was fixed; however, the training and validation samples were resampled across 200 model learning iterations. In both models, the predictive performances (i.e., RMSE and MAE) of the validation and test samples were similar. For example, the validation and test RMSEs for the 1-gram model were 1.148 and 1.187, respectively. The values for the 2-gram model are 1.171 and 1.213, respectively. Similarly, the MAEs in all cases were below 1 on a 5-point scale. These results imply that the validation sample performance (i.e., RMSE and MAE) closely approximated the predictive performance for unseen data (i.e., test RMSE and MAE). Although these results show that the 1-gram model performed slightly better on the validation and test samples, the 1- and 2-gram models exhibited similar performances, indicating that adding 2-gram terms to the input variables did not substantially impede performance.

Table 2. Mean RMSE and MAE (standard deviations in parentheses) for the 1- and 2-gram models

	RMSE		MAE	
	1-gram	2-gram	1-gram	2-gram
Training	0.743	0.717	0.522	0.494
	(0.006)	(0.007)	(0.008)	(0.009)
Validation	1.148	1.171	0.887	0.907
	(0.035)	(0.034)	(0.031)	(0.030)
Test	1.187	1.213	0.923	0.942
	(0.019)	(0.021)	(0.016)	(0.019)

In addition to predictive performance, the concurrent validity of the latent constructs is presented in Table 3. This table reports the mean correlation coefficients between the corresponding CFA- and text-based factor scores. Similar to the predictive performance, comparable results were obtained for both the 1- and 2-gram models, and the correlations tended to decrease in the validation and test samples compared with the training sample. However, the correlations for referral and feedback engagements in the validation and test samples were nearly zero, implying that it is not straightforward to predict these forms of engagement from customer experiences posted in review texts. In the subsequent sections, we discuss these reasons using additional analyses.

To test the predictive validity of the estimated factor scores, the average of the four customer loyalty item scores in Sect. 3.3 was used as the dependent variable, because customer loyalty is often assumed to be a goal of conceptual models involving customer engagement [e.g., 28–30]. Table 4 presents the results of linear regression analysis of customer engagement on customer loyalty. Unique training and validation samples were combined because they were used randomly for model learning. The text-based factor scores of these combined samples were estimated using mean term weights and their

Table 3. Mean correlation coefficients (standard deviations in parentheses) between CFA- and text-based factor scores

	Training		Validation		Test	
	1-gram	2-gram	1-gram	2-gram	1-gram	2-gram
Purchase	0.690	0.704	0.244	0.244	0.285	0.292
	(0.018)	(0.020)	(0.076)	(0.074)	(0.040)	(0.047)
Referral	0.635	0.658	0.042	0.041	0.010	−0.042
	(0.035)	(0.037)	(0.091)	(0.094)	(0.056)	(0.055)
Word-of-mouth	0.651	0.664	0.165	0.147	0.182	0.178
	(0.017)	(0.018)	(0.078)	(0.079)	(0.040)	(0.046)
Feedback	0.618	0.633	0.083	0.056	−0.044	−0.044
	(0.020)	(0.023)	(0.085)	(0.083)	(0.040)	(0.039)

corresponding text terms. Estimates of regression models with CFA-based factor scores were also reported.

In the combined learning sample (training and validation), the magnitudes and signs of the regression coefficients were similar across the three models. For example, purchase engagement was the most effective factor for customer loyalty, consistent across all three models. The R-squared values were 0.800, 0.423, and 0.445 for the CFA, 1-gram, and 2-gram models, respectively. The adjusted R-squared values were 0.799, 0.421, and 0.443 for the CFA, 1-gram, and 2-gram models. In the test sample with text-based factor scores, the R-squared values were 0.177 and 0.118 for the 1- and 2-gram models, respectively. The adjusted R-squared values are 0.151 and 0.161 for the 1- and 2-gram models, respectively. In contrast to model performance (Table 2) and concurrent validity (Table 3), these results show that the 2-gram model has better predictive validity. Although the R-squared and adjusted R-squared values of the test sample decreased substantially, these results indicate that text-based construct measurements can explain a portion of customer loyalty.

However, compared with the learning sample, the magnitudes of the regression coefficients for referral and word-of-mouth changed in the test sample. Specifically, the coefficient for referral has a negative sign, while the coefficient for word-of-mouth is statistically significant. One possible reason is that the survey scale for behavioral customer engagement may not accurately measure the intended theoretical constructs. While all forms of engagement were expected to have a significant relationship with customer loyalty, even in the CFA model, word-of-mouth and feedback had coefficients close to zero and were not significant. A similar result was observed in prior research [31] A more detailed discussion of this issue is provided in later sections.

4.2 Model Interpretation

The previous section discussed how similar performances and validities were observed for the 1- and 2-gram models. Next, to gain a more detailed and deeper understanding

Table 4. Regression coefficients (standard errors in parentheses) for the linear regression model for customer loyalty

	Training and validation (N = 1,160)			Test (N = 129)	
	CFA	1-gram	2-gram	1-gram	2-gram
Intercept	3.612***	3.521***	3.516***	3.560***	3.555***
	(0.011)	(0.019)	(0.019)	0.064	0.063
Purchase	1.019***	0.366***	0.388***	0.272***	0.267***
	(0.023)	(0.019)	(0.019)	0.084	0.084
Referral	0.132***	0.163***	0.161***	−0.190*	−0.201**
	(0.018)	(0.022)	(0.021)	0.100	0.090
Word-of-mouth	−0.046	0.017	0.016	0.198**	0.225**
	(0.036)	(0.021)	(0.021)	0.087	0.093
Feedback	−0.002	−0.009	0.001	−0.050	−0.023
	(0.020)	(0.019)	(0.019)	0.084	0.087
R^2	0.800	0.423	0.445	0.177	0.188
Adjusted-R^2	0.799	0.421	0.443	0.151	0.161

*** $p < 0.01$, ** $p < 0.05$, * $p < 0.1$

of the relationship between text terms (i.e., customer experience) and customer engagement, we present the results of the 2-gram model and its advantages. First, we examine the validity and stability of the interpretations of the latent factors (hidden units). Figures 1 and 2 show the means and SDs of the item weights and latent factor correlations, respectively. The upper matrix of Table 1 can be regarded as the factor loadings in the CFA. Therefore, the corresponding latent factors closely represent the theoretical constructs of behavioral customer engagement (i.e., purchase, referral, word-of-mouth, and feedback). The mean factor correlation matrix shown on the left side of Fig. 2 also supports a theoretical positive relationship among behavioral customer engagements. Additionally, their SDs, shown at the bottom and left sides of Figs. 1 and 2, respectively, indicate that the estimates of the term-weight matrix and correlation matrix are stable.

Figure 3 shows the means and SDs of the top 15 positive and negative term weights for each customer engagement level. The figures in the left and right columns show the results for 1- and 2-grams, respectively. For example, "performance" (1-gram term), which has the sixth largest weight in purchase engagement, can be understood as indicating a positive relationship with purchase engagement on average, and its instability is shown by the length of the error bar. This average and instability information for 1-gram is useful for understanding how a single term relates to the latent factor scores of the theoretical constructs. However, the meaning of a single term can be unclear regarding how it is used in a sentence; therefore, the context of the term cannot be readily identified. On the other hand, 2-gram term such as "customer service - good," which has the third largest weight in purchase engagement, provide a better understanding of how the text content relates to the latent factor scores of each customer engagement.

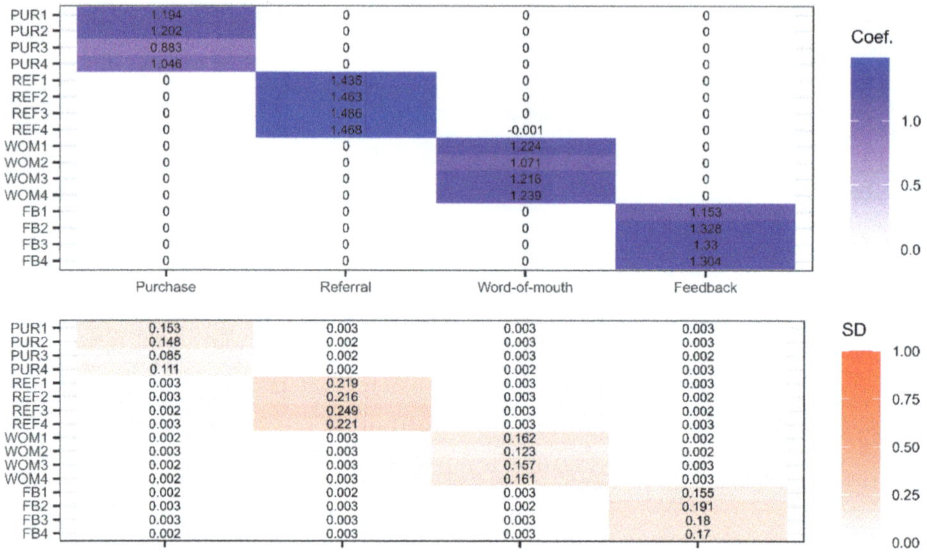

Fig. 1. Mean (upper) and SD (bottom) of term weight matrix

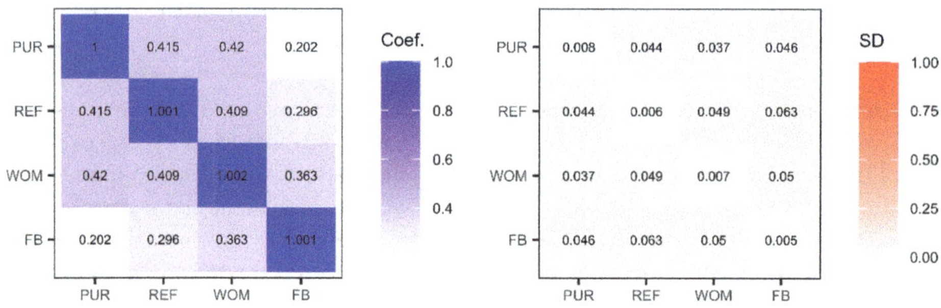

Fig. 2. Mean (left) and SD (right) of the factor correlation matrix

While several exceptions exist, terms with positive or negative meanings tend to be associated in the same direction as the latent scores for each customer engagement. However, when focusing on feedback engagement, several reverse relationships can be obviously found. For example, "hard," "old," and "cost-effectiveness - poor" have positive mean weights, whereas "good," "level - high," and "cost-effectiveness - high" have negative mean weights. These results suggest that negative rather than positive experiences drive the intention to provide feedback. This also implies that positive experiences decrease feedback behavioral intentions. Customers with negative experiences are motivated to communicate their dissatisfaction and suggestions to the company, whereas customers with positive experiences typically do not have the same need. Hence, the scale items for feedback engagement may not be suitable for identifying highly engaged customers with positive motivation. As per the validation of text-based factor scores in

Sect. 4.1, these are potentially the same reasons for obtaining different results for feedback engagement compared with other engagements. Similarly, the drivers of word-of-mouth engagement are likely to be multifaceted and may not exhibit a strong association with customers' positive or negative experiences.

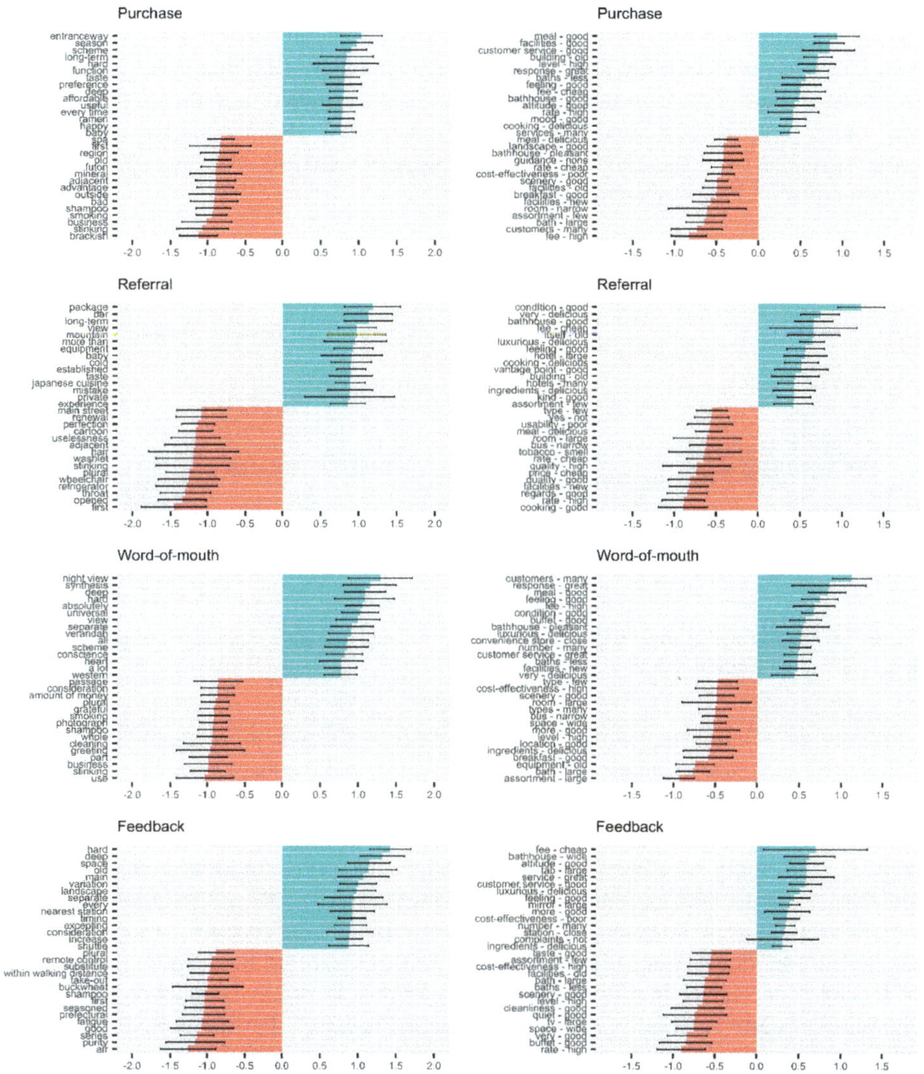

Fig. 3. Means (bars) and SDs (error bars) of the top 15 positive and negative term weights (1-gram: left, 2-gram: right) on each customer engagement

4.3 Further Scale Validation

Due to the unexpected results obtained from the aforementioned analysis, we conducted additional analyses to investigate the validity of the behavioral customer engagement scale. Figure 4 shows the means (bars) and SDs (error bars) of the term weights for "satisfaction" and "dissatisfaction" on each customer engagement, presented on the left and right sides, respectively. Because customer satisfaction or dissatisfaction is also an important direct or indirect driver of customer engagement [8, 10, 28], we focused on the relationship between these text terms and customer engagement. The mean term weights for satisfaction indicate that word-of-mouth and feedback engagements especially show a weaker relationship with the perception of satisfaction compared to purchase engagement. Even the mean term weight for referral engagement was less than 0.1. Alternatively, the extent of these engagements can be primarily determined by individual characteristics rather than the customer experience [7]. This may also explain the inconsistent results obtained from the concurrent and predictive validations in Sect. 4.1, with the exception of purchase engagement.

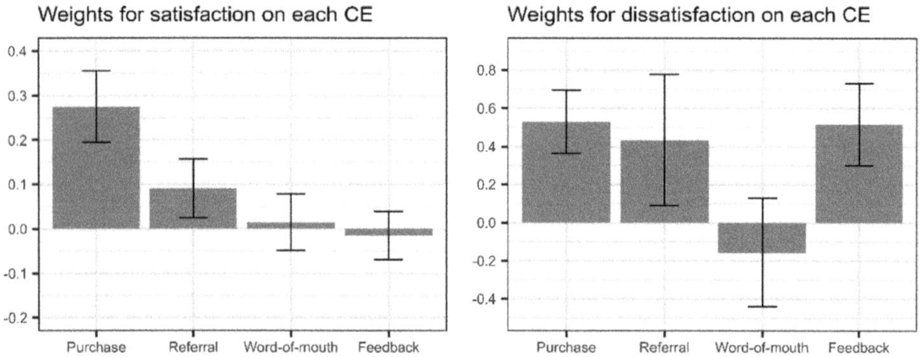

Fig. 4. Term weights for satisfaction (left) and dissatisfaction (right) on each customer engagement (CE) in the 2-gram model

Moreover, relationships with dissatisfaction provide even more anomalous results, such that dissatisfaction enhances purchase, referral, and feedback behaviors. Although we recognize the potential for model underspecification, we propose the following scenarios. Considering that referral engagement represents incentivized behavior (see Table 1), customers who wish to continue using the service despite their dissatisfaction may be motivated by potential referral incentives and the expectation that their complaints will reach the company. Alternatively, dissatisfied customers might continue purchasing a service if companies offer incentives through referrals and provide a channel for feedback on service improvements. These propositions may be valuable for testing with other research methods, such as experimental research. Furthermore, these results suggest the need for careful reconsideration of how customer engagement is measured and, potentially, a redefinition of what constitutes a highly engaged customer.

5 Conclusion

This study examined a well-established behavioral customer engagement scale [6] through nomological validation of the content of customer reviews and the scaled response scores. By applying a penalized neural network to analyze these relationships, our approach identified that positive (or negative) experiences, represented by 1- and 2-grams, potentially drive behavioral customer engagement. Additionally, the challenges of the behavioral engagement scale were discussed by comparing the obtained model interpretations with prior theory and empirical evidence. However, given the importance of measuring the behavioral aspects of customer engagement [32], we suggest continuous development and enhancement of the behavioral engagement scale.

The findings of this study and their discussion may also contribute to other customer engagement and related scales; however, there are three main limitations that should be mentioned. First, the customer reviews were collected through surveys. A more natural approach might involve surveying consumers who have posted genuine reviews on online platforms and combining the two datasets, potentially obtaining a more natural and public commentary. Another limitation is the potential influence of individual characteristics on customer engagement [7] beyond the impact of experience. While detailed personal characteristics are typically absent from public reviews and word-of-mouth data, future research should endeavor to collect and incorporate these variables into the analytical model. Furthermore, future research could explore the differences in outcomes resulting from the application of diverse analytical techniques because of the absence of a universally absolute text analysis method for comprehending customer experience.

Acknowledgements. The authors gratefully acknowledge the partial support from JSPS KAKENHI (Grant Number 22 K13496).

References

1. Hollebeek, L.D., Sharma, T.G., Pandey, R., Sanyal, P., Clark, M.K.: Fifteen years of customer engagement research: a bibliometric and network analysis. J. Prod. Brand Manag. **31**, 293–309 (2021). https://doi.org/10.1108/JPBM-01-2021-3301
2. Santos, Z.R., Cheung, C.M.K., Coelho, P.S., Rita, P.: Consumer engagement in social media brand communities: a literature review. Int. J. Inf. Manag. **63**, 102457 (2022). https://doi.org/10.1016/j.ijinfomgt.2021.102457
3. Hollebeek, L.D., Sarstedt, M., Menidjel, C., Sprott, D.E., Urbonavicius, S.: Hallmarks and potential pitfalls of customer- and consumer engagement scales: a systematic review. Psychol. Mark. **40**, 1074–1088 (2023). https://doi.org/10.1002/mar.21797
4. Lim, W.M., Rasul, T., Kumar, S., Ala, M.: Past, present, and future of customer engagement. J. Bus. Res. **140**, 439–458 (2022). https://doi.org/10.1016/j.jbusres.2021.11.014
5. Ferreira, M., Zambaldi, F., Guerra, D.d.S.: Consumer engagement in social media: scale comparison analysis. J. Prod. Brand Manag. **29**, 491–503 (2020). https://doi.org/10.1108/JPBM-10-2018-2095
6. Kumar, V., Pansari, A.: Competitive advantage through engagement. J. Mark. Res. **53**, 497–514 (2016). https://doi.org/10.1509/jmr.15.0044
7. van Doorn, J., et al.: Customer engagement behavior: theoretical foundations and research directions. J. Serv. Res. **13**, 253–266 (2010). https://doi.org/10.1177/1094670510375599
8. Pansari, A., Kumar, V.: Customer engagement: the construct, antecedents, and consequences. J. Acad. Mark. Sci. **45**, 294–311 (2017). https://doi.org/10.1007/s11747-016-0485-6
9. Brodie, R.J., Hollebeek, L.D., Jurić, B., Ilić, A.: Customer engagement: conceptual domain, fundamental propositions, and implications for research. J. Serv. Res. **14**, 252–271 (2011). https://doi.org/10.1177/1094670511411703

10. Kumar, V., Rajan, B., Gupta, S., Pozza, I.D.: Customer engagement in service. J. Acad. Mark. Sci. **47**, 138–160 (2019). https://doi.org/10.1007/s11747-017-0565-2
11. Ahn, J., Back, K.-J.: Antecedents and consequences of customer brand engagement in integrated resorts. Int. J. Hosp. Manag. **75**, 144–152 (2018). https://doi.org/10.1016/j.ijhm.2018.05.020
12. Rather, R.A., Rasul, T., Khan, H., Khan, I.: Unveiling the dynamics between consumer brand engagement, experience, and relationship quality towards luxury hotel brands: moderating investigation of brand reputation. Int. J. Hosp. Manag. **116**, 103572 (2024). https://doi.org/10.1016/j.ijhm.2023.103572
13. Rather, R.A., Hollebeek, L.D., Rasoolimanesh, S.M.: First-time versus repeat tourism customer engagement, experience, and value cocreation: an empirical investigation. J. Travel Res. **61**, 549–564 (2022). https://doi.org/10.1177/0047287521997572
14. Le, A.N.H., Khoi, N.H., Nguyen, D.P.: Unraveling the dynamic and contingency mechanism between service experience and customer engagement with luxury hotel brands. Int. J. Hosp. Manag. **99**, 103057 (2021). https://doi.org/10.1016/j.ijhm.2021.103057
15. NasarAmini Jeloudarlou, S., Aali, S., Faryabi, M., Bafandeh Zendeh, A.: The effect of servicescape on customer engagement: the mediating role of customer experience. J. Qual. Assur. Hosp. Tour. **23**, 318–344 (2022). https://doi.org/10.1080/1528008X.2020.1867696
16. Prentice, C., Wang, X., Loureiro, S.M.C.: The influence of brand experience and service quality on customer engagement. J. Retail. Consum. Serv. **50**, 50–59 (2019). https://doi.org/10.1016/j.jretconser.2019.04.020
17. Elwalda, A., Lü, K., Ali, M.: Perceived derived attributes of online customer reviews. Comput. Hum. Behav. **56**, 306–319 (2016). https://doi.org/10.1016/j.chb.2015.11.051
18. Yang, B., Liu, Y., Liang, Y., Tang, M.: Exploiting user experience from online customer reviews for product design. Int. J. Inf. Manag. **46**, 173–186 (2019). https://doi.org/10.1016/j.ijinfomgt.2018.12.006
19. Jia, M., Kim, H.-S., Tao, S.: B&B customer experience and satisfaction: evidence from online customer reviews. Serv. Sci. **16**, 42–54 (2024). https://doi.org/10.1287/serv.2022.0080
20. Mankad, S., "Spring" Han, H., Goh, J., Gavirneni, S.: Understanding online hotel reviews through automated text analysis. Serv. Sci. **8**, 124–138 (2016). https://doi.org/10.1287/serv.2016.0126
21. Zhang, J.: What's yours is mine: exploring customer voice on Airbnb using text-mining approaches. JCM. **36**, 655–665 (2019). https://doi.org/10.1108/JCM-02-2018-2581
22. Son, Y., Kim, W.: Development of methodology for classification of user experience (UX) in online customer review. J. Retail. Consum. Serv. **71**, 103210 (2023). https://doi.org/10.1016/j.jretconser.2022.103210
23. Sato, T.: Penalizing neural network and autoencoder for the analysis of marketing measurement scales in service marketing applications. In: Qiu, R., Lyons, K., Chen, W. (eds.) AI and Analytics for Smart Cities and Service Systems, pp. 31–42. Springer, Cham (2021). https://doi.org/10.1007/978-3-030-90275-9_3
24. Sato, T.: A penalized neural network model for predicting unobserved scores of construct indicators and reproducing latent scores of the theoretical constructs by using text information. In: Qiu, R., Chan, W.K.V., Chen, W., Badr, Y., Zhang, C. (eds.) City, Society, and Digital Transformation, pp. 1–14. Springer International Publishing, Cham (2022). https://doi.org/10.1007/978-3-031-15644-1_1
25. Huang, L., Qin, J., Zhou, Y., Zhu, F., Liu, L., Shao, L.: Normalization techniques in training DNNs: methodology, analysis and application. IEEE Trans. Pattern Anal. Mach. Intell. **45**, 10173–10196 (2023). https://doi.org/10.1109/TPAMI.2023.3250241
26. Sato, T., Kato, T.: Extracting unique keywords related to customer engagement from review text using uncorrelated weights estimation in neural networks. In: Lee, R. (ed.) Big Data and

Data Science Engineering, vol. 7, pp. 161–181. Springer Nature, Cham (2025). https://doi.org/10.1007/978-3-031-78373-9_13
27. Zeithaml, V.A., Berry, L.L., Parasuraman, A.: The behavioral consequences of service quality. J. Mark. **60**, 31–46 (1996). https://doi.org/10.1177/002224299606000203
28. Bergel, M., Frank, P., Brock, C.: The role of customer engagement facets on the formation of attitude, loyalty and price perception. J. Serv. Mark. **33**, 890–903 (2019). https://doi.org/10.1108/JSM-01-2019-0024
29. Islam, J.U., Rahman, Z., Hollebeek, L.D.: Consumer engagement in online brand communities: a solicitation of congruity theory. Internet Res. **28**, 23–45 (2018). https://doi.org/10.1108/IntR-09-2016-0279
30. Vivek, S.D., Beatty, S.E., Morgan, R.M.: Customer engagement: exploring customer relationships beyond purchase. J. Mark. Theory Pract. **20**, 127–145 (2012)
31. Parihar, P., Dawra, J.: The role of customer engagement in travel services. J. Prod. Brand Manag. **29**, 899–911 (2020). https://doi.org/10.1108/JPBM-11-2018-2097
32. Shin, H., Perdue, R.R.: Customer nontransactional value cocreation in an online hotel brand community: driving motivation, engagement behavior, and value beneficiary. J. Travel Res. **61**, 1088–1104 (2022). https://doi.org/10.1177/00472875211024752

RGP: Robust Goal Programming for Healthy Nutrition Tracking Using Patients' Dietary Image Predicted Data

Haiyan Yu[1], Zheng Zeng[1], Kun Zhang[1(✉)], Zhiqi Chen[2(✉)], Renying Xu[3], Senlin Li[1], Jianbin Chen[4], and Ehsan Hajizadeh[5]

[1] Key Laboratory of Big Data Intelligent Computing, Chongqing University of Posts and Telecommunications, Chongqing 404615, China
yhy188@pku.org.cn, zkun2022@126.com

[2] Shanghai Jiading District Central Hospital, Shanghai 201800, China
zhiqinicole@163.com

[3] Renji Hospital, School of Medicine, Shanghai Jiaotong University, Shanghai 200120, China

[4] School of Mathematics and Statistics, Beijing Institute of Technology, Beijing 100081, China
chenjianbin@bit.edu.cn

[5] Department of Industrial Engineering and Management Systems, Amirkabir University of Technology, Tehran 15875-4413, Iran
ehsanhajizadeh@aut.ac.ir

Abstract. Dietary nutrition tracking is crucial for personalized healthcare. Traditional methods, including weighing, dietary review, and food frequency, have limitations of low operational efficiency, availability for small samples, and noise from the patient's memory. Thus, it is essential to have an efficient and accurate method for tracking patients' nutrition. First, we utilized the U-net-based semantic segmentation method to complete the deep learning process with dietary food images, encompassing feature extraction, category recognition, volume estimation, and weight data collection of patient foods. Second, a data-driven goal optimization model is built to achieve the multiple nutrient intake goals and balance the body's energy requirements. Furthermore, considering the uncertainty set of dietary nutrition, a data-driven robust optimization algorithm is used to solve the optimization model under noisy data. Finally, we verified the effectiveness and robustness of the proposed method with real-world nutrition data, advancing the personalized services of nutrition tracking.

Keywords: Goal programming · diet problem · image convolution · robust optimization · nutrition tracking

1 Introduction

Imbalanced dietary intake has gradually become a significant global health challenge. It is well established that many chronic diseases [1], including diabetes,

X. Xie et al. (Eds.): ICSS 2025, LNOR, pp. 304–313, 2026.
https://doi.org/10.1007/978-3-032-13116-4_24

heart disease, and stroke, are closely associated with unhealthy eating habits. Consequently, the concept of a balanced diet has garnered unprecedented attention in recent years, being recognized as a crucial factor in preventing these chronic diseases and promoting a healthy lifestyle [2,3]. The objective of a balanced diet is to provide appropriate levels of essential nutrients, including carbohydrates, water, fats, proteins, vitamins, and minerals. Fats, carbohydrates, and proteins are particularly important as energy-providing nutrients and must be supplied to the body in specific proportions. However, the caloric and nutrient requirements of individuals are influenced by various factors, such as age, gender, height, and weight. Therefore, balancing caloric and nutrient intake is considered a complex multi-objective decision-making problem.

Robust optimization represents an algorithmic approach within the domain of mathematics that is specifically designed to address and mitigate the effects of uncertainty [4]. It enables the determination of optimal solutions within a framework that assumes the presence of the most adverse conditions, a scenario that arises when there is uncertainty surrounding the model's parameters. In the context of linear programming problems, robust optimization is often implemented to solve the issue of data uncertainty [5]. Variants of robust optimization were also proposed to alleviate the computational complexities associated with the data-driven optimizations [6,7]. A robust variant of the weighted goal programming method is also presented to provide robust solutions that exert control over the influence of estimation errors [8]. This method accounts for errors in data and is implemented to model the decision maker's risk preference.

In personalized dietary management [9,10], deep learning techniques are integrated in food image segmentation and nutrient tracking systems [11]. A variety of neural network models are used in image recognition [12]. However, the accuracy of those methods seldom satisfied the requirements of food image segmentation in personalized diets. Thus, it is essential to have an efficient and accurate method for tracking patients' nutrition.

In this study, using food weight data predicted from patient diet images, we developed a data-driven goal optimization model to achieve multiple nutrient intake targets while maintaining the body's energy balance. Additionally, we propose a robust goal programming method specifically designed for dietary plan optimization.

2 Robust Optimization of the Diet Problem

The system is defined as a six-tuple, (X, A, B, Y, F, C). The decision variable, X, represents the quantity of a specific food intake selected by the decision-maker, such as a patient. The constant A denotes the attributes of the food, particularly the associated nutrients. The constant B represents the reference value for nutrient intake or the nutrition intake value recommended by medical guidelines, commonly referred to as the recommended nutrient intake (RNI). The observed data Y indicates the actual food categories consumed by the user, which is a categorical variable. The observed data F represents the actual food intake data, or the actual nutrient intake (referred to as actual nutrient intake (ANI)). The covariate C includes pertinent information or preferences of the decision-maker, such as gender, age, physical activity level, and other relevant factors.

The observed data of the system can be expressed as a set of values, (x, a, b, y, f, c), where the category y is the value converted into a categorical data type. The decision variable x is the quantity that needs to be modeled and solved for (in units of grams). The constants a and b are derived from health management domain knowledge or clinical guidelines (in weight units). The data y is obtained through food image classification to determine food categories, data f is derived from food image segmentation to ascertain actual nutrient intake values (in grams), and data c is acquired from the user system. Moreover, we assumed that each nutrient content intake has a linear relationship with the weight of the food that needs to be consumed.

In a balanced diet, the nutrient intake derived from the food consumed by an individual aligns with the recommended dietary intake, $\sum_{j=1}^{n} x_j \times a_{ji} = b_i$, where x_j is the quantity of food item j, a_{ji} is the nutritional content of food item j for nutrient i (per unit of x_j), b_i is the total daily requirement for nutrient i, $j = 1, \ldots, n$ the index over food items (e.g., chicken, rice, vegetables), and i the index over nutrients (e.g., protein, carbohydrates, vitamin C). In practical scenarios, a discrepancy invariably exists between the ANI and the RNI.

Let a deviation d_i characterize the variance between actual and recommended intakes. Here, d_i^+ quantifies the excess intake beyond the recommended threshold, signifying the nutrients that should be decreased. In the context of goal programming, this is represented as $d_i^+ \geq 0$. Conversely, d_i^- encapsulates the deficiency of actual intake relative to the recommended level, $d_i^- \geq 0$. We express the actual intake of the i-th nutrient as f_i. The positive deviation d_i^+ and negative deviation d_i^- are defined as:

$$d_i^+ = \max\{0, f_i - b_i\}, \quad d_i^- = \max\{0, b_i - f_i\}. \tag{1}$$

where $d_i^- \times d_i^+ = 0$, meaning that both deviations cannot be positive simultaneously. Considering the balance constraints of daily nutrient intake, we assume $f_i(x) \in [f_i^l(x), f_i^u(x)]$, where $f_i(x)$ represents the total intake of the i-th nutrient per day; $f_i^l(x)$ denotes its lower interval; $f_i^u(x)$ denotes its upper interval.

In the goal programming model, the primary objective is to approximate predefined target values by minimizing the corresponding deviation variables. This minimization ensures that nutrient intake deviations converge toward recommended targets. Notably, in dietary optimization using goal programming, different deviation variables may have distinct units or scales. To reconcile this, we normalize deviations by converting them into percentage values, thereby enabling consistent measurement across all variables. This normalization establishes a mathematical foundation for solving the multi-objective optimization problem in nutritional diet planning.

$$\min_{x_j} V(d_i^+, d_i^-) = \sum_{i=1}^{p} \left(\frac{d_i^- + d_i^+}{b_i}\right)$$
$$s.t. \begin{cases} \sum_{j=1}^{n} a_{ij} x_j + d_i^- - d_i^+ = b_i, \ i = 1, 2, 3..., I \\ f_i^u(x) \geq f_i(x) \geq f_i^l(x), \ i = 1, 2, 3..., I \\ \sum_{j=1}^{n} x_j \leq B \\ \sum_{j=1}^{n} l_{\{x_j > 0\}} = \lambda \\ x_j \geq 0, \ j = 1, 2, 3..., n \\ d_i^-, d_i^+ \geq 0, \ i = 1, 2, 3..., I \end{cases} \tag{2}$$

where $l_{\{x_j>0\}}$ denotes the limitations in food variety choices. Specifically, when $x_j > 0$, $l_{\{x_j>0\}} = 1$; otherwise, it is 0. The parameter λ represents the number of food types selected; B is a constant representing the upper bound of food selection. Here, $\lambda = 0$ indicates that the i-th food item is not consumed by a specific group, while $\lambda = 1$ indicates that the j-th food item is consumed.

This optimization model considers the equilibrium of nutrients, food weight, and dietary diversity within daily meals, utilizing recommended intake guidelines from medical advice as the benchmark for optimal nutritional intake. The constraints of the model are as follows: the initial constraint is the goal constraint, which ensures that the dietary improvement objectives are met; the second constraint governs the intake bounds for nutrient i; the third constraint imposes an upper limit B on the total food weight; additional constraints delineate the permissible ranges for the decision and deviation variables.

3 Robust Goal Optimization for Healthy Nutrition

Although deterministic dietary optimization models are amenable to direct resolution, the indeterminacy inherent in food information obtained through food image recognition may render the resulting solutions infeasible. Considering that such uncertainty is an intrinsic characteristic of deep learning, it is prudent to preprocess this data rather than depend solely on post hoc sensitivity analyses. Consequently, this investigation expands the initial problem to incorporate a robust optimization model, thereby accounting for this indeterminacy.

Bertsimas and Sim [13] proposed the use of an adjusted non-negative integer to modulate the level of conservativeness in the robust solution. Acknowledging the dimensional disparities between food energy and nutrient content, we differentiate between energy and nutritional information in our treatment. Within this model, we consider the matrices representing food nutritional information and energy as uncertain parameters, and we constrain their values within the specified budget uncertainty sets μ_c and μ_A, respectively.

$$\mu_c = \left\{ C \in \mathbf{R}^{p \times n} : \begin{array}{l} \bar{c}_{ij} - \hat{c}_{ij}\eta_{ij} \leq c_{ij} \leq \bar{c}_{ij} + \hat{c}_{ij}\eta_{ij}, \\ \sum_{j=1}^{n} \eta_{ij} \leq \Gamma_i, \quad 0 \leq \eta_{ij} \leq 1, \qquad \forall i, \ \forall j \end{array} \right\}, \tag{3}$$

$$\mu_A = \left\{ A \in \mathbf{R}^{m \times n} : \begin{array}{l} \bar{a}_{kj} - \hat{a}_{kj}\varsigma_{kj} \leq a_{kj} \leq \bar{a}_{kj} + \hat{a}_{kj}\varsigma_{kj}, \\ \sum_{j=1}^{n} \varsigma_{kj} \leq \Gamma'_k, \quad 0 \leq \varsigma_{kj} \leq 1, \quad k = 1, \ \forall j \end{array} \right\}, \tag{4}$$

where parameter Γ_i locates in the interval $[0, |J_i|]$, J_i and $J_i^{'}$ in the interval $[0, |J_k^{'}|]$, representing the sets of coefficients c_{ij} and a_{kj}. Indices $j \in J_i$ and $j \in J_k'$, c_{kj} takes values in the interval $[\bar{c}_{ij} - \hat{c}_{ij}, \bar{c}_{ij} + \hat{c}_{ij}]$, and $a_{kj} \in [\bar{a}_{kj} - \hat{a}_{kj}, \bar{a}_{kj} + \hat{a}_{kj}]$.

The uncertainty set μ_c allows for at most changes $\lfloor \Gamma_i \rfloor$ in these coefficients, with one coefficient c_{ij} changing by $(\Gamma_i - \lfloor \Gamma_i \rfloor)\, \bar{c}_{ij}$. For the chosen Γ_i, we consider a subset that satisfies the conditions $S_i \subseteq J_i$ and $\lfloor S_i \rfloor = \lfloor \Gamma_i \rfloor$. Similarly, the uncertainty set μ_A allows for at most $\lfloor \Gamma_k' \rfloor$ changes in these coefficients, with one coefficient a_{kj} changing by $(\Gamma_k' - \lfloor \Gamma_k' \rfloor)\, \hat{a}_{kj'}$. For the chosen Γ_k', we consider a subset $S_k' \subseteq J'$ that meets the conditions $|S_k'| = \lfloor \Gamma_k' \rfloor$. Thus, the optimization problem in nutritional diet planning can be converted into

$$
\begin{aligned}
&\min_{x_j,\, d_i^-, d_i^+} \quad \sum_{i=1}^{p} \left(\frac{w_i^- d_i^-}{g_i} + \frac{w_i^+ d_i^+}{g_i} \right) \\
&\text{subject to} \\
&\quad \sum_{j=1}^{n} \bar{c}_{ij} x_j + \max_{\substack{S_i \subseteq J_i, |S_i| = \lfloor \Gamma_i \rfloor \\ t_i \in J_i \setminus S_i}} \left(\sum_{j \in S_i} \hat{c}_{ij} |x_j| + (\Gamma_i - \lfloor \Gamma_i \rfloor) \hat{c}_{it_i} |x_{t_i}| \right) + d_i^- - d_i^+ = g_i, \qquad \forall i, \\
&\quad \sum_{j=1}^{n} \bar{a}_{kj} x_j + \max_{\substack{S_k' \subseteq J_k', |S_k'| = \lfloor \Gamma_k' \rfloor \\ j_k' \in J_k' \setminus S_k'}} \left(\sum_{j \in S_k'} \hat{a}_{kj} |x_j| + (\Gamma_k' - \lfloor \Gamma_k' \rfloor) \hat{a}_{kj_k'} |x_{j_k'}| \right) \leq b_k, \qquad \forall k, \\
&\quad d_i^- \geq 0, \quad d_i^+ \geq 0, \qquad \forall i.
\end{aligned}
\tag{5}
$$

where $S_i \subseteq J_i$ is a subset of food items j (with $|S_i| = \lfloor \Gamma_i \rfloor$) for which the cost deviations $\hat{c}_{ij}$ are fully activated, $S_k' \subseteq J_k'$ is a subset of food items j (with $|S_k'| = \lfloor \Gamma_k' \rfloor$) for which nutritional deviations $\hat{a}_{kj}$ are fully activated, Γ_i and Γ_k' is the budget of uncertainty parameters controlling the conservatism of the model, $\lfloor \Gamma_i \rfloor$ and $\lfloor \Gamma_k' \rfloor$ are the integer parts of Γ_i and Γ_k', w_i^- and w_i^+ are the weights for negative/positive deviations d_i^- and d_i^+, and g_i is the adjusted nutritional target.

If Γ_i is an integer, the i-th robust constraint simplifies to the following optimization model:

$$
\beta_i(x, \Gamma_i) = \max_{S_i \subseteq J_i, |S_i| = \Gamma_i} \left\{ \sum_{j \in S_i} \hat{c}_{ij} |x_j| \right\}, \tag{6}
$$

The model $\beta_i(x, \Gamma_i)$ computes the worst-case cost deviation protected against. It is noted that when $\Gamma_i = 0$, the constraint condition is equivalent to the nominal problem. Furthermore, we can adjust the model to accommodate the conservativeness level of the solution.

$$
\beta_i(x^*, \Gamma_i) = \max_{\substack{S_i \subseteq J_i, \\ |S_i| = \lfloor \Gamma_i \rfloor, \\ t_i \in J_i \setminus S_i}} \left\{ \sum_{j \in S_i} \hat{c}_{ij} |x_j| + (\Gamma_i - \lfloor \Gamma_i \rfloor) \hat{c}_{it_i} |x_{t_i}| \right\}, \tag{7}
$$

Model (7) is equivalent to the following linear optimization problem:

$$\begin{aligned} \min_{\eta_{ij}} \quad & \sum_{j \in J_i} \hat{c}_{ij} |x_j^*| \eta_{ij} \\ \text{s.t.} \quad & 0 \leq \eta_{ij} \leq 1, \qquad \forall j \in J_i, \\ & \sum_{j \in J_i} \eta_{ij} \leq \Gamma_i, \end{aligned} \tag{8}$$

where the continuous variable $\eta_{ij} \in [0, 1]$ represents the fraction of maximum deviation considered for each coefficient. When $\eta_{ij}= 1$, the model fully protects against deviations in item j, while $\eta_{ij}= 0$ provides no protection. The constraint $\sum \eta_{ij} \leq \Gamma_i$ ensures the total protection level remains within the specified uncertainty budget.

The model (8) can be transferred into the dual problem (9), which aims to minimize the total penalty ($\sum p_{ij}$) plus the budget cost ($\Gamma_i z_i$).

$$\begin{aligned} \min_{z_i, p_{ij}} \quad & \sum_{j \in J_i} p_{ij} + \Gamma_i z_i \\ \text{s.t.} \quad & z_i + p_{ij} \geq \hat{c}_{ij} |x_j^*|, \qquad \forall j \in J_i, \\ & p_{ij} \geq 0, \qquad \forall j \in J_i, \\ & z_i \geq 0. \end{aligned} \tag{9}$$

where variable z_i is the dual variable associated with the budget constraint $\sum_{j \in J_i} \eta_{ij} \leq \Gamma_i$, representing the marginal cost of increasing the budget Γ_i (the maximum allowed allocation for category i). Variable p_{ij} is the dual variable associated with the box constraint $0 \leq \eta_{ij} \leq 1$, representing the penalty for activating the deviation η_{ij} (i.e., setting $\eta_{ij} > 0$) for item j in category i.

Since problem (8) is feasible and bounded for $\Gamma_i \in [0, |\Gamma_i|]$, the strong duality theorem indicates that the dual problem (9) is also feasible and bounded. Thus, the optimal objective value of problem (9) is equal to the optimal objective value of the original problem (8). Thus, the constraint in model (5) is equivalent to

$$\begin{aligned} & \textstyle\sum_{j=1}^{n} \bar{c}_{ij} x_j + z_i \Gamma_i + \sum_{j \in J_i} p_{ij} + d_i^- - d_i^+ = g_i, \ z_i \geq 0, \ \ \forall i, \\ & z_i + p_{ij} \geq \hat{c}_{ij} \nu_j, \ p_{ij} \geq 0, \ \ \forall i, \ \ \forall j \in J_i, \\ & -\nu_j \leq x_j \leq \nu_j, \ \ \forall j \in J_i. \end{aligned} \tag{10}$$

where ν_j is the absolute value auxiliary variable, enforcing $|x_j|$ through constraints $-\nu_j \leq x_j \leq \nu_j$ and converting the nonlinear $|x_j|$ terms to linear form. This is essential for maintaining convexity while handling absolute deviations.

Similarly, we have

$$\begin{aligned} & \textstyle\sum_{j=1}^{n} \bar{a}_{kj} x_j + \Gamma_k' z_k' + \sum_{j \in J_k'} p_{kj}' \geq b_k, \ z_k' \geq 0, \ \ \forall k, \\ & z_k' + p_{kj}' \geq \hat{a}_{kj} \nu_j, \ p_{kj}' \geq 0, \ \ \forall k, \ \ \forall j \in J_k'. \end{aligned} \tag{11}$$

where z_k is the nutritional constraint dual and analogous to z_i but for nutritional uncertainty; p_{kj}' is the nutritional protection variable and measures excess nutritional deviations beyond z_k', $p_{kj}' = max(0, \hat{a}_{kj}|x_j| - z_k')$.

Thus, the robust optimization model for the diet problem can be equivalently expressed in the dual form:

$$
\begin{aligned}
& \min G = \sum_{i=1}^{I} \left(\frac{d_i^-}{g_i} + \frac{d_i^+}{g_i} \right) \\
\text{s.t.} & \sum_{j=1}^{n} \bar{c}_{ij} x_j + z_i \Gamma_i + \sum_{j \in J_i} p_{ij} + d_i^- - d_i^+ = g_i, \quad z_i \geq 0, \qquad \forall i, \\
& \sum_{j=1}^{n} \bar{a}_{kj} x_j + \Gamma_k' z_k' + \sum_{j \in J_k'} p_{kj}' \geq b_k, \quad z_k' \geq 0, \qquad \forall k, \\
& z_i + p_{ij} \geq \hat{c}_{ij} \nu_j, \quad p_{ij} \geq 0, \qquad \forall i, \ \forall j \in J_i, \\
& z_k' + p_{kj}' \geq \hat{a}_{kj} \nu_j, \quad p_{kj}' \geq 0, \qquad \forall k, \ \forall j \in J_k', \\
& -\nu_j \leq x_j \leq \nu_j, \qquad \forall j, \\
& z_i \geq 0, \quad d_i^- \geq 0, \quad d_i^+ \geq 0, \qquad \forall i, \\
& z_k' \geq 0, \quad d_k'^- \geq 0, \quad d_k'^+ \geq 0, \qquad \forall k.
\end{aligned}
\tag{12}
$$

The evaluation of meal planning can be assessed by the root mean square error (RMSE):

$$
\delta(f) = \sqrt{\frac{1}{I} \sum_{i=1}^{I} \left(\frac{\hat{f}_i - f_i}{f_i} \right)^2} \times 100\%, \tag{13}
$$

where RMSE denotes the root mean squared error of the relative distance between the nutrient composition of the ith type in the meal planning and the benchmark nutrient levels. A lower RMSE value signifies that the meal planning is more aligned with the recommended benchmark nutrient intake.

4 Numerical Studies

In this experimental study, a representative male individual was chosen for analysis, using data obtained from the China Health and Nutrition Survey (CHNS) dataset. The subject's demographic information is as follows: 28 years of age, 174 cm in height, 70 kg in weight, and engaged in light physical activity.

The daily caloric intake for individuals is calculated based on their Basal Metabolic Rate (BMR), which is determined through factors including body weight, height, age, and gender. The Schofield equation, which was endorsed by the World Health Organization (WHO), is deemed the most appropriate for approximating BMR reference values for the Chinese population [11]. The total energy expenditure (TEE) of the human body over a 24-hour period can be computed by integrating BMR with the Physical Activity Level (PAL), as formulated by $TEE = PAL \times BMR$. The subject's nutrient intake, including energy, was documented utilizing the 24-hour dietary recall method (24HR) for a designated day. The subject's necessary energy intake can be computed by

$TEE = 1.6 \times 15.3 \times 70 + 679 = 2800.00(kcal)$. The recommended protein intake is $2800 \times 15\% \div 4 = 105.00(g)$. The recommended fat intake is $2800 \times 25\% \div 9 \approx 77.78(g)$. The carbohydrate intake is $2450 \times 60\% \div 4 \approx 420.00(g)$.

Figure 1 illustrates the food weight data prediction process using patient diet images. This process entails inputting standardized meal images, followed by data preprocessing and image segmentation. Subsequently, a deep learning-trained food recognition model is employed to classify foods and to ascertain their volumes and weights. This process facilitates the computation of the nutrient content within various food images, thereby enabling the determination of the user's nutrient intake for the respective meal.

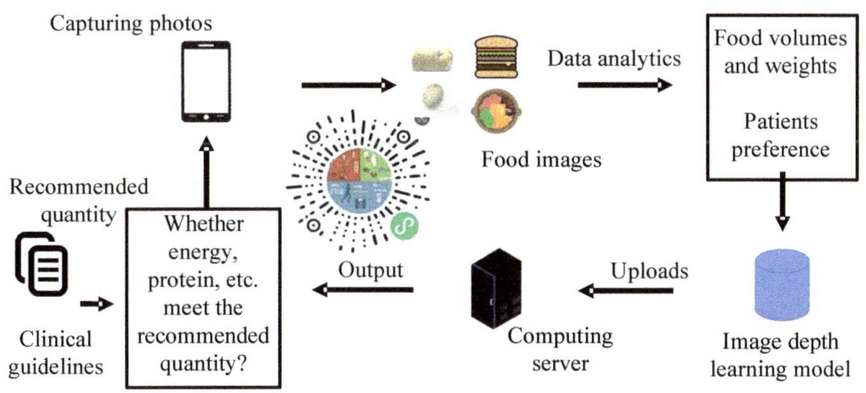

Fig. 1. Food weight data prediction process using patient diet images.

Optimization and solution procedures were executed for both the original GP model and the robust optimization model, utilizing the RSOME package [4]. These computations were performed on a Gourbi (11.0) platform hosted on a PC equipped with an Intel CPU operating at 2.40 GHz. A collection of foods was selected, and this same set of foods was solved using both the original model and the robust optimization model, with varying budget levels controlled.

When $\Gamma = 0$, the problem corresponds to the original model, which does not account for the uncertainty in food data. By manipulating different budget levels, the variations in the objective's optimal value in response to different λ values are compared. Table 1 shows that for conventional multi-objective programming (MOP), Pareto solutions can be obtained within a specific time frame through the application of heuristic algorithms.

We are striving to identify the optimal meal planning strategy by reducing the RSME. As the quantity of food items increases, the RMSE decreases. Unlike deterministic approaches, accounting for food content uncertainty reduces RMSE as more food items are included. These findings corroborate the intuitive understanding of dietary balance in everyday life: a more diverse range of foods must be consumed to achieve a balanced diet, a principle that holds even when

Table 1. Solutions and related methods comparison

	Ingredients	MOP			RGP without λ	RGP with λ		
λ		3	4	5	NA	3	4	5
RMSE		0.1619	0.0979	0.0716	0.2108	0.1841	0.0967	0.0561*
1	Rice	486.70	283.58	0	173.29	0	0	100.00
2	Noodles	0	296.62	417.85	172.62	500.00	500.00	412.85
3	Pineapple	161.01	0	0	13.66	0	0	0
4	Carrot	0	0	0	7.26	500.00	500	269.89
5	Egg	0	0	103.14	160.13	0	234.86	250.09
6	Beef	0	0	90.18	112.36	0	0	0
7	Soy Milk	0	0	0	29.19	0	0	0
8	Milk	0	0	0	396.21	0	0	0
9	Chicken	231.47	131.61	102.12	42.53	215.48	0	0
10	Pork	0	83.51	95.14	40.69	0	149.03	144.43

Notes: MOP refers to Multi-Objective Programming (MOP) with NSGA-II, while not considering lambda; RGP refers to the robust goal programming.

accounting for the uncertainty in food nutritional data. Nevertheless, this approach also entails an increased burden, as incorporating a wider variety of foods necessitates additional expenses.

5 Conclusions

This study proposes an intelligent dietary recommendation method that employs multi-objective programming to optimize nutrient intake. The method incorporates user physiological data and a budget-constrained robust optimization model to select suitable food components and mitigate discrepancies in food image recognition data, thereby ensuring nutritional requirements are fulfilled in the presence of uncertainty. Future research will investigate the model's optimization constraints (such as patients' preferences) to better conform to specific nutritional benchmarks and individual needs, with ingredient quantities adjusted to align with daily recommendations, thereby enhancing user satisfaction and the practicality of the method.

Acknowledgements. This work was supported in part by Chongqing Municipal Education Commission (No. KJZD-M202500605; KJQN202200608), the Ministry of Education of China (No. 23XJAZH004) and Chongqing Municipal Health Commission (No. 2023DBXM008; 2024ZYZD004), Chongqing Science and Technology Commission (No. CSTB2023NSCQ-MSX0073, CSTB2022TIAD-KPX0155), Shanghai Jiading District Health Commission (2024-KY-ZD-06), and National Natural Science Foundation of China (No. 62272077).

References

1. Yu, H., Yang, C.-C., Yu, P.: Constrained optimization for stratified treatment rules in reducing hospital readmission rates of diabetic patients. Eur. J. Oper. Res. **308**(3), 1355–1364 (2023)
2. Yu, H., Piri, S., Qiu, H., Xu, R., Miao, H.: Personalized algorithmic pricing decision support tool for health insurance: the case of stratifying gestational diabetes mellitus into two groups. Inf. Manag. **61**(3), 103945 (2024)
3. Yu, H., Zuo, X., Tang, J., Fu, Y.: Identifying causal effects of the clinical sentiment of patients' nursing notes on anticipated fall risk stratification. Inf. Process. Manag. **60**(6), 103481 (2023)
4. Chen, Z., Sim, M., Xiong, P.: Robust stochastic optimization made easy with RSOME. Manag. Sci. **66**(8), 3329–3339 (2020)
5. Selvi, A., Ben-Tal, A., Brekelmans, R., den Hertog, D.: Convex maximization via adjustable robust optimization. INFORMS J. Comput. **34**(4), 2091–2105 (2022)
6. Qi, Z., Pang, J.-S., Liu, Y.: On robustness of individualized decision rules. J. Am. Stat. Assoc. **118**(543), 2143–2157 (2023)
7. Baringo, L., Baringo, A.: A stochastic adaptive robust optimization approach for the generation and transmission expansion planning. IEEE Trans. Power Syst. **33**(1), 792–802 (2017)
8. Hanks, R.W., Weir, J.D., Lunday, B.J.: Robust goal programming using different robustness echelons via norm-based and ellipsoidal uncertainty sets. Eur. J. Oper. Res. **262**(2), 636–646 (2017)
9. Carroll, R.J.: Estimating the distribution of dietary consumption patterns. Stat. Sci. Rev. J. Inst. Math. Stat. **29**(1), 2 (2014)
10. Camirand Lemyre, F., Carroll, R.J., Delaigle, A.: Semiparametric estimation of the distribution of episodically consumed foods measured with error. J. Am. Stat. Assoc. **117**(537), 469–481 (2022)
11. Zhang, Y., et al.: Deep learning in food category recognition. Inf. Fusion **98**, 101859 (2023)
12. Falk, T., Mai, D., Bensch, R., Çiçek, Ö., Abdulkadir, A., Marrakchi, Y., et al.: U-Net: deep learning for cell counting, detection, and morphometry. Nat. Methods **16**(1), 67–70 (2019)
13. Bertsimas, D., Sim, M.: The price of robustness. Oper. Res. **52**(1), 35–53 (2004)

Author Index

X. Xie et al. (Eds.): ICSS 2025, LNOR, pp. 315–316, 2026.
https://doi.org/10.1007/978-3-032-13116-4

GPSR Compliance
The European Union's (EU) General Product Safety Regulation (GPSR) is a set of rules that requires consumer products to be safe and our obligations to ensure this.

If you have any concerns about our products, you can contact us on

ProductSafety@springernature.com

In case Publisher is established outside the EU, the EU authorized representative is:

Springer Nature Customer Service Center GmbH
Europaplatz 3
69115 Heidelberg, Germany

www.ingramcontent.com/pod-product-compliance
Ingram Content Group UK Ltd.
Pitfield, Milton Keynes, MK11 3LW, UK
UKHW020632260926
475829UK00003BA/37
* 9 7 8 3 0 3 2 1 3 1 1 5 7 *